LOCAL MODELS FOR SPATIAL ANALYSIS

SECOND EDITION

LOCAL MODELS FOR SPATIAL ANALYSIS

SECOND EDITION

CHRISTOPHER D. LLOYD

CRC Press
Taylor & Francis Group
Boca Raton London New York

CRC Press is an imprint of the
Taylor & Francis Group, an **informa** business

CRC Press
Taylor & Francis Group
6000 Broken Sound Parkway NW, Suite 300
Boca Raton, FL 33487-2742

First issued in paperback 2019

ISBN-13: 978-1-4398-2919-6 (hbk)
ISBN-13: 978-0-367-86493-4 (pbk)

Library of Congress Cataloging-in-Publication Data

Lloyd, Christopher D.
    Local models for spatial analysis / Christopher D. Lloyd. -- 2nd ed.
        p. cm.
    Includes bibliographical references and index.
    ISBN 978-1-4398-2919-6 (hardcover : alk. paper)
    1. Geographic information systems--Mathematical models. 2. Spatial analysis (Statistics) I. Title.

G70.212.L59 2011
519.5--dc22
                                                            2010027816

Visit the Taylor & Francis Web site at
http://www.taylorandfrancis.com

and the CRC Press Web site at
http://www.crcpress.com

For my parents

Contents

Preface to the second edition xiii

Acknowledgements xv

1 Introduction   **1**

1.1   Remit of this book 2

1.2   Local models and methods 2

1.3   What is local? 4

1.4   Spatial dependence and autocorrelation 5

1.5   Spatial scale 7

     1.5.1   Spatial scale in geographical applications 8

1.6   Stationarity 9

1.7   Spatial data models 9

     1.7.1   Grid data 9

     1.7.2   Areal data 10

     1.7.3   Geostatistical data 10

     1.7.4   Point patterns 10

1.8   Datasets used for illustrative purposes 11

     1.8.1   Monthly precipitation in Great Britain in 1999 11

     1.8.2   A digital elevation model (DEM) of Great Britain . . 13

     1.8.3   Positions of minerals in a slab of granite 13

     1.8.4   Landcover in Turkey 13

     1.8.5   Digital orthophoto of Newark, New Jersey 17

     1.8.6   Population of Northern Ireland in 2001 17

     1.8.7   Other datasets 19

1.9   A note on notation 20

1.10   Overview 21

2 Local Modelling   **23**

2.1   Standard methods and local variations 23

2.2   Approaches to local adaptation 23

2.3   Stratification or segmentation of spatial data 27

2.4   Moving window/kernel methods 28

2.5   Locally-varying model parameters 30

2.6   Transforming and detrending spatial data 30

2.7   Categorising local statistical models 30

     2.7.1   The nature of spatial subsets 31

	2.7.2	The relationship of spatial subsets to the complete dataset	32
	2.7.3	The relationship between a global statistic and the corresponding local statistic	32
2.8		Local models and methods and the structure of the book	34
2.9		Overview	35

3 Grid Data — **37**

3.1		Exploring spatial variation in gridded variables	37
3.2		Global univariate statistics	38
3.3		Local univariate statistics	38
3.4		Analysis of grid data	38
3.5		Moving windows for grid analysis	39
	3.5.1	Image smoothing: Low pass filtering	41
	3.5.2	High pass filters	42
	3.5.3	Edge detectors	44
	3.5.4	Texture	44
	3.5.5	Other approaches	47
3.6		Wavelets	48
	3.6.1	Fourier transforms and wavelet transforms	48
	3.6.2	Continuous wavelet transform	50
	3.6.3	Discrete wavelet transform (DWT)	50
	3.6.4	Wavelet basis functions	51
	3.6.5	Implementation of the DWT	53
	3.6.6	Fast wavelet transform: Illustrated example	54
	3.6.7	Two-dimensional (2D) wavelet transforms	56
	3.6.8	Other issues	60
	3.6.9	Applications of wavelets	61
3.7		Segmentation	62
3.8		Analysis of digital elevation models	65
3.9		Overview	71

4 Spatial Patterning in Single Variables — **73**

4.1		Local summary statistics	73
4.2		Geographically weighted statistics	74
4.3		Spatial autocorrelation: Global measures	80
	4.3.1	Testing for spatial autocorrelation	83
4.4		Spatial autocorrelation: Local measures	85
	4.4.1	Local indicators of spatial association	86
	4.4.2	Use of different approaches	97
4.5		Spatial association and categorical data	98
4.6		Other issues	98
4.7		Overview	98

5   Spatial Relations **101**
    5.1   Global regression 101
    5.2   Spatial and local regression 103
    5.3   Regression and spatial data 105
    5.4   Spatial autoregressive models 106
    5.5   Multilevel modelling 108
    5.6   Allowing for local variation in model parameters 109
          5.6.1   Spatial expansion method 111
          5.6.2   Spatially varying coefficient processes 116
    5.7   Moving window regression (MWR) 122
    5.8   Geographically weighted regression (GWR) 122
          5.8.1   Illustrated application of MWR and GWR 124
          5.8.2   Selecting a spatial bandwidth 126
          5.8.3   Testing the significance of the GWR model 127
          5.8.4   Geographically weighted regression and collinearity . . 128
          5.8.5   Case study: MWR and GWR 135
          5.8.6   Other geographically weighted statistics 138
    5.9   Spatially weighted classification 138
    5.10 Local regression methods: Some pros and cons 142
    5.11 Overview 142

**6   Spatial Prediction 1: Deterministic Methods, Curve Fitting,
    and Smoothing** **145**
    6.1   Point interpolation 146
    6.2   Global methods 147
    6.3   Local methods 148
          6.3.1   Thiessen polygons: Nearest neighbours 148
          6.3.2   Triangulation 151
          6.3.3   Natural neighbours 151
          6.3.4   Trend surface analysis and local polynomials 153
          6.3.5   Linear regression 153
          6.3.6   Inverse distance weighting (IDW) 154
          6.3.7   Thin plate splines 158
          6.3.8   Thin plate splines case study 167
          6.3.9   Generalised additive models and penalised regression
                  splines 168
          6.3.10 Finite difference methods 171
          6.3.11 Locally adaptive approaches for constructing digital
                  elevation models 171
    6.4   Areal interpolation 173
    6.5   General approaches: Overlay 175
    6.6   Local models and local data 177
          6.6.1   Generating surface models from areal data 177
          6.6.2   Population surface case study 180
          6.6.3   Local volume preservation 181

	6.6.4	Making use of prior knowledge	184
	6.6.5	Uncertainty in areal interpolation	189
6.7	Limitations: Point and areal interpolation		189
6.8	Overview		190

7   Spatial Prediction 2: Geostatistics — **191**

7.1	Random function models		192
7.2	Stationarity		193
	7.2.1	Strict stationarity	193
	7.2.2	Second-order stationarity	193
	7.2.3	Intrinsic stationarity	194
	7.2.4	Quasi-intrinsic stationarity	194
	7.2.5	Nonstationarity	194
7.3	Global models		194
7.4	Exploring spatial variation		195
	7.4.1	The covariance and correlogram	195
	7.4.2	The variogram	197
	7.4.3	The cross-variogram	198
	7.4.4	Variogram models	198
7.5	Kriging		204
7.6	Globally constant mean: Simple kriging		206
7.7	Locally constant mean models		208
7.8	Ordinary kriging		208
7.9	Cokriging		212
	7.9.1	Linear model of coregionalisation	212
	7.9.2	Applying cokriging	213
7.10	Equivalence of splines and kriging		215
7.11	Conditional simulation		215
7.12	The change of support problem		216
7.13	Other approaches		216
7.14	Local approaches: Nonstationary models		217
7.15	Nonstationary mean		220
	7.15.1	Trend or drift?	220
	7.15.2	Modelling and removing large scale trends	220
7.16	Nonstationary models for prediction		222
	7.16.1	Median polish kriging	222
	7.16.2	Kriging with a trend model	222
	7.16.3	Intrinsic model of order k	225
	7.16.4	Use of secondary data	226
	7.16.5	Locally varying anisotropy	233
7.17	Nonstationary variogram		233
7.18	Variograms in texture analysis		240
7.19	Summary		241

8 Point Patterns and Cluster Detection **243**

   8.1   Point patterns 243

   8.2   Visual examination of point patterns 244

   8.3   Measuring event intensity and distance methods 245

   8.4   Statistical tests of point patterns 246

   8.5   Global methods 246

   8.6   Measuring event intensity 247

       8.6.1   Quadrat count methods 247

       8.6.2   Quadrat counts and testing for complete spatial randomness (CSR) 248

   8.7   Distance methods 249

       8.7.1   Nearest neighbour methods 250

       8.7.2   The K function 251

   8.8   Other issues 255

   8.9   Local methods 256

   8.10 Measuring event intensity locally 256

       8.10.1 Quadrat count methods 256

       8.10.2 Kernel estimation 256

   8.11 Accounting for the population at risk 261

   8.12 The local K function 261

   8.13 Point patterns and detection of clusters 264

   8.14 Overview 270

9 Summary: Local Models for Spatial Analysis **271**

   9.1   Review 271

   9.2   Key issues 272

   9.3   Software 272

   9.4   Future developments 273

   9.5   Summary 274

A Software **275**

References **277**

Index **307**

Preface to the second edition

The first edition of this book was written in recognition of recent developments in spatial data analysis that focussed on differences between places. The second edition reflects continued growth in this research topic, but is also a function of feedback and reflection on the first edition. One aim of the second edition has been to provide more detailed critical accounts of key methods that were introduced in the first edition.

Difference is at the core of geography and cognate disciplines, and it provides the rationale behind this book. One focus of these disciplines is to describe differences or similarities between places. In quantitative terms, characterising spatial dependence — the degree to which neighbouring places are similar — is often important. In social geography, for example, researchers may be interested in the degree of homogeneity in areas of a city. Some areas of a city may have mixed population characteristics, while others may be internally very similar. In geomorphology, different processes operate in different places and at different spatial scales. Therefore, a model which accounts for this spatial variation is necessary. However, in practice, global models have often been applied in analyses of such properties. Use of global models in such contexts will mask underlying spatial variation, and the data may contain information which is not revealed by global approaches. One simple definition of a global model is that which makes use of all available data, while a local model makes use of some subset of the complete dataset. In this book, a variety of different definitions are given — a local model may, for example, make use of all available data and account for local variations in some way. This book is intended to enhance appreciation and understanding of local models in spatial analysis. The text includes discussions on key concepts such as spatial scale, nonstationarity, and definitions of local models. A key focus is on the description and illustration of a variety of approaches which account for local variation in both univariate and multivariate spatial relations.

The development and application of local models is a major research focus in a variety of disciplines in the sciences, social sciences, and the humanities, and recent work has been described in a number of journal papers and books. This book provides an overview of a range of different approaches that have been developed and employed within Geographical Information Science (GIScience). Examples of approaches included are methods for point pattern analysis, the measurement of spatial autocorrelation, and spatial prediction. The intended audience for this book is geographers and those concerned with the analysis of spatial data in the physical or social sciences. It is assumed that readers will have prior experience of Geographical Information Systems. Knowledge of GIScience is, as in any discipline, often developed through examination of examples. As such, key components of this book are worked examples and case studies. It is intended that the worked examples will help develop an understanding of how the algorithms work, while the case studies will demonstrate their practical utility and range of application. The applications used to illustrate the methods discussed are based on data representing various physical properties and a set of case studies using data about human populations are also included.

While many readers of this text have been exposed to some of the ideas outlined, it is hoped that readers will find it useful as an introduction to a range of concepts which are new to them, or as a means of expanding their knowledge of familiar concepts. The detailed reference list will enable readers to explore the ideas discussed in more depth, or seek additional case studies.

Updates, including errata, references to new research, and links to other relevant material, are provided on the book web site (http://www.crcpress.com/ product/isbn/9781439829196).

Acknowledgements

A large number of people have contributed in some way to the writing of this book. I would like to acknowledge the organisations who supplied the data used in the case studies. Thanks are due to the British Atmospheric Data Centre (BADC) for providing access to the United Kingdom Meteorological Office (UKMO) Land Surface Observation Stations Data. The UKMO is acknowledged as the originator of these data. The Northern Ireland Statistics and Research Agency (NISRA) is thanked for providing data from the 2001 Census of Population of Northern Ireland. The United States Geological Survey (USGS) is also thanked for allowing the use of their data. Suha Berberoglu and Alastair Ruffell provided datasets, and their help is acknowledged. Other freely-available datasets are used in the book for illustrative purposes, and the originators and providers of these data are also thanked.

I would like to thank Irma Shagla at CRC Press for her support during the book proposal and production processes. Several people kindly took the time to read the whole or parts of the manuscript of the first edition on which the present book builds. Peter Atkinson provided advice and suggestions at various stages of the process of writing the first edition, and I am very grateful for his time, expertise, and encouragement. In addition, Chris Brunsdon, Gemma Catney, Ian Gregory, David Martin, Jennifer McKinley, Ian Shuttleworth, and Nick Tate are all are thanked for their helpful comments on parts of the text of the first edition. I also wish to thank Stewart Fotheringham who commented on the structure of the book. Various individuals who wrote reviews of the first edition for academic journals are thanked for their comments, all of which were considered in the revision of the text. Chris Brunsdon kindly conducted a review of the manuscript of the second edition, and his help is gratefully acknowledged. Gemma Catney provided helpful comments on various chapters. Peter Congdon is thanked for his advice on Bayesian statistical modelling and for his comments on Chapter 5. I am grateful to Doug Nychka for his comments on thin plate splines. Any errors or omissions are, of course, the responsibility of the author. I am grateful to various colleagues at Queen's University, Belfast, for their support in various ways. My parents and wider family provided support in a less direct although no less important way. Finally, as for the first edition, my thanks go to Gemma who provided input and encouragement throughout.

1

Introduction

A key concern of geography and other disciplines which make use of spatially-referenced data is with differences between places. Whether the object of study is human populations or geomorphology, space is often of fundamental importance. We may be concerned with, for example, factors that effect unemployment or factors that influence soil erosion; traditionally, global methods have often been employed in quantitative analyses of datasets that represent such properties. The implicit assumption behind such methods is that properties do not vary as a function of space. In many cases, such approaches mask spatial variation and the data are under-used. The need for methods which do allow for spatial variation in the properties of interest has been recognised in many contexts. In geography and cognate disciplines, there is a large and growing body of research into local methods for spatial analysis, whereby differences between places are allowed for. This book is intended to introduce a range of such methods and their underlying concepts. Some widely-used methods are illustrated through worked examples and case studies to demonstrate their operation and potential benefits. To aid implementation of methods, relevant software packages are mentioned in the text. In addition, a summary list of selected software packages is provided in Appendix A.

The book is intended for researchers, postgraduate students, and professionals, although parts of the text may be appropriate in undergraduate contexts. Some prior knowledge of methods for spatial analysis, and of Geographical Information Systems (GISystems), is assumed. Background to some basic concepts in spatial data analysis, including elements of statistics and matrix algebra, is provided by O'Sullivan and Unwin (304), de Smith et al. (104), and Lloyd (245). There is a variety of published reviews of local models for spatial analysis (23), (127), (128), (367), but each has particular focuses. One concern here is to bring together discussion of techniques that could be termed 'local' into one book. A second concern is to discuss developments of (relatively) new techniques.

This chapter describes the remit of the book before introducing local models and methods. Then, the discussion moves on to issues of spatial dependence, spatial autocorrelation, and spatial scale. The concept of stationarity, which is key in the analysis of spatially or temporally referenced variables, is also outlined. Finally, key spatial data models are described, and the datasets used for illustrative purposes are detailed.

1.1   Remit of this book

The development and application of local models is a major research focus
in a variety of disciplines in the sciences, social sciences, and the humanities.
Recent work has been described in a number of journal papers and books.
This book provides an overview of a range of different approaches that
have been developed and employed within Geographical Information Science
(GIScience). This book is not intended to be an introduction to spatial
statistics in general. The aim is to start from first principles, to introduce
users of GISystems to the principles and application of some widely used
local models for the analysis of spatial data. The range of material covered
is intended to be representative of methods being developed and employed
in geography and cognate disciplines. Work is presented from a range of
disciplines in an attempt to show that local models are important for all who
make use of spatial data. Some of the techniques discussed are unlikely to
enter widespread use in the GISystems community, and the main stress is
on those approaches that are, or seem likely to be in the future, of most use
to geographers. Some topics addressed, such as image processing, are not
covered in detail. Rather, the principles of some key local approaches are
outlined, and references to more detailed texts provided.

The applications used to illustrate the methods discussed are based on data
representing various physical properties (e.g., precipitation and topography)
and on human populations. These applications serve to highlight the breadth
of potential uses of the methods and models discussed.

1.2   Local models and methods

Broadly, this chapter will stress the distinction between global and local
methods. With a global model, the assumption is that variation is the
same everywhere. However, it may be the case that a global model does
not represent well variation at any individual location. Global methods make
use of all available data, whereas local methods are often defined as those that
make use of some subset of the data. But there are also approaches whereby
the data are transformed in some way. For example, removal of a global trend
(representing the spatially-varying mean of the property) may be conducted
to remove large scale variation — the aim would be to obtain residuals from a
regular trend (i.e., a relatively constant increase or decrease in values in some
direction) across the region of interest, allowing a focus on local variation.

Local models have been used widely in some disciplines for several decades. For example, in image processing local filters have long been used to smooth or sharpen images. However, in geography a focus on the development of methods that account for local variation has been comparatively recent. Some methods, by definition, work locally. For example, many methods for analysing gridded data are always employed on a moving window basis (for example, methods for drainage network derivation and spatial filters). This book will discuss such techniques, although the main focus will be on reviewing models and methods of which there are global versions and local versions, the latter adapting in some way to local spatial variation.

That properties often vary spatially is recognised by Unwin and Unwin (367) who, in a review of the development of local statistics, outline some key concerns of geographical analysis. In particular, they note that:

1. Most spatial data exhibit spatial dependence (see Section 1.4).

2. Many analyses are subject to the modifiable areal unit problem (MAUP — results of an analysis depend on the division of space; see Section 6.4).

3. It is difficult to assume stationarity (see below) in any process observed over geographical space (for example, the mean and variance may vary markedly from place to place and thus the process can be called nonstationary).

The development of GISystems and the increased availability of spatial data has led to both the creation of problems and the development of solutions. Availability of datasets covering large areas (in particular, remotely-sensed images) increased the probability that regions with different properties would be encountered. As such, the need for local models that account for these differences increased. In addition, the capacity to collect data at very fine spatial resolutions meant that concern with spatial variation and its relation to spatial scale would increase (367).

Against this background, a key change in geography has been from a focus on *similarities* between places to *differences* across space (127). Fotheringham and colleagues (125), (128) include within this movement approaches for dissecting global statistics into their local components. Related concerns include concentration on local exceptions rather than the search for global regularities and production of local mappable statistics rather than global summaries.

Central to the theme of this section is the idea of spatial nonstationarity. That is, if the property of interest (for example, precipitation, elevation, or human population) varies from place to place, for some scale of analysis, then a nonstationary model is appropriate in the analysis of this property. Stationarity is discussed in more detail in Section 1.6. A model with constant parameters may not be appropriate in various situations. Fotheringham (125) gives three possible reasons:

1. There are spatial variations in observed relationships due to random sampling variations.

2. Some relationships are intrinsically different across space.

3. The model used to measure relationships is a gross misspecification of reality — relevant variables are missing, or the form of model adopted is inappropriate.

In practice, it may be difficult to distinguish between these reasons, but the methods described in this book constitute ways to explore these issues and, hopefully, to enhance our understanding of spatial processes. Fotheringham and Brunsdon (127) divide local methods into those approaches for analysis of univariate data, methods for analysis of multivariate data, and methods for analysis of movement patterns (spatial interaction models). The first two areas are concerns within this book, but the latter is largely outside its remit (although spatial interaction modelling approaches may be based on regression, a topic which is a concern here).

In the last decade, several important developments have taken place in quantitative geography and in allied disciplines. Such developments include methods for exploring local spatial autocorrelation (see Section 4.4) and methods for exploring variation in spatial relations between multiple variables (geographically weighted regression, discussed in Section 5.8, is an example of this). That is, models have been developed to allow for differences in properties at different locations. For example, the relationship between two properties may be markedly different in one region than in another and a local model that allows for these differences may be more appropriate than a model for which the parameters are fixed. In other areas major developments have taken place. For example, wavelets provide a powerful means of decomposing and analysing imagery (see Section 3.6). Such methods are receiving widespread attention, and there is a large range of sophisticated software packages to implement such methods. This book is intended to bring together discussions of such methods and to provide pointers to material about these methods which will enable their exploration further. Allied to these developments is a number of important summaries of recent developments, written by various authors, which are cited in the text.

1.3   What is local?

The term local can have a multitude of meanings in different contexts. In physical geography, for example, a local space may be some area over which a particular process has an obvious effect. A watershed might also be considered, in some sense, a local space. In geomorphology, a landscape may be classified into discrete spatial areas which could be regarded as local spaces. In socioeconomic contexts, a local space may be the neighbourhood which an individual is familiar with, or the wider set of areas with which they interact on a regular basis. In terms of spatial data analysis, a local space is often expressed in terms of distance from some point or area (the locality or neighbourhood of that point or area).

A study area can only be local in the context of a global dataset, or a larger subset of the dataset. Of course, a locally-based approach is not necessarily beneficial. If a dataset is transformed or partitioned this may, for example, provide a better model fit but the results may not be meaningful or interpretable. Application of local models may be more problematic than the application of global models because of the additional complexity — factors such as the size of a moving window or the type of transform applied may have a major impact on the results obtained from an analysis. Indeed, the division of geographical space is important in any analyses of spatial data. For example, the statistics computed from an image are a function of the spatial resolution of the image (as discussed below). Similarly, results from analyses based on one set of administrative zones will be different than those obtained when another set of zones is used. Unwin and Unwin (367) outline the need to (i) define which areas to include in an operation, and (ii) decide how to treat non-zero entries. That is, which data are included in the analysis, and how much influence (weight) should each observation have? An example of the latter is the weight assigned to an observation using a spatial kernel, as discussed in, for example, Sections 2.2, 2.4, 5.8, and 8.10.2, as well as throughout Chapter 4.

1.4 Spatial dependence and autocorrelation

The core principle behind many local methods is the concept of spatial dependence. That is, objects close together in space tend to be more similar than objects which are farther apart. This principle was termed the "First Law of Geography" (as outlined by Tobler (361)). In cases where data values are not spatially dependent many forms of geographical analysis are pointless. Figure 1.1 shows synthetic examples of strong and weak spatial dependence (note that the data are treated as ratio variables in this example). The term spatial autocorrelation refers to the correlation of a variable with itself and where neighbouring values tend to be similar this is termed positive spatial autocorrelation — the data are spatially dependent. Where neighbouring values tend to be dissimilar this is termed negative spatial autocorrelation. This topic is developed further in Section 4.3.

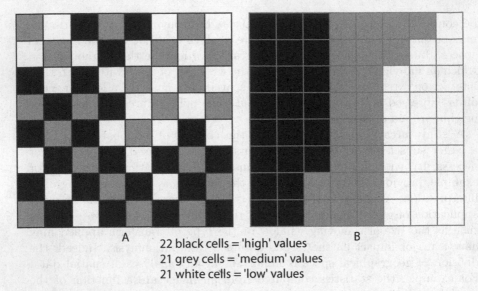

A   B

22 black cells = 'high' values
21 grey cells = 'medium' values
21 white cells = 'low' values

FIGURE 1.1: (A) Weak spatial dependence and (B) Strong spatial dependence.

Spatial dependence is often accounted for explicitly through the use of geographical weighting functions. This can be illustrated using the example of spatial interpolation. Suppose that there is the need to predict the value of some property at a location where no data are available. One sensible way to proceed is to take a weighted average of the observations surrounding the location at which we wish to make a prediction. That is, observations close to the prediction location will be given larger weights (more influence) than observations that are more distant from the prediction location. A procedure which operates on this principle is inverse distance weighting (IDW; see Section 2.2). The spatial structure of a process, and thus the degree of spatial dependence, may vary from place to place. Standard IDW accounts for local variation using a geographical weighting scheme of a given form, but it may be that a different weighting function should be employed in different areas. This book introduces methods that allow for variation in spatial structure. A property may be spatially structured at one scale, but spatially unstructured at another scale. For example, an image may appear 'noisy' at a fine spatial scale and structured at a coarser spatial scale, and this structure may vary from place to place.

1.5 Spatial scale

The concept of scale is central to all disciplines concerned with the spatial arrangement of properties. The term has been defined in many different ways. For example, a map is defined by its scale (we can talk of a large scale map or a small scale map). In the context of this discussion, and in common with the account of Atkinson and Tate (29), scale is taken to refer to the size or extent of a process, phenomenon, or investigation. Bian (41) uses the term 'operational scale' to refer to the scale over which a process operates. The availability of a wide range and type of data sources for locations around the globe means that users of spatial data are faced with working with multiscale representations — for example, a user may have, for one region, several remotely-sensed images that have different spatial resolutions. Users of such data usually have little choice about the scale of measurement (i.e., in this context, spatial resolution). As such, it is necessary to develop ways to work at a range of spatial scales (152).

Atkinson and Tate (29) state that spatial scale comprises (i) scales of spatial measurement, and (ii) scale of spatial variation (see Section 7.14 for relevant discussion). There are two scales of measurement — the support (geometrical size, shape, and orientation of the measurement units) and the spatial coverage of the sample (29). A variety of approaches exist for characterising scales of spatial variation and some approaches are mentioned below. There are many reviews of spatial scale problems; Atkinson and Tate (29) provide such a review in a geostatistical framework (see Chapter 7 for more information on related approaches).

The scale of spatial variation may change with location. That is, the dominant scale of spatial variation at one location may be quite different from that at another location. Hence, there is a need for approaches that allow variation with respect to (i) spatial scale, and (ii) spatial location. Many spatial processes, both (for example) physical and socioeconomic, may appear homogeneous at one scale and heterogeneous at another (Lam (228) discusses this idea with respect to ecological diversity). Clearly, locally-adaptive approaches are only necessary if the property of interest is spatially heterogenous at the scale of measurement. This book outlines a variety of approaches that allow exploration of local differences in scales of spatial variation.

Clearly, the shape and size of the area over which a property is recorded affects directly the results obtained through analyses of those data. Each level of a hierarchy of data presented at different spatial scales has unique properties that are not necessarily a simple sum of the component (disaggregated) parts (41). As Lam (228) states, the spatial resolution of an image changes fundamental biophysical relationships (known as the ecological fallacy), and the same is true in other contexts. That is, spatial models are frequently scale dependent — models that are applicable at one scale may not be appropriate at another scale (41). The modifiable areal unit problem (as defined on page 3 and discussed further in Section 6.4) reflects the fact that areal units can be changed and observed spatial variation altered. For example, the degree of spatial dependence is likely to change as the areal units are changed (see, for example, Lloyd (242)).

In the context of physical geography, Atkinson and Tate (29) note that nearly all environmental processes are scale dependent. So, the observed spatial variation is likely to vary at different scales of measurement. This means that there is a need to construct a sampling strategy that enables identification of spatial variation of interest. To facilitate acquisition of suitable data and integration of data at different spatial scales or different variables, the scaling properties of spatial variables should be used (29). However, spatial dependence may be unknown or may differ markedly in form from place to place, and there is the added problem that patterns at a given scale may be a function of interactions amongst lower-level systems (29). This has been referred to as the dichotomy of scale. As such, it is often necessary to downscale (starting at a coarse resolution relative to the spatial scale of interest — an increase in the spatial resolution) or upscale (starting with fine resolution components and constructing outputs over a coarser resolution — a decrease in the spatial resolution).

Many different methods have been developed to allow analyses of scales of spatial variation. These include fractal analysis, analysis of spatial structure using variograms, and wavelets (228). Throughout this book, spatial scale is a central concern. Chapter 7, in particular, deals with the characterisation of dominant scales of spatial variation while methods such as geographically weighted regression (Chapter 5), kernel estimation, and the K function (Chapter 8), for example, allow exploration of scales of spatial variation. In Section 6.4, some methods for changing from one set of areal units to another are discussed.

1.5.1   Spatial scale in geographical applications

There have been many published studies which detail attempts to characterise spatial variation in some geographical property. In a social context, an individual's perception of an area is a function of their knowledge of the neighbourhood and such perceptions have, therefore, inherent scales (149). Likewise, to model appropriately some physical process it is necessary to obtain measurements that capture spatial variation at the scale of interest. For example, if the sample spacing is larger than the scale of spatial variation that is of interest, then models derived from these data may not be fit for the task in hand (149).

As noted previously, the concern here is with approaches that enable exploration of local differences in scales of spatial variation. A relevant study is that by Lloyd and Shuttleworth (253), who show that the relations between mean commuting distance (as represented in the 1991 Northern Ireland Census of Population) and other variables differ markedly from place to place. In addition, the size of the areas over which these relations were assessed (that is, the size of the spatial kernel (see Section 2.4)) was varied. This study demonstrated regional variation in the spatial scale of the relations between these variables (see Section 5.8 for more details of this kind of approach).

1.6   Stationarity

The concept of stationarity is central in the analysis of spatial or temporal variation. In order to utilise the literature on local models for spatial analysis, an understanding of the key concepts is essential. The term stationarity is often taken to refer to the outcome of some process that has similar properties at all locations in the region of interest — it is a stationary process. In other words, the statistical properties (e.g., mean and variance) of the variable or variables do not change over the area of interest. A stationary model has the same parameters at all locations, whereas with a nonstationary model the parameters are allowed to vary locally. So, the focus of this book is on nonstationary models. There is little point in employing a nonstationary model if it offers, for example, no increased ability to characterise spatial variation or to map accurately a particular property. As such, it would be useful to be able to test for stationarity. However, testing for stationarity is not strictly possible, as discussed in Section 7.2. In the case of spatial prediction, for example, the performance of a stationary and a nonstationary model could be compared through assessment of the accuracy of predictions and thus the utility of a nonstationary approach considered.

1.7   Spatial data models

This book discusses local models that can be used in the analysis of properties that are represented by different kinds of data models. The key data models (or data types) are defined below. However, many analytical tools can be applied to properties represented using a range of different data models.

1.7.1   Grid data

Many operations used in the analysis of grid (or raster) data are, by definition, local. In particular, there is a wide range of methods used to analyse image data that are based on the idea of a moving window. Some key classes of operations are outlined in Chapter 3. Given the importance of remotely-sensed imagery in many applications areas, grid operations are a particular concern in this book.

1.7.2   Areal data

A frequent concern with areal data (e.g., areas dominated by a certain soil type, or population counts over particular zones) is to ascertain the neighbours of an area. That is, with what other areas does a particular area share boundaries, and what are the properties of these areas? Analysis of areal data is discussed in Chapters 4 and 5, while reassigning values across different areal units is discussed in Chapters 6 and 7. The centroids of areas may, in some contexts, be analysed in the same way as geostatistical data, as outlined next.

1.7.3   Geostatistical data

A typical geostatistical problem is where there are samples at discrete locations and there is a need to predict the value of the property at other, unsampled, locations. An example is an airborne pollutant sampled at a set of measurement stations. The basis of geostatistical analysis is the characterisation of spatial variation, and this information can be used to inform spatial prediction or spatial simulation. Parts of Chapters 4, 5, 6, and 7 discuss methods for the analysis of these kinds of data (Chapter 7 is concerned with geostatistical methods specifically).

1.7.4   Point patterns

Most of the models described in this book are applied to examine spatial variation in the values of properties. With point pattern analysis the concern is usually to analyse the spatial configuration of the data (events), rather than the values attached to them. For example, the concern may be to assess the spatial distribution of disease cases with respect to the total population. The population density is greater in urban areas than elsewhere and the population at risk is spatially varying — this must, therefore, be taken into account. The focus of Chapter 8 is on methods for assessing local variations in event intensity and spatial structure.

1.8   Datasets used for illustrative purposes

A variety of applications are mentioned to help explain different techniques, and specific case studies are also given to illustrate some of the methods. The application of a range of techniques is illustrated, throughout the book, using six main sets of data with supplementary datasets used in particular contexts. These are (with data model in parenthesis):

1. Monthly precipitation in Great Britain in 1999 (geostatistical).

2. A digital elevation model (DEM) of Great Britain (grid).

3. Positions of minerals in a slab of granite (point pattern).

4. A Landsat Thematic Mapper (TM) image and vector field boundaries for a region in south eastern Turkey (grid, areal).

5. A digital orthophoto of Newark, New Jersey (grid).

6. Population of Northern Ireland in 2001 (two datasets; areal).

The datasets are described below. In addition, two other areal datasets used in particular contexts are detailed in Section 1.8.7.

1.8.1   Monthly precipitation in Great Britain in 1999

The data are ground data measured across Great Britain under the auspices of the UK Meteorological Office as part of the national rain gauge network. The data were obtained from the British Atmospheric Data Centre (BADC) Web site*. Daily and monthly data for July 1999 were obtained and combined into a single monthly dataset. Only data at locations at which measurements were made for every day of the month of July were used. The locations of observations made during July 1999 are shown in Figure 1.2, and summary statistics are as follows: number of observations = 3037, mean = 38.727 mm, standard deviation = 37.157 mm, skewness = 2.269, minimum = 0.0 mm, and maximum = 319.0 mm. The smallest values were two zeros and the next smallest value was 0.5 mm. Elevation measurements are also available for each of the monitoring stations and, since precipitation and elevation tend to be related over periods of weeks or more, these data are used to demonstrate the application of multivariate techniques.

In parts of the book, a subset of the data is used to illustrate the application of individual techniques. Two subsets were extracted: one containing five observations (used to illustrate spatial prediction) and one containing 17

---

*www.badc.rl.ac.uk

FIGURE 1.2: Measurements of precipitation for July 1999.

observations (used to illustrate local regression techniques). The full dataset is then used to demonstrate differences in results obtained using alternative approaches. The data are used in Chapters 5, 6, and 7.

1.8.2 A digital elevation model (DEM) of Great Britain

The relevant section of the global 30 arc-second (GTOPO 30) digital elevation model (DEM)[†] was used. After conversion from geographic coordinates to British National Grid using a nearest-neighbour algorithm (see Mather (264) for a summary), the spatial resolution of the DEM (Figure 1.3) was 661.1 metres. The data are used in Chapter 3 and to inform analyses in Chapters 5, 6, and 7.

1.8.3 Positions of minerals in a slab of granite

A regular grid with a 2 mm spacing was placed over a granite slab and the presence of quartz, feldspars, and hornblende was recorded at each node of the grid. The dominant mineral in each grid cell was recorded. Feldspar and mafic minerals such as hornblende occur in clusters and give an indication of magma chamber crystal settling. Therefore, the degree of clustering or dispersion aids interpretation of a rock section. The slab is illustrated in Figure 1.4. The dataset comprised 1326 point locations (37 columns by 36 rows, but there were no observations at 6 locations since none of the three selected minerals were present at those 6 locations). The data cover an arbitrarily selected part of the surface of a granite slab and, therefore, the study region is an arbitrary rectangle. The locations identified as comprising mafic minerals are shown in Figure 1.5. The data are used in Chapter 8 to illustrate methods for the analysis of point patterns[‡].

1.8.4 Landcover in Turkey

The dataset is a Landsat Thematic Mapper (TM) image for 3rd September 1999. It covers an area in the south-eastern coastal region of Turkey called Cukurova Deltas. The area has three deltas that are formed by the rivers Seyhan, Ceyhan, and Berdan. The study area lies in the centre of this region, and covers an area of approximately 19.5 km by 15 km (29,250 hectares). The image was geometrically corrected and geocoded to the Universal Transverse Mercator (UTM) coordinate system using 1:25,000 scale topographic maps. The image was then spatially resampled to a spatial resolution of 25 m using a nearest-neighbour algorithm. The dataset is described in more detail in Berberoglu et al. (39). In that paper, the focus was on classification

---

[†]edcdaac.usgs.gov/gtopo30/gtopo30.html
[‡]The data were provided by Dr. Alastair Ruffell of Queen's University, Belfast.

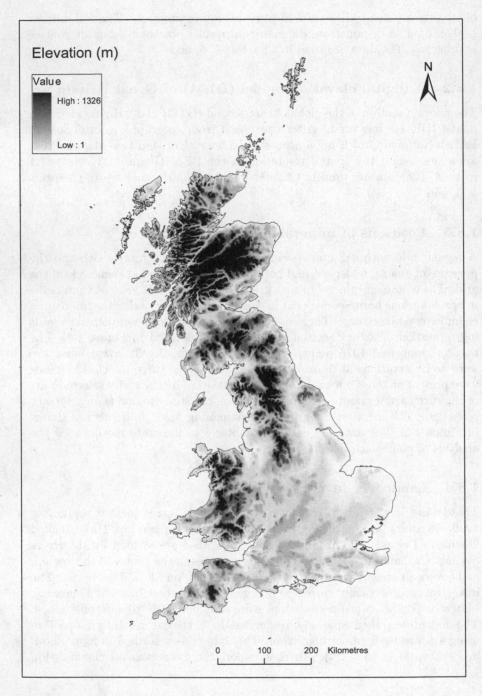

FIGURE 1.3: DEM of Great Britain; spatial resolution of 661.1 m.

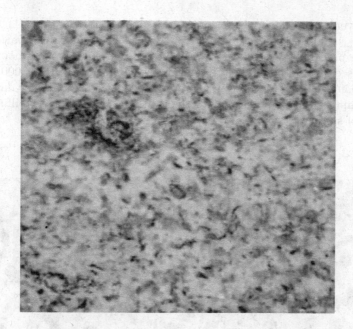

FIGURE 1.4: Slab of granite, width = 76mm.

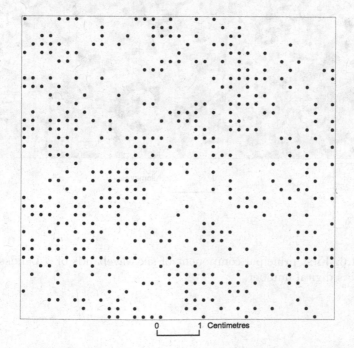

FIGURE 1.5: Mafic mineral locations in a slab of granite.

of land covers. In the book, the first principal component (PC1) of six wavebands (bands 1–5 and 7; Figure 1.6) is used to illustrate a variety of local image processing procedures. A related dataset is vector field boundary data digitised from Government Irrigation Department (DSI) 1:5000 scale maps. The image of PC1 is used in Chapter 3. The vector boundary data are used in Chapter 4, in conjunction with the values from the image, to illustrate measures of spatial autocorrelation[§].

FIGURE 1.6: First principal component of six wavebands of a Landsat TM image. DN is digital number.

[§]Dr. Suha Berberoglu, of the University of Cukurova, provided access to the processed data.

1.8.5 Digital orthophoto of Newark, New Jersey

A digital orthophoto quadrangle (DOQ) [¶] of part of Newark, New Jersey, was acquired and a subset was extracted for the illustration of the discrete wavelet transform in Chapter 3. The image subset is shown in Figure 1.7.

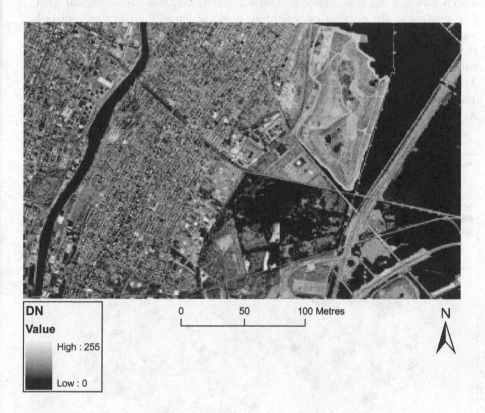

FIGURE 1.7: Digital orthophoto of Newark, New Jersey. Image courtesy of the US Geological Survey.

1.8.6 Population of Northern Ireland in 2001

The datasets detailed in this section represent the human population of an area. The data are population counts made as a part of the 2001 Census of Population of Northern Ireland. Of the two datasets used in the book, the first had counts within zones called Output Areas. There are 5022

[¶]See http://online.wr.usgs.gov/ngpo/doq/

Output Areas, with populations ranging between 109 and 2582 and a mean average population of 336. Population counts are shown in Figure 1.8. Population densities, as shown in Figure 1.9, are a sensible way of visualising such data and urban areas like Belfast obviously become more visible using such an approach. The data are used in Chapter 6 to illustrate areal interpolation from zone (that is, Output Area) centroids to a regular grid. The centroids are population-weighted, and the large majority of centroids were positioned using household counts (using COMPASS [COMputerised Point Address Service], a database of spatially-referenced postal addresses) with some manual adjustments where centroids fell outside of their Output Area because the zone was unusually shaped (for example, a crescent shape). These data are available through the Northern Ireland Statistics and Research Agency[||].

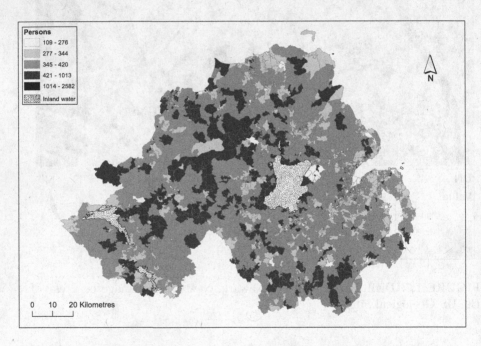

FIGURE 1.8: Population of Northern Ireland in 2001 by Output Areas. Northern Ireland Census of Population data — © Crown Copyright. Reproduced under the terms of the Click-Use Licence.

The second population dataset utilised in this book relates to the percentages of people in Northern Ireland who, in 2001, were Catholic or

[||]www.nisra.gov.uk

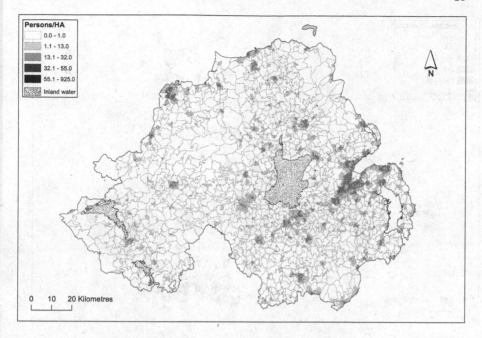

FIGURE 1.9: Population of Northern Ireland per hectare in 2001 by Output Areas. Northern Ireland Census of Population data — © Crown Copyright. Reproduced under the terms of the Click-Use Licence.

non-Catholic (i.e., mostly Protestant) by community background ('religion or religion brought up in'). The percentages were converted to log-ratios given by $1\sqrt{2}\times\ln(\text{CathCB/NonCathCB})$, where CathCB is the percentage of Catholics by community background and NonCathCB is the percentage of non-Catholics. The mapped values are shown in Figure 1.10. The rationale behind this approach is detailed by Lloyd (242). The data are for 1km grid squares, one of several sets of zones for which counts were provided from the Census (see Shuttleworth and Lloyd (331) for more details).

1.8.7 Other datasets

Two other datasets are used to illustrate particular methods. Both of these datasets are well-known and have been described in detail elsewhere. The first dataset represents residential and vehicle thefts, mean household income and mean housing value in Columbus, Ohio, in 1980. The data are presented by Anselin (9), and the zones used to report counts are shown in Figure 5.4 (page 112). The second dataset comprises births and cases of sudden infant death syndrome (SIDS) in North Carolina. The data were used by Symons et al. (355) and Bivand (45) provides a description. These data are used in Section 8.13 to demonstrate a method for detection of clusters of events.

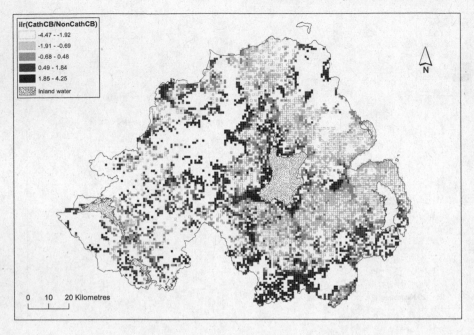

FIGURE 1.10: Isometric log-ratio (ilr) for Catholics/Non-Catholics in Northern Ireland in 2001 by 1km grid squares. Northern Ireland Census of Population data — © Crown Copyright.

1.9   A note on notation

Some symbols are used in the text to mean different things, but consistency has been the aim between chapters where possible where this does not conflict with well-known use of symbols in particular situations. Examples of symbols defined differently include the use of h to mean a scaling filter in the context of wavelets (Chapter 3), and as a separation distance in the case of geostatistics (Chapter 7). In Chapter 5 spatial coordinates are given with u, v as x is used to denote an independent variable. Elsewhere the more conventional x, y is used. Two forms of notation are use to represent location i. These are the subscript i (e.g., z_i) and the vector notation $\mathbf{s}_i$ (e.g., $z(\mathbf{s}_i)$). The selection of one or the other is based on clarity and convenience. It is hoped that readers will find the meaning to be clear in each case following definitions given in individual chapters, although in most cases notation is consistent between chapters.

1.10   Overview

At a very broad level the book discusses methods that can be used to analyse data in two key ways. That is, when the concern is with the analysis of (i) spatial variation in the properties of observations of one or more variables, and (ii) spatial variation in the configuration of observations (for example, are observations more clustered in some areas than in others?). Most of the book focuses on (i) but (ii), in the form of point pattern analysis, is discussed in Chapter 8.

Chapter 2 discusses some ways of adapting to local variation. In Chapter 3, the focus is on local models for analysing spatial variation in single variables on grids, Chapter 4 is concerned with analysis of single variables represented in a variety of ways, while in Chapter 5 the concern is with local models that can be used to explore spatial variation in multivariate relations between variables. Chapter 6 outlines some methods for the prediction of the values of properties at unsampled locations; techniques which enable transfer of values between different zonal systems and from zones to points are also discussed. Chapter 7 illustrates geostatistical methods for analysing spatial structure and for spatial prediction. Chapter 8 is concerned with the analysis of spatial point patterns. Chapter 9 summarises the main issues raised in the previous chapters and brings together some key issues explored in this book.

2

Local Modelling

In this chapter, the basic principles of locally-adaptive methods are elucidated. There is a wide variety of methods which adapt in different ways to spatial variation in the property or properties of interest, and the latter part of this chapter looks at some key types of locally-adaptive methods. These include stratification or segmentation of data, moving window/kernel methods, as well as various data transforms. Finally, some ways of categorising local models are outlined, and links are made to the models and methods described in each of the chapters.

2.1   Standard methods and local variations

This section briefly outlines some ways in which standard approaches can allow assessment of local variations. The most obvious way of identifying local variations is simply to map values or events and such an approach often provides the first hint that a local model might be of value in a given case. Residuals from a global regression analysis provide an indication either that additional variables might be necessary or that the relations between variables vary spatially (see Section 5.1). Tools such as the Moran scatterplot (see Section 4.3) and the variogram cloud (Section 7.4) allow assessment of local departures in terms of spatial dependence between observations. The remainder of this chapter focuses on ways of exploring or coping with local variations.

2.2   Approaches to local adaptation

Some definitions of global models state that change in any one observation affects all results (e.g., a global polynomial trend model is dependent on all data values), whereas with local models change in one observation only affects results locally. In this book, a broader definition is accepted and a range of methods that either adapt to local spatial variation, or which can be used to transform data, such that the transformed data have similar characteristics (e.g., mean or variance) at all locations, are discussed. In other words, the concern is with nonstationary models and with methods that can be used to transform, or otherwise modify, data so that a stationary model can be applied to the transformed data.

A widely-used approach to accounting for spatial variation is a geographical weighting scheme. A distance matrix can be used to assign geographical weights in any standard operation:

$$\mathbf{W}(\mathbf{s}_i) = \begin{bmatrix} w_{i1} & 0 & \cdots & 0 \\ 0 & w_{i2} & \cdots & 0 \\ \vdots & \vdots & \cdots & \vdots \\ 0 & 0 & \cdots & w_{in} \end{bmatrix}$$

where $\mathbf{s}_i$ is the ith location $\mathbf{s}$, with coordinates x, y, and w_{in} is the weight with respect to locations i and n.

There are various other options in addition to weighting schemes based on distances. Getis and Aldstadt (142) list some common geographical weighting schemes:

1. Spatially contiguous neighbours.

2. Inverse distances raised to some power.

3. Lengths of shared borders divided by the perimeter.

4. Bandwidth (see Section 2.4) as the nth nearest neighbour distance.

5. Ranked distances.

6. Constrained weights for an observation equal to some constant.

7. All centroids within distance d.

8. n nearest neighbours.

and the authors list more recently-developed schemes including bandwidth distance decay, as discussed in Section 2.4. This chapter (and the rest of the book) expands on most of these schemes.

A widely used example of distance weighting is provided by the inverse distance weighting (IDW) interpolation algorithm. In that case, the objective is to use measurements $z(\mathbf{s}_i)$, $i = 1, 2, ..., n$, made at point locations, to make a prediction of the value of the sampled property at a location, $\mathbf{s}_0$, where no observation is available. The weights assigned to samples are a function of the distance of the sample from the prediction location. The weights are usually obtained by taking the inverse squared distance (signified by the exponent -2), and the prediction can be given by:

$$\hat{z}(\mathbf{s}_0) = \frac{\sum_{i=1}^{n} z(\mathbf{s}_i) d_{i0}^{-2}}{\sum_{i=1}^{n} d_{i0}^{-2}} \tag{2.1}$$

where d_{i0} is the distance by which the location $\mathbf{s}_0$ and the location $\mathbf{s}_i$ are separated. Changing the value of the exponent alters the influence of observations at a given distance from the prediction location. Inverse distance squared weighting is depicted in Figure 2.1, and an example of distance and inverse distance squared weights, w, for prediction location $\mathbf{s}_0$ are given in Figure 2.2. This method is discussed in more detail in Section 6.3.6; spatial interpolation is the concern of Chapters 6 and 7. O'Sullivan and Unwin (304), Dubin (112), and Lloyd (245) provide further discussions about weight matrices. Note that, as stated in Section 1.9, spatial locations are indicated in two different ways in this book. The notation of the form $z(\mathbf{s}_i)$, as outlined above, is one and the other uses the form z_i (and w_{ij} can be used to indicate the weight given locations i and j). The two forms of notation are used following common convention. The first form is used primarily in the case of interpolation and for geographically weighted regression (see Section 5.8), while the second is used in a variety of other contexts.

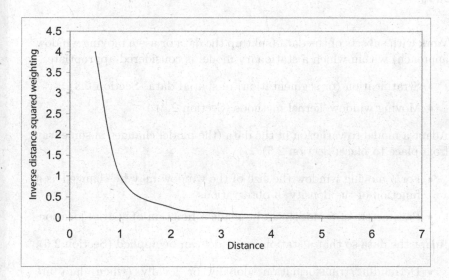

FIGURE 2.1: Inverse squared distance weights against distance.

The remainder of this section outlines some general approaches for local modelling. Local models can be categorised in various ways. One possible division is into three broad approaches that can be used to deal with data that have locally-varying properties. Some examples are given for each type of approach:

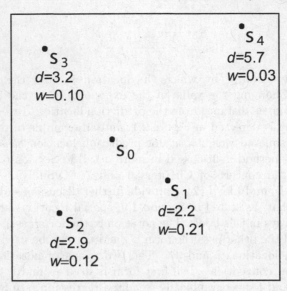

FIGURE 2.2: Distance, d, and inverse squared distance weights, w, for location s_0.

1. Work with subsets of the data (split up the data or use a moving window approach) within which a stationary model is considered appropriate.

   - Stratification (or segmentation) of spatial data (Section 2.3)
   - Moving window/kernel methods (Section 2.4)

2. Adapt a model to variation in the data (the model changes in some way from place to place; Section 2.5).

   - For a moving window the size of the window may be changed as a function of the density of observations
   - Parameters of a model can be made a function of spatial location

3. Adapt the data so that a stationary model can be applied (Section 2.6).

   - Detrending/transformation globally or locally (when data are detrended globally the intention is usually to remove global trends and enable a focus on local variation)
   - Spatial deformation (deform geographical space such that the deformed space has similar properties at all locations)

In practice, these approaches are often used in combination. For example, data within a moving window (type 1 approach) of locally varying size as a function of data density (type 2) may be detrended (type 3).

Another distinction is between the application of (i) the same model locally, and (ii) a different model locally. For example, in the case of (i) a kernel (see Section 2.4) of fixed form and size might be used at all locations, while for (ii) the form and size of the kernel may be varied from place to place. Throughout the following chapters the type of model which a given approach represents will be made clear.

In the following sections some more details about each of the main methods are provided, starting with the most conceptually simple approaches.

2.3 Stratification or segmentation of spatial data

If there is evidence that a dataset may be divided into two or more distinct populations then it may be possible to divide the dataset and treat each subset as a separate population. The data may be divided using existing data, or by using some classification or segmentation algorithm. Figure 2.3 shows a raster grid on which is superimposed the boundaries of segments generated using region-growing segmentation (see Section 3.7). Note that there are some very small segments in this example and, in practice, some minimum segment size constraint is likely to be applied. In an application making use of existing data, Berberoglu et al. (39) classified Mediterranean land cover on a per-field basis using imagery that was subdivided using vector field boundary data.

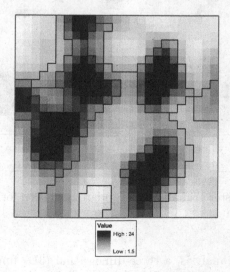

FIGURE 2.3: Raster grid with segment boundaries superimposed.

2.4   Moving window/kernel methods

The most widely used approach to local adaptation in spatial analysis is the
moving window. A moving window may be used to estimate statistics based
on an equal weighting of data within the moving window. Alternatively, a
geographical weighting scheme may be used (as illustrated in Section 2.2 and
below) whereby observations are weighted according to their distance from the
location of interest (e.g., the centre of the moving window). Moving windows
are used very widely in image processing. Such operations are known as focal
operators. With a focal operator, the output values are a function of the
neighbouring cells. In other words, if we use a three-by-three pixel mean
filter, a form of focal operator, then the mean of the cells in the windows is
calculated and the mean value is written to the location in the output grid
that corresponds to the location of the central cell in the window. Figure
2.4 shows a raster grid and the standard deviation for a three-by-three pixel
moving window. Clearly, the local standard deviation highlights areas of
contrasting values. Focal operators are discussed further in Section 3.5.

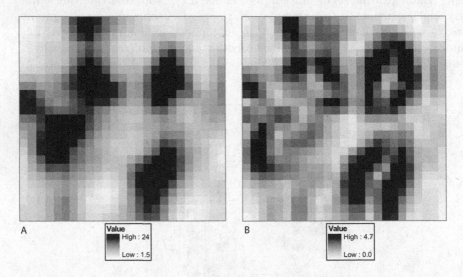

FIGURE 2.4: (A) Raster grid and (B) standard deviation for a three-by-three
pixel moving window.

With kernel based analysis, a three-dimensional (3D) function (kernel) is
moved across the study area and the kernel weights observations within its
sphere of influence. A nonuniform kernel function can be selected so that the

weight assigned to observations is a function of the distance from the centre of the kernel. A kernel function is depicted in Figure 2.5. The kernel function depicted (the quartic kernel) is often used in kernel estimation as discussed in Section 8.10.2; the size of the kernel is determined by its bandwidth, τ.

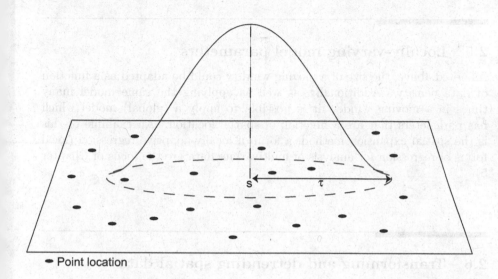

● Point location

FIGURE 2.5: The quartic kernel, with bandwidth τ, centred on location **s**.

As discussed elsewhere, selecting an appropriate window size is problematic. Different window sizes will capture spatial variation at different scales. It is possible to select a window size using some kind of optimisation procedure, but such an approach does not necessarily enable identification of the most meaningful scale of analysis. Another approach is to use a locally-adaptive window — that is, the size of the window varies from place to place according to some specific criterion such as, for irregularly (spatially) distributed data, observation density.

Kernels, whether of constant size or locally-adaptive, are usually isotropic — they do not vary with direction. However, various strategies exist for adapting the form of the kernel in different directions. A simple approach is to use an anisotropy ratio whereby the kernel is stretched or compressed with respect to a particular direction. Similar approaches are described in the book.

Hengl (181), in a discussion about spatial interpolation, makes a distinction between *localised* and *local* models. Localised models are defined as those which make use of local subsets of the data for prediction largely to increase the speed of the prediction process, while with local models the assumption is that the properties of the variable change from place to place and model parameters are estimated locally. In this book, both model types are considered relevant.

2.5   Locally-varying model parameters

As noted above, the size of a moving window could be adapted as a function of data density. Additionally, as well as applying the same model many times in a moving window it is possible to apply a (global) model which has parameters that are a function of spatial location. An example of this is the spatial expansion method, a form of locally-adaptive regression (local forms of regression for analysis of multivariate data are the focus of Chapter 5).

2.6   Transforming and detrending spatial data

A standard way of removing large scale trends in a dataset is to fit a global polynomial and work with the residuals from the trend. For example, rainfall in Britain tends to be more intense in the north and west of the country than elsewhere. Therefore, the mean precipitation amount varies across Britain. If we fit a polynomial trend model to, say, monthly precipitation values in Great Britain and subtract this from the data, then the resulting residuals will, if the polynomial fits the data well, have a more similar mean at all locations than did the original data. So, we can perhaps apply to the detrended data a model which does not have locally varying parameters (at least as far as the mean average of the property is concerned). If there are local trends in the data then the data can be detrended in a moving window. Detrending is discussed in Section 6.3.4. The subject of deformation of geographical space, such that a global model can be applied to the data in the deformed space, is discussed in Section 7.17.

2.7   Categorising local statistical models

The various forms of local models and methods outlined in this book are connected in different ways, and they can be grouped using particular characteristics or properties (for example, as in Section 2.2). Unwin (368) divides local statistics into three groups comprising those statistics based on (i) non-overlapping sets of calculations, (ii) calculations with respect to an observation and its neighbours, and (iii) local values standardised by global values derived from the entire dataset. A recent important contribution to the characterisation and development of local statistical methods is the paper by Boots and Okabe (54). The authors present local spatial statistical analysis (LoSSA) as an integrative structure and a framework that facilitates the development of new local and global statistics. In their paper, Boots and Okabe (54) aim to formalise what is involved in the implementation of a local statistic and to explicitly consider the main components and limitations of the procedures employed. In particular, the paper focuses on: (i) the nature of spatial subsets, (ii) their relationship to the complete dataset, and (iii) the relationship between a global statistic and the corresponding local statistic, and each of these is summarised below.

2.7.1 The nature of spatial subsets

Different forms of local spaces exist and Boots and Okabe (54) define focused local spaces and unfocused local spaces:

- Focused local space: determined with respect to a given location, an example being a circle drawn around the location.

- Unfocused local space: determined with respect to attributes, rather than locations. An example is a subset of zones with a population density greater than some particular threshold.

A local space may be connected (e.g., sharing boundaries with another local space) or disjointed.

The n data values at sites s_i can be given by $z_{s_i}, i = 1, ..., n$. Given a local space $S_{\mathrm{L}j}$ with data values of sites in the local space $z_{\mathrm{L}j1}, ..., z_{\mathrm{L}jn_{\mathrm{L}j}}$, Boots and Okabe (54) term the global and local statistic *similar* in the case where the former is given by $F(n; z_{s_1}, ..., z_{s_n})$ and the latter by $F(n_{\mathrm{L}j}; z_{\mathrm{L}j_1}, ..., z_{\mathrm{L}jn_{\mathrm{L}j}})$. In such cases, the application of a common statistical test may be possible.

An important distinction made by Boots and Okabe (54), and other authors, is between global statistics that are decomposable into local statistics (see the local indicators of spatial association defined in Section 4.4 as an example) and those that are not. Furthermore, in the case of a global statistic which is a linear combination of the local statistics, the authors use the term linearly decomposable. Boots and Okabe (54) demonstrate their framework by developing a local variant of the cross K function (see Sections 8.8 and 8.12 for relevant material).

2.7.2   The relationship of spatial subsets to the complete dataset

In terms of relationships between global space and local spaces, Boots and Okabe (54) define four conditions:

- Local spaces exhaustively covering global space.

- Local spaces not exhaustively covering global space.

- Overlap between local spaces.

- No overlap between local spaces.

These conditions can be structured and defined as follows:

1. Exhaustive:

   - Overlapped — covering
   - Non-overlapped — tessellation (zones filling space with, as indicated, no gaps and no overlaps)

2. Non-exhaustive

   - Overlapped — clumps
   - Non-overlapped — incomplete tessellation (e.g., only zones with more than a particular population amount)
   - Non-overlapped — islands

Figure 2.6 shows local spaces defined with respect to a set of point locations (variable 1). These include local spaces exhaustively covering global space (local space 1) and local spaces which do not exhaustively cover global space (local space 2). The latter local spaces overlap (in some cases), while the former do not. In the example, variable 1 could represent the location of a school and variable 2 the location of pupils.

As Boots and Okabe (54) note, the relationship between global and local spaces is an important one and has implications for significance testing. Where there are overlaps between local spaces it will be necessary to adjust significance levels, and this topic is discussed in various contexts in the book.

2.7.3   The relationship between a global statistic and the corresponding local statistic

Boots and Okabe (54) consider different ways of evaluating the local statistic relative to the global statistic. It is possible to evaluate the global and local statistics or just the local statistics. The possibilities can be outlined as shown in Figure 2.7.

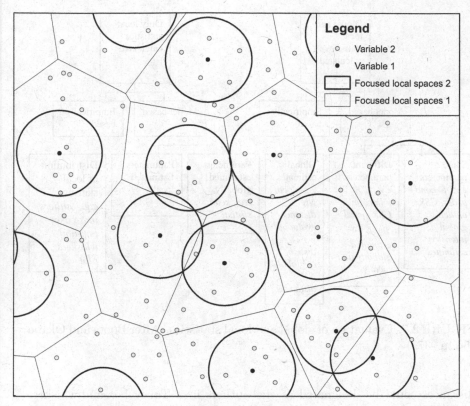

Legend

○ Variable 2

● Variable 1

▢ Focused local spaces 2

▢ Focused local spaces 1

FIGURE 2.6: Local spaces.

A model-based case of computation of both the global and local statistics using the same parameters is a second-order point pattern analysis whereby the global estimate of the intensity is used as a null model of complete spatial randomness (CSR; see Section 8.4) (54). Boots and Okabe (54) note that, in the case of black/white cell join counts in a binary map assessed against a CSR model, separate global and local estimates of the probability, p_b, of observing a black cell may be made (lower line, second box from left). Section 4.3 summarises the joins count approach, and Section 4.5 provides relevant discussion in a local context. Cases where only the local statistics are calculated include scan statistics (outlined in Section 8.13) and the geographical analysis machine (GAM; again, see Section 8.13). Geographically weighted regression (Section 5.8) provides another example of a case where local statistics only are computed.

Boots and Okabe (54) give the empirically-based example of computing the global mean and variance of nearest-neighbour distances and using these to evaluate the probability of occurrence of a local nearest-neighbour distance (lower line, third box from left), or the significant local nearest-neighbour

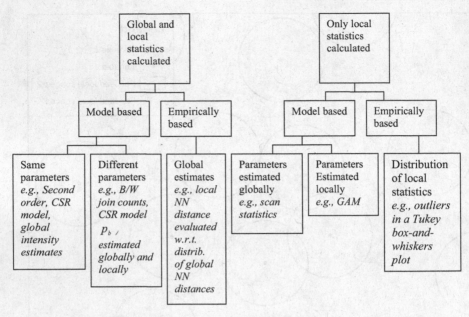

FIGURE 2.7: Evaluation of global and local statistics. After Boots and Okabe (54, p. 372).

distances could be designated as the outliers in a Tukey box-and-whiskers plot (365) of distances (lower line, first box from right).

   The contribution of Boots and Okabe (54) is important in providing a means to formally assess the components and implementation of local statistics. LoSSA also facilitates the development of new procedures.

2.8   Local models and methods and the structure of the book

To provide a link between this chapter and the remainder of the book, the main classes of model or method utilised in each chapter are detailed:

- Chapter 3 Grid data

  - Moving windows, segmentation (moving window based), geographical weights

- Chapter 4 Spatial patterning in single variables

– Moving windows, geographical weights, global summary depicting outliers (Moran scatterplot)

- Chapter 5: Spatial relations

  – Moving windows, geographical weights, spatially varying processes (Bayesian spatially varying regression model), model as a function of location (spatial expansion model)

- Chapter 6: Spatial prediction 1: deterministic methods, curve fitting and smoothing

  – Moving windows, geographical weights, transformation (global and local detrending), use of local data (areal interpolation)

- Chapter 7: Spatial prediction 2: geostatistics

  – Moving windows, geographical weights, global summary depicting outliers (variogram cloud), transformations (in terms of mean and variogram)

- Chapter 8: Point patterns and cluster detection

  – Moving windows, geographical weights, quadrat-based analyses

Moving windows and geographical weighting schemes appear throughout the book while other approaches appear in specific contexts. The above list is not exhaustive, but it serves to highlight the range of approaches which are represented.

2.9 Overview

To recap, the following chapters outline a range of approaches that fall within the broad headings given in this chapter. The book is structured according to either data type or the problem (e.g., analysing spatial relations or spatial prediction), rather than the approach used for adapting to spatial variation; each of the chapters includes several broad approaches to local adaptation, as summarised above. The next chapter deals with the analysis of gridded data, and the main approach employed in this context, as noted above, is the moving window.

3

Grid Data

In this chapter, exploration of spatial variation in single variables available on grids is the focus. Methods outlined would be appropriate in, for example, analysis of single properties such as elevation or digital number in a remotely-sensed image. While univariate methods are detailed here, it should be noted that there are multivariate versions of some of the methods described in this chapter. The methods described in Chapter 4 can also be applied to gridded data, as can those detailed in Chapter 5. The present chapter focuses on methods specifically designed for, or adapted for use with, gridded data. Some of the approaches outlined have a relatively long history, and the potential scope of this chapter is vast. Inevitably, the focus is on a restricted range of approaches that are used to highlight key concepts.

3.1 Exploring spatial variation in gridded variables

Methods which are discussed widely in the GISystems literature (for example, standard moving window statistics) are discussed only quite briefly in this chapter. Other methods that are covered in little detail in the GISystems literature, but widely in other literature (for example, wavelets) have more space devoted to them. Section 3.2 provides a summary of some global statistics. In Section 3.3, the purpose of local univariate statistics for gridded data is outlined. Section 3.4 provides some context for the following sections, Section 3.5 is concerned with moving window operations, Section 3.6 outlines wavelets, and the subject of Section 3.7 is segmentation. Various derivatives of altitude are described in Section 3.8 and such approaches are used widely in GISystems contexts. In each section in the book, where appropriate, standard global methods are described first and then local methods are introduced, which enable assessment of variation across the region of interest. However, in this chapter, aside from a short section below, the focus is exclusively on local models. Section 3.9 summarises the chapter.

3.2   Global univariate statistics

Summaries of single variables are often global, for example the arithmetic mean or histogram of a dataset may be estimated. Such measures may provide useful summaries, but they obscure any spatial variation in the data. Point operations (75) are global, in that the output pixel depends only on the value of the corresponding input pixel. Point operations include contrast enhancement, contrast stretching, and grey-scale transformations (75). In practice, local features are often of interest — especially in terms of the neighbours of a pixel, and various related approaches are described below.

3.3   Local univariate statistics

Local (for example, moving window) statistics provide a means of exploring spatial variation in the property of interest; the moving average is an obvious case. Kernel methods (see Section 2.4) provide a means of accounting for spatial structure in the estimation of statistics locally. This chapter describes a variety of approaches that are based on moving windows. The particular focus is on methods for local analysis of single variables represented by grids. Applications that will be discussed include data smoothing, analysis of spatial structure, and analysis of topographic surfaces.

3.4   Analysis of grid data

Many standard image processing procedures are local, by definition. In this chapter, the principles of local grid-based analysis are detailed. In addition, approaches to segmenting grids (such that the segments are, in some sense, homogeneous) and wavelet analysis are discussed. The following section is intended to review a range of approaches that are important in GISystems contexts; it does not provide an introduction to the principles of image processing and the focus is intentionally biased. For a detailed review of image processing and analysis, readers can consult a range of texts (for example, (311), (338)). In the context of remote sensing, the book by Mather (264) provides a good introduction to the handling of images.

3.5   Moving windows for grid analysis

The moving window is applied for a range of local image (grid) analysis procedures. Local processing of image data includes smoothing and gradient operators (these are termed convolution filters) and both are outlined below. As discussed in Chapter 2, the choice of window size will affect any analysis. For grids, the immediate neighbours of a pixel are often used to derive some measure or function locally. That is, a three-by-three pixel moving window is used. The mean and variance are commonly computed in this way. In many applications, all eight neighbours are used (using the analogy with chess, this is sometimes referred to as queen's case) but in other applications only four neighbours (excluding the diagonals; referred to as rook's case) may be used. Queen's and rook's case for a grid are illustrated in Figure 3.1. Some operators may utilise orientation of neighbours as well as their values. An example of this is edge detection, discussed in Section 3.5.3. One position of a three-by-three pixel moving window is shown in Figure 3.2 superimposed on a four-by-four pixel image. The central pixel is shaded, and the window is shown in heavy black lines.

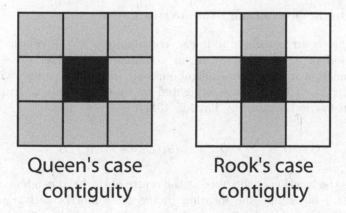

Queen's case
contiguity   Rook's case
contiguity

FIGURE 3.1: Queen's case and rook's case contiguity for a grid: central cell is black and neighbouring cells used in calculations are in light grey.

A spatial filter is a form of focal operator, whereby the results for a given cell are a function of the value of the neighbouring cells (broader discussions about spatial filters are provided by Chou (81), Mather (264), Pavlidis (311), and Sonka et al. (338)). Spatial filters provide the basis of smoothing and gradient operators and edge detectors. A spatial filter works as follows:

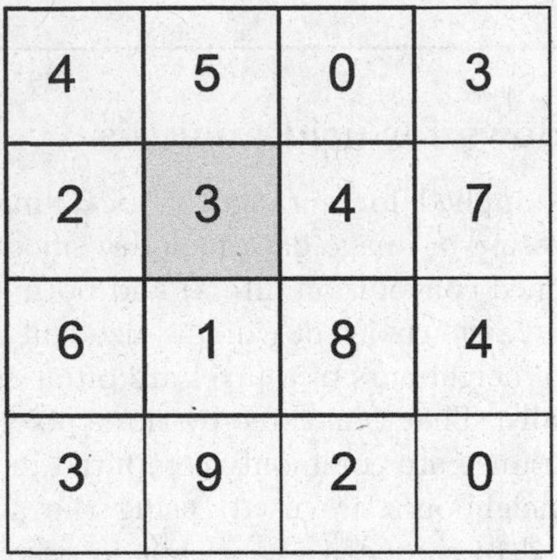

4	5	0	3
2	3	4	7
6	1	8	4
3	9	2	0

FIGURE 3.2: One position of a three-by-three pixel moving window on a four-by-four pixel grid.

1. Visit the first cell in the grid (e.g., the top left cell).

2. Calculate some measure, which is a function of the neighbouring cell values.

3. Move to the next cell and return to step 2.

One option is to calculate a linear combination, c, of values in the neighbourhood of a pixel, z. A linear combination of pixel values with weighting function w, in a local neighbourhood (usually a square window) of the pixel z in an input image, weighted by the coefficients w (usually a three-by-three array) is given by Pavlidis (311):

$$c(x,y) = \sum_{i=-m}^{m} \sum_{j=-n}^{n} w(x,y,i,j) z(x+i, y+j) \qquad (3.1)$$

where x, y is the location of the pixel at the centre of the local neighbourhood, the indices i, j determine the location in the filter matrix with a moving window of $2m + 1$ pixels by $2n + 1$ pixels (so, if the window is square, m and n are equal). Where the weights are equal at each location, the weight is given as $w(x, y)$ (that is, the index i, j is dropped). The positions of cells in a moving window are given in Figure 3.3.

Methods for local (pre-)processing of images can be divided into two groups: (i) smoothing operators, and (ii) gradient operators (338). Smoothing operators are used to reduce noise or outliers, while with gradient operators locations in the images where there are rapid changes are highlighted. These two classes of operators are outlined below.

Most spatial filters are employed with fixed windows. That is, the same model is applied everywhere, but it is applied locally and individual estimates

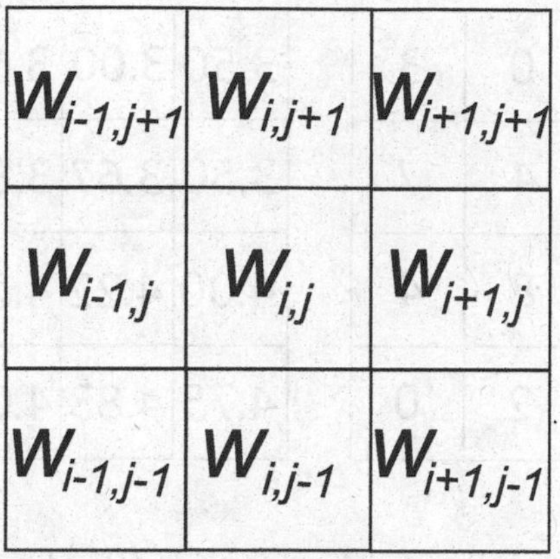

$W_{i-1,j+1}$	$W_{i,j+1}$	$W_{i+1,j+1}$
$W_{i-1,j}$	$W_{i,j}$	$W_{i+1,j}$
$W_{i-1,j-1}$	$W_{i,j-1}$	$W_{i+1,j-1}$

FIGURE 3.3: Position of cells in a three-by-three pixel moving window.

of, for example, the mean are obtained at each location. As Section 3.5.5 details, there are alternatives to the use of a standard fixed moving window.

3.5.1 Image smoothing: Low pass filtering

Image smoothing is used to reduce the effect of extreme values (for example, noise) in a grid (Section 3.6 also discusses approaches that can be used for smoothing). The mean filter is the most widely used smoothing filter. The convolution mask with equal weighting for all pixels in a three-by-three pixel neighbourhood is given as:

$$w = \frac{1}{9} \begin{pmatrix} 1 & 1 & 1 \\ 1 & 1 & 1 \\ 1 & 1 & 1 \end{pmatrix} \tag{3.2}$$

Put simply, using this weighting scheme, the central pixel in the output image is the mean of the pixels in this neighbourhood of the input image.

The example in Figure 3.4 illustrates the calculation of the mean for a three-by-three pixel moving window in which all pixels are weighted equally. In this example, mean values are calculated at the edges where there are less than nine neighbouring cells — in practice, the edges are often ignored such that all statistics are estimated from nine pixels (assuming a three-by-three pixel moving window) — so the output (convolved) image is often smaller than the input image (see Lloyd (245) for an example). In Figure 3.5, the mean in a three-by-three pixel moving window is shown for the landcover data described in Chapter 1.

4	5	0	3
2	3	4	7
6	1	8	4
3	9	2	0

A.

3.50	3.00	3.67	3.50
3.50	3.67	3.89	4.33
4.00	4.22	4.22	4.17
4.75	4.83	4.00	3.50

B.

FIGURE 3.4: (A) Numerical example of a grid. (B) Mean calculated over a three-by-three pixel moving window.

The median (middle of ranked values) and mode (most frequent value) are also sometimes computed in a moving window. The key advantage of the median is that it is not sensitive to outliers, unlike the mean. The mode can be derived from categorical data. There are various approaches that aim to smooth without blurring sharp edges. Using such an approach, the average is computed only from pixels in the neighbourhood that have properties similar to the pixel at the centre of the window. A summary of edge preserving methods is provided by Sonka et al. (338).

An alternative approach would, as discussed in Chapter 2, be to use a kernel which weights pixels according to their proximity to the pixel at the centre of the window. Some examples are given by:

$$w = \frac{1}{10} \begin{pmatrix} 1\ 1\ 1 \\ 1\ 2\ 1 \\ 1\ 1\ 1 \end{pmatrix} \tag{3.3}$$

$$w = \frac{1}{16} \begin{pmatrix} 1\ 2\ 1 \\ 2\ 4\ 2 \\ 1\ 2\ 1 \end{pmatrix} \tag{3.4}$$

For larger convolution masks the weights can be assigned according to the Gaussian distribution formula, for example. The mask coefficients are then assigned to have a unit sum. A Gaussian kernel is illustrated in Section 4.2, where a Gaussian weighting scheme is used to assign weights to irregularly distributed points.

3.5.2   High pass filters

High pass filters produce images that have sharper edges, but they also amplify noise (311). An image can be viewed as the sum of its low and high frequency

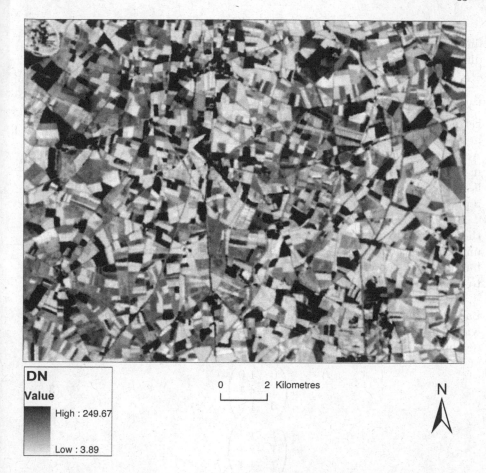

DN
Value

High : 249.67

Low : 3.89

0   2 Kilometres

N

FIGURE 3.5: Digital Number (DN): Mean calculated over a three-by-three pixel moving window.

parts. The high frequency component can be obtained by subtracting the low pass image from the original image (264). Thomas et al. (358) use the relation:

$$z' = z - f\overline{z} + c \qquad (3.5)$$

where z' is the filtered pixel value at the centre of a three-by-three matrix, z is the original pixel value, f is the fraction of the surrounding pixel values that is subtracted from the value of the central pixel, $\overline{z}$ is the average of pixel values in the neighbourhood, and c is a constant (added to ensure that z' is positive). Thomas et al. (358) used a value of 0.8 for f and a value of 20.0 for c.

Edge detection methods, like the approaches for high pass filtering outlined above, enhance areas within which neighbouring pixels have markedly different values. Edge detection is the focus of the following section.

3.5.3   Edge detectors

Edge detectors (or gradient operators) are used to identify significant changes in neighbouring pixels (that is, changes in the intensity function (338)). An edge is a vector variable — it has magnitude and direction. The edge magnitude represents the magnitude of the gradient, and the edge direction is rotated by 90° from the gradient direction (338). The gradient direction represents the direction with maximum growth of the function.

A common objective is to identify boundaries that represent cells where there is the maximum rate of change. Another objective may be to highlight uniform areas in images. In images containing discrete features, such as buildings in remotely-sensed images, edge detection algorithms may be used to identify edges of buildings and extract lines for conversion to vector arcs.

Where the concern is only with edge magnitude and not direction, a Laplacian differential operator may be used. The Laplacian operator, used widely in image processing, is given by the following weights for 4 and 8 pixel neighbourhoods respectively (338):

$$\begin{pmatrix} 0 & 1 & 0 \\ 1 & -4 & 1 \\ 0 & 1 & 0 \end{pmatrix} \tag{3.6}$$

$$\begin{pmatrix} 1 & 1 & 1 \\ 1 & -8 & 1 \\ 1 & 1 & 1 \end{pmatrix} \tag{3.7}$$

Figure 3.6 shows the output of the 8 neighbour Laplacian operator applied to the landcover data described in Section 1.8.4. Other edge detection algorithms, including the Roberts operator, the Prewitt operator and the Sobel operator, are discussed by Sonka et al. (338). Edge detection is the basis for some segmentation procedures, and segmentation is discussed in Section 3.7. A related procedure is edge sharpening, where the objective is to make edges steeper (that is, increasing gradients at edges (338)). The Laplacian operator is often used to measure gradient as a part of an image sharpening procedure, the other component of which is a coefficient which gives the strength of the sharpening.

3.5.4   Texture

In the context of image analysis, texture can be defined as referring to the structure or spatial arrangement of grey levels of pixels (75). Texture may be divided into tone (pixel intensity) and structure (the spatial relationship

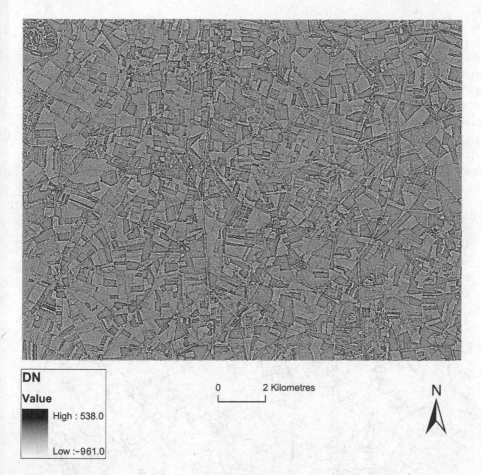

DN
Value

High : 538.0

Low :−961.0

0   2 Kilometres

N

FIGURE 3.6: The output of the 8 neighbour Laplacian operator applied to the landcover data.

between primitives) — the latter is most relevant here. In this context, measures of spatial autocorrelation (discussed in Section 4.3) are used in image processing to aid characterisation of spatial structure. Measures of variation between pixels are widely used in spatial analysis (for example, see LaGro (226)).

Image texture is calculated using a moving window, and there is a wide range of methods used to characterise texture (summaries are provided by Haralick et al. (174) and Sonka et al. (338)). Simple measures include the local standard deviation. Figure 3.7 illustrates the standard deviation calculated in a three-by-three pixel moving window for the Turkish landcover data. The local variance has been used to ascertain an appropriate spatial resolution of remotely-sensed imagery. With this approach, the local variance

is computed at a range of spatial resolutions and the average local variance
for each spatial resolution is plotted against the spatial resolution; the plot
usually rises to a peak (corresponding to the maximum variance for which the
spatial resolution may be selected) and declines for coarser spatial resolutions
(25). However, such an approach is problematic where a scene contains
a variety of landcovers. Lloyd et al. (250) outline an approach whereby
the map of local variances is retained, and the spatial resolution at which
the maximum local variance is reached is mapped. Also, a map of spatial
resolutions corresponding to the maximum local variance is obtained. In this
way, the limitations of identifying a single appropriate spatial resolution can
be assessed.

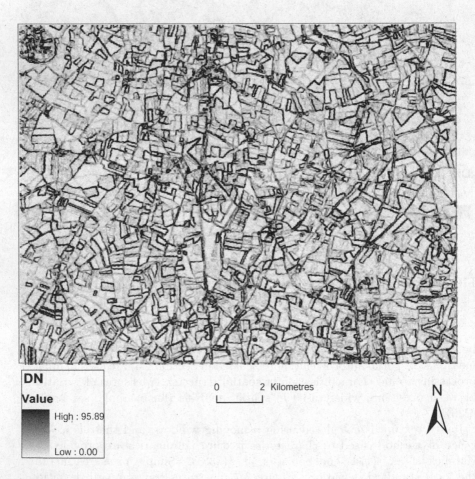

FIGURE 3.7: DN: Standard deviation calculated over a three-by-three pixel
moving window.

3.5.4.1 Texture in predefined areas

As well as moving window approaches, texture has been computed over areas defined using existing boundary data (that is, a per-field approach). In the analysis of remotely-sensed images per-field approaches are used widely. In some cases, prior information is available (for example, vectorised field boundaries as for the agricultural landcover image described in Chapter 1). Berberoglu et al. (39) and Lloyd et al. (251) present studies where Landsat Thematic Mapper (TM) imagery of agricultural landcover is classified using maximum likelihood and artificial neural networks. Inputs to the classifiers included a range of texture measures. Lloyd et al. (251) used per-field statistics derived from the grey level cooccurrence matrix and various geostatistical structure functions (see Chapter 7). A review of methods for linking spectral and textural information from remotely-sensed images is provided by Berberoglu and Curran (38). An alternative to the use of existing boundary data is to use some form of segmentation procedure to identify areas with similar properties (see Section 3.7).

3.5.5 Other approaches

In addition to simple fixed size windows with isotropic weights, weights can be modified as a function of direction, or window size could be made adaptive (for example, small when there is a large amount of variation, and large when variation is minimal). The weights assigned to the pixels could be adapted at each location as a function of the mean and the variance of the grey levels (pixel values) covered by the window (264). Other approaches include rotating windows (338). Such approaches are different to the those discussed in this chapter so far, in that local variation in model parameters may be allowed. In contrast, the main focus in this chapter has been on methods where the same model is used at all locations (for example, computation of the mean in a three-by-three pixel moving window).

There is a wide range of discrete image transforms available; the filters described above comprise some transformation approaches. Other transforms include the Fourier transform, which is used to decompose an image into features that correspond to different spatial frequencies. Introductions to the topic of discrete image transforms include Castleman (75) and Sonka et al. (338). The following discussion introduces the wavelet transform which has been the focus of much interest in recent years in image processing and, increasingly, in GISystems and remote sensing. The Fourier transform is also introduced to provide context for the discussion of wavelets.

3.6 Wavelets

Wavelets are a major area of research in the analysis of gridded data (for example, they are applied for the analysis of remotely-sensed imagery, as discussed below); they provide a means of examining local variations in frequency. It is for these reasons that they are discussed here. This section provides only a brief overview of the basic ideas behind wavelet analysis. There is a wide variety of texts that could be consulted for more details and proofs. A review of linear discrete image transforms, including wavelets, is provided by Sonka et al. (338) and other introductions to wavelets for image analysis are given by Castleman (75) and Starck et al. (343). An introduction to wavelets for the non-mathematician is provided by Hubbard (190) and introductions are provided by Abramovich et al. (1), Graps (160), Strang (348), and Walker (374). The book by Addison (3) includes a very clear introduction to the concept of the wavelet transform. Key texts about wavelets include the books by Daubechies (103) and Mallat (257). In addition, there are some good introductions to wavelets on the internet, including the text by Clemens Valens (369).

Wavelets are not yet widely used in GISystems contexts, although there is a range of applications of wavelets for the analysis of remotely-sensed imagery (for example, (57), (295)) and simplification of digital terrain models (49). The increasing use of wavelets in such fields makes a discussion here pertinent.

3.6.1 Fourier transforms and wavelet transforms

A wavelet is simply a localised waveform. Wavelets are mathematical functions used to split data into different frequency components. Each component is then analysed with a resolution matched to its scale (160). In essence, the wavelet transform (WT) is used to transform a signal (or image) into another form that facilitates understanding of spatial (or temporal) variation in that signal. The subject of wavelets is best introduced with reference to Fourier analysis. The Fourier transform (FT) provides a way of decomposing a function into sinusoids of different frequency that sum to the original function.

The Fourier transform translates a function f which depends on space to a function $\hat{f}$ which depends on frequency. The transform enables the identification of the different frequency sinusoids and their amplitudes. The Fourier coefficients represent the contribution of each sine and cosine for a given frequency. An inverse Fourier transform can then be used to transform back from the frequency domain to the time (or space) domain. With Fourier analysis, the basis functions are sines and cosines, and a function is constructed using combinations of sines and cosines of different amplitudes and frequencies. In contrast, there is not a single set of wavelet basis functions.

An account of the FT and wavelets is provided by Kumar and Foufoula-Georgiou (223).

Where sampling at discrete locations on a function, the discrete Fourier transform (DFT) can be used. The computation of the FT can be speeded up through the use of a fast FT (FFT). With a conventional DFT, large values of n lead to heavy computational loads, as multiplying an $n \times n$ matrix by a vector results in n^2 arithmetic operations. Where the discrete samples are spaced evenly, the Fourier matrix can be factored into a product of a few sparse matrices. The resulting factors can then be applied to a vector with a total of $n \log_2 n$ arithmetic operations. The inverse transform matrix is the transpose of the original matrix.

Press et al. (319, p.498) note the large difference between n^2 and $n \log_2 n$ arithmetic operations, and they present algorithms for the FFT. The algorithm of Cooley and Tukey (92) is widely used for FFT. A description of the FFT is provided by Castleman (75, pp. 177–178), and a more extensive account is the book by Brigham (59). The excellent paper by Graps (160) introduces the Fourier transform (and FFT) as background for developing an account of wavelets.

Fourier functions are localised in frequency, but not in space. In other words, the whole function is affected by change in any part. Wavelets, however, are localised in frequency (via dilation) and time or space (via translations). Ordinarily, sharp spikes and other discontinuities can be represented using fewer wavelet bases than sine-cosine basis functions. That is, wavelets are better suited to capturing sharp discontinuities. It is the adaptation to local variation in frequencies that makes wavelets of interest in this chapter. However, wavelets are not a replacement for Fourier analysis, rather they constitute an additional set of tools that may be more appropriate in some instances.

As for Fourier analysis, those engaging in wavelet analysis can make a choice between a continuous wavelet transform (CWT) and a discrete wavelet transform (DWT). The main focus here is on the DWT, which can be applied readily for the analysis of images. A short introduction to the CWT is provided for context.

Fourier analysis can be conducted in a moving window, but in this case a single window is used for all frequencies, and the resolution of the analysis is the same at all locations. That is, a windowed FT (WFT) is appropriate where an assumption is made of stationarity at the scale of the window. If this assumption is unreasonable, then wavelets may provide a more suitable option. In windowed FT, the size of the window is fixed, whereas the Mother wavelet (note that in DWT, as described below, there is also a Father wavelet (the scaling function)) is stretched or compressed (dilated), changing the size of the window as shown in Figure 3.8 (illustrating the Daubechies 12 wavelet; the Daubechies wavelets are discussed below).

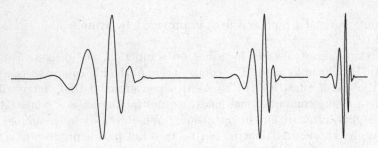

FIGURE 3.8: Dilating the Mother wavelet (Daubechies 12 wavelet).

3.6.2   Continuous wavelet transform

A set of wavelet basis functions can be generated through translating and scaling the basic wavelet $\psi(x)$:

$$\psi_{a,b}(x) = \frac{1}{\sqrt{a}} \psi \left(\frac{x-b}{a} \right) \tag{3.8}$$

where a is the dilation (that is, scaling) parameter, and b is the translation (position) parameter. The continuous wavelet transform (CWT; also called the integral wavelet transform) of a function $f(x)$ for a given analysing wavelet ψ is given as:

$$W(a,b) = \int_{-\infty}^{\infty} f(x)\psi_{a,b}(x) \, \mathrm{d}x \tag{3.9}$$

where $W(a,b)$ is the wavelet coefficient of the function $f(x)$. So, the wavelet is moved along the signal (it is translated) and it is stretched (dilated). The CWT equation shows that a function $f(x)$ can be decomposed in basis functions $\psi(x)$. Where the wavelet matches the signal well a large WT value is returned and, conversely, where the wavelet does not match the signal well a small WT value is returned. The WT can be viewed as the cross-correlation of a signal with a set of wavelets of various widths (3). Note that the wavelet basis functions are not specified here — different wavelet bases can be selected to suit particular purposes.

3.6.3   Discrete wavelet transform (DWT)

With the CWT, there is smooth continuous movement across locations, while for the discrete wavelet transform (DWT), the movement is in discrete steps (3). The analysis of gridded data is the focus of this section, and in this context the wavelet transform in its discrete form is widely used. To compute the DWT only the function at the values of time (or space) where the wavelet is non-zero is needed. For the DWT, the dilation (scaling) parameter, a is usually set to 2^j where j is an integer value, and the translation (position) parameter, b, is also an integer.

With CWT, there are an infinite number of wavelets in the transform. This problem was overcome by Mallat (256), with the introduction of the scaling function, ϕ (see Section 3.6.4 for an example). The scaling function contains the information in a signal up to a given scale while the detail in the signal is accounted for by the wavelet function. By using a scaling function and wavelets it is possible to decompose a signal and then reconstruct it without loss of information. Analysis of a signal using DWT results in a set of wavelet and scaling function transform coefficients. The work of Mallat (256)(in conjunction with Yves Meyer), in developing multiresolution analysis, marked a turning point in the practical application of wavelets. Mallat demonstrated the links between wavelets and a range of other approaches used in image processing, subband coding, and digital speech processing (190). Mallat showed that in multiresolution analysis, the WT of a signal can be conducted using simple digital filters (the wavelet function and scaling function themselves are not needed), and this is the subject of Section 3.6.5.

3.6.4 Wavelet basis functions

The most simple Mother function is the Haar. The Haar wavelet is a simple step. The Haar scaling function (ϕ; corresponding to smoothing of the signal) is given by:

$$\phi_i^j(x) = \phi(2^j x - i) \quad i = 0, .., 2^j - 1 \tag{3.10}$$

for the resolution j, where

$$\phi(x) = \begin{cases} 1 & 0 \leq x < 1 \\ 0 & \text{otherwise} \end{cases}$$

The Haar wavelet (ψ; corresponding to detail in the signal) is given by:

$$\psi_i^j(x) = \psi(2^j x - i) \quad i = 0, .., 2^j - 1 \tag{3.11}$$

where

$$\psi(x) = \begin{cases} 1 & 0 \leq x < 0.5 \\ -1 & 0.5 \leq x < 1 \\ 0 & \text{otherwise} \end{cases}$$

The Haar scaling function and wavelet are illustrated in Figure 3.9. Further illustrations are provided by Burrus et al. (71). The Haar wavelet can be applied using the fast wavelet transform described in the following section. A more in-depth introduction to the Haar wavelet is provided in the book by Stollnitz et al. (347).

The wide range of existing basis functions enables the user to select a basis function that is well suited to the task in hand. Choice of a basis function

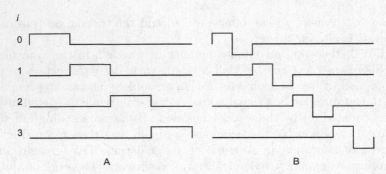

FIGURE 3.9: A. Haar scaling function $(\phi(4x-i))$ and B. Haar wavelet $(\psi(4x-i))$ for $j=2$.

represents a trade-off between how compactly the basis functions are localised in space and the degree of smoothness (160). Other basis functions (that is, wavelet families) include the Meyer (271) and the Daubechies (102) Mother functions. The Daubechies basis functions are widely used in wavelet analyses, and WT based on one member of this wavelet family (Daubechies 4, shown in Figure 3.10) is demonstrated in the illustrated example in Section 3.6.6.

An important property that some wavelet bases possess is orthogonality (346). In an orthogonal basis, all of the basis functions are orthogonal to each other. That is, with an orthogonal transform, information is coded once only (Hubbard (190) provides a clear summary of orthogonality). So, where the concern is to prevent redundancy (if, for example, the objective is compression), use of an orthogonal basis is important.

Where the elements of the basis are normalised to have length of one, the basis is termed orthonormal (short for orthogonal and normalised). Ingrid Daubechies introduced a family of basis functions that are orthonormal (102), and a case study is given below using one member of this family (see Section 3.6.6). There are several advantages of orthonormality, including ease of computation. The normalised Haar basis is given by:

$$\phi_i^j(x) = \sqrt{2^j}\phi(2^j x - i) \tag{3.12}$$

$$\psi_i^j(x) = \sqrt{2^j}\psi(2^j x - i) \tag{3.13}$$

The new normalised coefficients are derived by taking each of the old coefficients with superscript j and dividing them by $\sqrt{2^j}$ (347). Stollnitz et al. (347) provide a more in-depth account of the Haar basis with examples and pseudocode.

Each family of wavelets comprises subclasses that are distinguished by the number of coefficients and the number of iterations (160). The number of vanishing moments constitutes the primary means of classification; this is

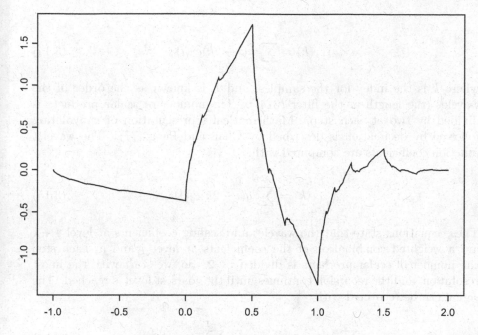

FIGURE 3.10: Daubechies 4 wavelet.

related to the number of wavelet coefficients (for example, Daubechies 4 indicates four coefficients).

3.6.5 Implementation of the DWT

In multiresolution analysis, the wavelet coefficients, c_j, may be thought of as a filter (160). The coefficients are placed in a transformation matrix that is applied to the raw data vector. The coefficients are ordered using two dominant patterns: one works as a smoothing filter, while the other extracts information on local variation (detail). In signal processing these orderings are called a quadrature mirror filter pair. With multiresolution analysis, the coefficients of the scaling function and wavelet coefficients at one scale can be derived from the scaling function coefficients at the next scale. This approach provides the basis for most applications of wavelets.

The fast wavelet transform (FWT), defined by its $i = 0, 1, ..., L - 1$ filter coefficients, is achieved by successive low-pass filtering and high-pass filtering of the data vector. The outputs of the low-pass (scaling) filter, h, are the "smooth" components, c, and the outputs of the high-pass (wavelet) filter, g, are the "detail" components, w.

It is possible to compute the scaling function coefficients at one (coarser) scale, $c_{j+1}(k)$, from those at the next (finer) scale, $c_j(k)$:

$$c_{j+1}(k) = \sum_{i=0}^{L-1} h(i - 2k)c_j(i) \tag{3.14}$$

where k is the index for the samples, and L is known as the order of the wavelet (the length of the filter (314)). The number of scalar products is divided by two at each step. Mathematical representation of convolution followed by decimation is described by Chan and Peng (77). The wavelet function coefficients are computed with:

$$w_{j+1}(k) = \sum_{i=0}^{L-1} g(i - 2k)c_j(i) \tag{3.15}$$

These equations state that the wavelet and scaling coefficients at level $j + 1$ are a weighted combination of the coefficients at level j and at each step the number of scalar products is divided by 2. So, we start with the finest resolution, and the recursion continues until the coarsest level is reached. The signal can be recovered with:

$$c_j(k) = \sum_{i=0}^{L-1} h(k - 2i)c_{j+1}(i) + \sum_{i=0}^{L-1} g(k - 2i)w_{j+1}(i) \tag{3.16}$$

To recap, with FWT the wavelet coefficient matrix is applied to the data vector in a hierarchical algorithm (also known as a pyramidal algorithm). The coefficients are ordered such that odd rows contain an ordering of wavelet coefficients that act as the smoothing filter, while even rows contain an ordering of wavelet coefficients with different signs that extract local detail. Firstly, the matrix is applied to the original full-length vector. The vector is then smoothed and decimated by half, and the matrix is applied again. The smoothed and halved vector is smoothed and halved again, and the matrix applied again. The process continues until a trivial number of "smooth-smooth-smooth..." data remain. In other words, each time the wavelet coefficient matrix is applied a higher (spatial) resolution of the data is extracted, while the remaining data are smoothed. The DWT output comprises the accumulated detail components and the remaining smooth components (160). This concept is illustrated below.

3.6.6   Fast wavelet transform: Illustrated example

Press (319) provide code for FWT. They discuss the Daubechies wavelet class, and code is given to implement one member of this class, DAUB4. The DAUB4 wavelet was illustrated in Figure 3.10. This code was used to conduct the following illustrated example. The computation of the FWT, as detailed below, follows the structure of the code. DAUB4 has four wavelet coefficients as follows:

$$h_0 = \frac{1+\sqrt{3}}{4\sqrt{2}} = 0.4829629131445341$$

$$h_1 = \frac{3+\sqrt{3}}{4\sqrt{2}} = 0.8365163037378079$$

$$h_2 = \frac{3-\sqrt{3}}{4\sqrt{2}} = 0.2241438680420134$$

$$h_3 = \frac{1-\sqrt{3}}{4\sqrt{2}} = -0.1294095225512604$$

and

$$g_0 = h_3; g_1 = -h_2; g_2 = h_1; g_3 = -h_0$$

For inputs c_j the DAUB4 scaling function can be given as:

$$c_{j+1}(k) = h_0 c_j(2k) + h_1 c_j(2k+1) + h_2 c_j(2k+2) + h_3 c_j(2k+3)$$

and the DAUB4 wavelet function can be given as:

$$w_{j+1}(k) = g_0 c_j(2k) + g_1 c_j(2k+1) + g_2 c_j(2k+2) + g_3 c_j(2k+3)$$

Scaling and wavelet coefficients are computed at each iteration and the index, k, is incremented by two with each iteration (215). Kaplan (215) provides a good introduction to this wavelet function. The four coefficients are arranged in a transformation matrix (note the wrap-around of the last two rows) (319):

$$
\begin{bmatrix}
h_0 & h_1 & h_2 & h_3 & & & & & \\
h_3 & -h_2 & h_1 & -h_0 & & & & & \\
& & h_0 & h_1 & h_2 & h_3 & & & \\
& & h_3 & -h_2 & h_1 & -h_0 & & & \\
\vdots & \vdots & & & & & \ddots & & \\
& & & & & & h_0 & h_1 & h_2 & h_3 \\
& & & & & & h_3 & -h_2 & h_1 & -h_0 \\
h_2 & h_3 & & & & & & & h_0 & h_1 \\
h_1 & -h_0 & & & & & & & h_3 & -h_2
\end{bmatrix}
$$

A row of 16 values was extracted from the landcover image outlined in Chapter 1. The transformation matrix above was then applied to this data vector to obtain the wavelet transform. The data values extracted were: 17, 26, 29, 43, 59, 80, 71, 59, 61, 60, 58, 58, 61, 75, 69, 63.

Multiplication of the transformation matrix by the matrix vector gives "smooth" components and "detailed" components as illustrated below. First, the matrix is applied to the original full-length vector as given in Table 3.1. The output alternates between "smooth" components and "detail" components and after permutation the eight "smooth" components precede the eight "detail" components. The matrix is applied again to the "smooth" vector, giving Table 3.2. The matrix is applied to the "smooth-smooth" vector, giving the output in Table 3.3. There are only two "smooth-smooth-smooth" components, and this is the last stage.

TABLE 3.1
"Smooth" components and
"detail" components: 1.

Output	Permute
30.895	30.895
−4.536	52.848
52.848	103.695
−2.674	89.553
103.695	85.146
5.331	80.497
89.553	99.513
−0.363	86.471
85.146	−4.536
−0.837	−2.674
80.497	5.331
−5.701	−0.363
99.513	−0.837
2.588	−5.701
86.471	2.588
−21.387	−21.387

3.6.7   Two-dimensional (2D) wavelet transforms

There are various approaches to applying the WT to two (or higher) dimensional data. This section outlines the horizontal and vertical analyses of Mallat (256); (Starck et al. (343), provide a summary).

Where the data are two dimensional, analysis can proceed by constructing 2D wavelets, ψ, together with the 2D scaling function ϕ:

$$\phi(x, y) = \phi(x)\phi(y) \tag{3.17}$$

$$\psi^1(x, y) = \phi(x)\psi(y) \tag{3.18}$$

TABLE 3.2
"Smooth" components and
"detail" components: 2.

Input	Output	Permute
30.895	70.783	70.783
52.848	27.648	133.661
103.695	133.661	119.574
89.553	−1.143	120.481
85.146	119.574	27.648
80.497	12.420	−1.143
99.513	120.481	12.420
86.471	−31.940	−31.939

TABLE 3.3
"Smooth" components and
"detail" components: 3.

Input	Output	Permute
70.783	157.206	157.206
133.661	2.719	157.103
119.574	157.103	2.719
120.481	−47.822	−47.822

$$\psi^2(x,y) = \psi(x)\phi(y) \tag{3.19}$$

$$\psi^3(x,y) = \psi(x)\psi(y) \tag{3.20}$$

where ψ^1 relates to vertical high frequencies (horizontal edges), ψ^2 relates to horizontal high frequencies (vertical edges) and ψ^3 represents high frequencies in both directions (the diagonals) (256). The wavelets are dilated by scaling factors 2^j along x and y translated on an infinite rectangular grid $(2^j_n, 2^j_m)$ (190).

In essence, the 2D algorithm entails the application of several one-dimensional (1D) filters. The steps followed are as given below:

1. Convolve the rows of the image with a 1D filter.

2. Discard the odd numbered columns (where the left most column is numbered zero).

3. Convolve the columns of the resulting signals with another 1D filter.

4. Discard the odd numbered rows (where the top row is numbered zero) (256); (75).

This process is conducted with both the h filter and the g filter. The result is four images (two obtained using h and two with g). Three of these images gg, gh, and hg represent the highest resolution wavelet coefficients (they represent "detail" components in the image). The image hh is a smoothed representation of the original image and the filters can be applied to it in the same way as to the original image, leading to four new images $gg(hh)$, $gh(hh)$, $hg(hh)$, and $hh(hh)$. The filters are then applied to the twice-smoothed image $hh(hh)$ and so on. The decomposition of an image is illustrated in Figure 3.11 (where the down-arrow indicates columns or rows being discarded) and the 2D DWT coefficient matrices are illustrated in Figure 3.12 (for example, s1-d1 corresponds to the hg image, where s = smooth and d = detail).

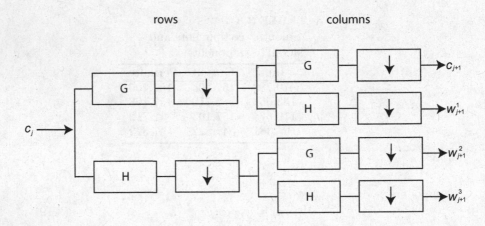

FIGURE 3.11: 2D DWT decomposition.

The scaling function at resolution $j+1$ is obtained from that at resolution j with:

$$c_{j+1}(k_x, k_y) = \sum_{i_x=0}^{L-1} \sum_{i_y=0}^{L-1} h(i_x - 2k_x)h(i_y - 2k_y)c_j(i_x, i_y) \qquad (3.21)$$

and the "detail" components are obtained with (343):

$$w_{j+1}^1(k_x, k_y) = \sum_{i_x=0}^{L-1} \sum_{i_y=0}^{L-1} g(i_x - 2k_x)h(i_y - 2k_y)c_j(i_x, i_y) \qquad (3.22)$$

$$w_{j+1}^2(k_x, k_y) = \sum_{i_x=0}^{L-1} \sum_{i_y=0}^{L-1} h(i_x - 2k_x)g(i_y - 2k_y)c_j(i_x, i_y) \qquad (3.23)$$

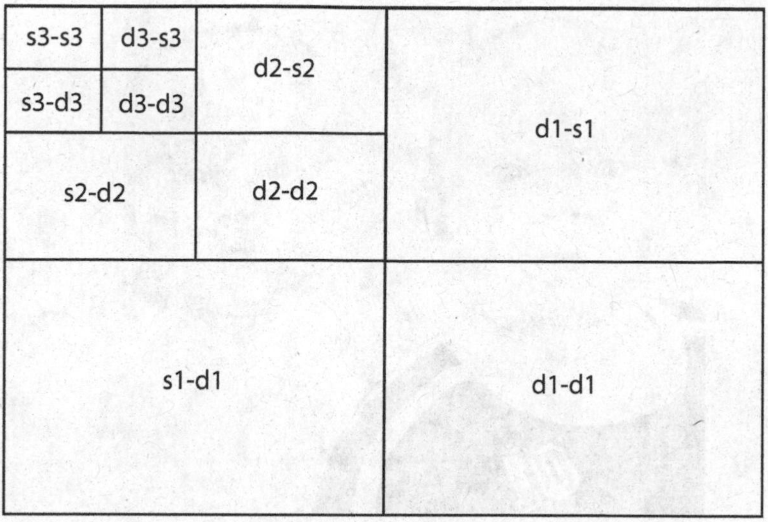

s3-s3	d3-s3	d2-s2	d1-s1
s3-d3	d3-d3		
s2-d2		d2-d2	
s1-d1			d1-d1

FIGURE 3.12: 2D discrete wavelet transform matrix.

$$w_{j+1}^3(k_x, k_y) = \sum_{i_x=0}^{L-1} \sum_{i_y=0}^{L-1} g(i_x - 2k_x)g(i_y - 2k_y)c_j(i_x, i_y) \qquad (3.24)$$

There are various kinds of orientable wavelets and these include steerable wavelets. Using this method, a filter is applied in a few (e.g., three) directions, and the filter at other directions is then interpolated.

3.6.7.1 Two-dimensional wavelet transform case study

The basic principles of wavelets are illustrated here using two images: the babies image shown in Figure 3.13 and the Newark, New Jersey, image (introduced in Section 1.8.5). The analyses were both conducted using the S+ Wavelets® software (61). The DAUB4 Mother function was used. In practice, appropriate Mother functions are often sought as outlined above.

The DWT of the image shown in Figure 3.13 is given (to one level) in Figure 3.14 to show clearly how the DWT works. The DWT image was displayed using the standard deviation contrast stretch (two standard deviations) to highlight the wavelet coefficients. The visibility of different horizontal, vertical, and diagonal features is clear in the different subimages. The lower-left image relates to vertical high frequencies (horizontal edges), the upper-right image relates to horizontal high frequencies (vertical edges) while the lower right image indicates high frequencies in both directions (the diagonals). This is evident in, for example, the horizontal edges picked out in the lower-left image and the vertical edges picked out in the upper-right image (for example, the characters '01').

FIGURE 3.13: Babies image.

The DWT of the Newark, New Jersey, image (given in Chapter 1, but also given in Figure 3.15) to one level is shown in Figure 3.16. The output is more difficult to interpret than that for the previous case, but it is clear that linear features aligned in particular directions are indicated by the relevant subimages. The standard deviation contrast stretch was used, as for the babies DWT images, to highlight the wavelet coefficients. In the lower left image, horizontal features are apparent while in the upper right image several vertical features are visible, particularly in the east of the region. Finally, the diagonal features in the east of the region can be seen in the lower right image. Many applications could follow the decomposition — the objective may be, for example, texture analysis or image compression. In the former case, the sub-images give information about variation at different spatial resolutions which could be used as an input to some classification scheme. In the latter case, the most significant coefficients could be identified, and only those retained. Some applications of wavelets are outlined in Section 3.6.9.

3.6.8 Other issues

Wavelet packets are an extension of the wavelet transform. Alternatively, the wavelet transform may be seen as a subset of the wavelet packet transform (160). Wavelet packets are linear combinations of wavelets. In essence, a wavelet packet combines a wavelet, which adapts to abrupt changes, and an oscillating function, that adapts to regular fluctuations. Wavelet packets are suitable for signals that combine nonstationary and stationary characteristics (Hubbard (190), gives fingerprints as an example).

FIGURE 3.14: DWT of babies image.

Approaches have been developed for selecting, on an automated basis, the most suitable basis function in a given situation. A basis of adapted waveform is the best basis function for a given signal representation (160). The basis selected should contain substantial information about the signal. If the basis description is efficient, then the signal information can be compressed.

3.6.9   Applications of wavelets

Applications of wavelets include image compression, noise suppression, human vision, and earthquake prediction. Graps (160) provides a brief summary of some applications. Image compression is based on sparse coding if an appropriate wavelet basis is selected. That is, only the most significant coefficients are retained. There are a variety of applications of wavelets in remote sensing. Lloyd et al. (250) use the DWT to quantify changes in energy with change in spatial resolution. There is now a wide range of WT software available. Zatelli and Antonello (402) outline WT routines for use in the GIS GRASS. The authors also outline several applications, including

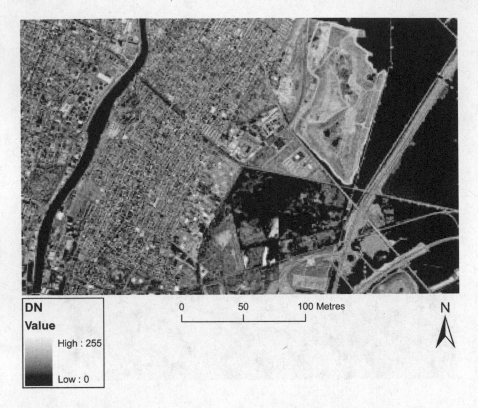

FIGURE 3.15: Digital orthophoto of Newark, New Jersey. Image courtesy of the US Geological Survey.

use of the WT to remove errors (in the form of local 'peaks') from a DEM generated by laser scanning.

3.7   Segmentation

Segmentation is discussed in this book for two reasons: (i) the objective of segmentation is to identify regions which are, in some sense, homogeneous (such that a stationary model can be applied to the segments if further processing or analysis is required — the first of the key approaches discussed in Section 2.2), and (ii) in image segmentation (as distinct from cluster analysis), new segments are formed by comparing neighbouring pixels or existing segments and assessing their similarity. This section outlines briefly some key approaches to image segmentation. More detailed accounts of such

FIGURE 3.16: DWT of Newark, New Jersey, image.

techniques are provided by Haralick and Shapiro (175) and Sonka et al. (338). Haralick and Shapiro (175) consider that the aim with any segmentation technique must be to form homogeneous segments that are simple and with few holes. Furthermore, adjacent regions should have significantly different properties or values while the boundaries of segments should be simple and spatially accurate. The most simple approach would be to apply a clumping procedure, whereby all neighbouring pixels with the same value comprise a discrete group. Thresholding provides another simple approach. Using this approach, each pixel is compared to some threshold value, and it is assigned to one of two classes depending on whether the threshold value is exceeded or not (311). The threshold is usually selected by examining the histogram and in the case of, for example, a bimodal distribution the mid-point between the two peaks may represent an appropriate threshold, but in most cases threshold selection is more problematic. However, this is a global (nonspatial) approach, and the focus below is on local methods for segmenting images.

Haralick and Shapiro (175) have divided image segmentation techniques into six groups, each of which are defined below:

1. Measurement space guided spatial clustering.

2. Single linkage region-growing.

3. Hybrid linkage region-growing.

4. Centroid linkage region-growing.

5. Spatial clustering.

6. Split and merge.

Clustering forms segments in measurement space whereas image segmentation works in the spatial domain and the latter is the focus here.
However, various spatially-weighted classification (clustering) schemes have
been developed and such approaches are considered below. There are spatial
clustering techniques that combine clustering in measurement space with
region-growing segmentation (defined below). The focus in this section is on
univariate segmentation approaches, while multivariate classification methods
are discussed briefly in Chapter 5.

Region-growing approaches all work by comparing neighbouring pixels or
segments, and joining them if they are similar in some sense. With single
linkage region-growing, neighbouring pixels with similar values are joined.
Such an approach is simple, but a key problem is the possibility of unwanted
region chaining, whereby regions are incorrectly merged because of one join
between neighbouring pixels. This is particularly likely with complex or noisy
data (175). Using hybrid linkage region-growing, pixels are joined if their
neighbourhoods are similar in some sense. Such approaches tend to be more
robust than single linkage methods when the data are noisy. One approach
entails joining neighbouring pixels if neither pixel is on an edge, as defined
by an edge operator. With centroid linkage region-growing, each pixel is
compared to existing neighbouring segments; such an approach is defined
below. Split and merge segmentation entails splitting the image into quarters
and successive quartering of the new segments until a segment is considered
sufficiently homogeneous. It is then sometimes necessary to merge adjacent
split segments. A range of region-growing segmentation approaches is detailed
by Haralick and Shapiro (175).

Centroid linkage segmentation algorithms scan an image on some predefined
basis, such as from top to bottom and left to right, and each pixel value
is compared to the mean of an existing neighbouring region. Where the
difference between the mean and a given pixel is less than a specific threshold,
T (selection of a meaningful threshold is problematic, as Pavlidis (311)
discusses), the pixel is merged with the segment (175). One criterion for
uniformity entails comparison of the difference between the value of a pixel
and the average over a region against some threshold. Haralick and Shapiro
(175) apply a T test to assess if a given pixel should join a segment.

Centroid linkage segmentation could be conducted as follows: after the first
comparison of a pixel and a region (or another sole pixel), subsequent pixels
are compared with their neighbours that already have segment labels. Where
a neighbouring segment is sufficiently similar to the pixel, the pixel is added
to the segment and the mean of the segment updated. Where a region is
sufficiently similar to more than one neighbouring segment, such that it could
be merged with more than one segment, the pixel is merged with the segment
with which it is most similar. Such an algorithm has been outlined by Levine

and Shaheen (236). A single scan may fail to join segments where, for instance, the region of analysis is partially divided by some feature running against the direction of the scan (175). Therefore, the data may be scanned a second time in a different direction. Thus, if the first loop was from top to bottom and left to right the second could be from bottom to top and right to left.

A centroid linkage region-growing segmentation algorithm was applied to the agricultural landcover image described in Chapter 1. The vectorised output from this procedure is illustrated in Figure 3.17. Segmentation was conducted through a single scan of the image using a purpose-written Fortran computer program. Where differences between pixels and segments were less than the mean difference between all neighbouring pixels (a completely arbitrary threshold), then the pixel was added to the segment. The segments follow many major field boundaries. However, the effect of the presence of pixels with markedly different spectral values within one field is clear in that many of the segments do not match the vectorised field boundaries illustrated in Chapter 1. Therefore, the segments are, not surprisingly, more complex than the field boundaries.

3.8 Analysis of digital elevation models

Digital elevation models (DEMs) are one of the most widely used forms of grid in GISystems contexts. Derivatives (for example, gradient and aspect) are usually approximated by computing differences between values in a square filter, or by fitting polynomials to the data in the moving window. Typically, a three-by-three cell moving window is used to obtain derivatives. Alternatively, a triangulated irregular network (TIN) model may be used, and slopes derived from the triangular facets. Only a brief review is given here, as extensive summaries of derivatives of altitude can be found in several publications (for example, (117), (119), (122), (281), (289)). With the methods discussed in this section, the same model is applied at all locations, and the moving window is of fixed size.

Gradient and aspect are the two first-order derivatives of altitude. Slope comprises gradient (the maximum rate of change in altitude) and aspect (the direction of the maximum rate of change; also called the azimuth). The terms slope and gradient are sometimes used interchangeably (70). Slope has been used directly in modelling water runoff, soil erosion, and for cost surface analysis, as well as other applications. Profile convexity and plan convexity (118) are the two second-order derivatives of altitude.

The gradient in the x direction at location i, j can be derived using a simple finite difference estimator given as (334):

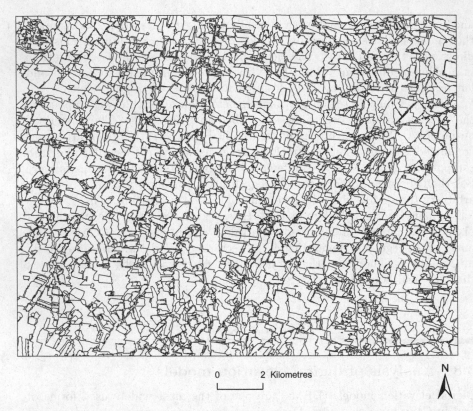

0   2 Kilometres

N

FIGURE 3.17: Region-growing segmentation output from the landcover image.

$$[\partial z/\partial x]_{i,j} = [(z_{i+1,j} - z_{i-1,j})/2h_x] \tag{3.25}$$

and for the y direction:

$$[\partial z/\partial y]_{i,j} = [(z_{i,j+1} - z_{i,j-1})/2h_y] \tag{3.26}$$

where h_x and h_y are the distances between cell centroids along the rows and columns, respectively. Gradient is calculated with:

$$\tan G = [(\partial z/\partial x)^2 + (\partial z/\partial y)^2]^{0.5} \tag{3.27}$$

and aspect is obtained with:

$$\tan A = (\partial z/\partial x)/(\partial z/\partial y) \tag{3.28}$$

Local errors in terrain may contribute to large errors in slope estimated using this approach. A range of other estimators of slope have been developed which are less sensitive to this problem.

The derivatives can be obtained through using a quadratic surface, the parameters of which are determined by elevation values in a three-by-three cell submatrix (118), (403):

$$z = Ax^2 + By^2 + Cxy + Dx + Ey + F \qquad (3.29)$$

This surface is not constrained to fit the data points. A surface can be selected that does fit exactly through the elevation values using the following quadratic equation (403):

$$z = Ax^2y^2 + Bx^2y + Cxy^2 + Dx^2 + Ey^2 + Fxy + Gx + Hy + I \qquad (3.30)$$

Given a 3 by 3 submatrix coded as follows:

$$z_1 \ z_2 \ z_3$$
$$z_4 \ z_5 \ z_6$$
$$z_7 \ z_8 \ z_9$$

The nine parameters of the equation are related to the nine submatrix elevations by:

$$A = [(z_1 + z_3 + z_7 + z_9)/4 - (z_2 + z_4 + z_6 + z_8)/2 + z_5]/h^4$$
$$B = [(z_1 + z_3 - z_7 - z_9)/4 - (z_2 - z_8)/2]/h^3$$
$$C = [(-z_1 + z_3 - z_7 + z_9)/4 + (z_4 - z_6)/2]/h^3$$
$$D = [(z_4 + z_6)/2 - z_5]/h^2$$
$$E = [(z_2 + z_8)/2 - z_5]/h^2$$
$$F = [(-z_1 + z_3 + z_7 - z_9)/4h^2$$
$$G = [(-z_4 + z_6)/2h$$
$$H = [(z_2 - z_8)/2h$$
$$I = z_5$$

Gradient can then be calculated with:

$$\text{Gradient} = (G^2 + H^2)^{1/2} \qquad (3.31)$$

Aspect is computed with:

$$\text{Aspect} = \arctan(-H/-G) \qquad (3.32)$$

An alternative approach is presented by Horn (188) whereby $\partial z/\partial x$ (weighted average of central differences in the west–east direction) and $\partial z/\partial y$ (weighted average of central differences in the north–south direction) are calculated with:

$$\partial z/\partial x = ((z_3 + 2z_6 + z_9) - (z_1 + 2z_4 + z_7))/8h_x \qquad (3.33)$$

$$\partial z/\partial y = ((z_3 + 2z_2 + z_1) - (z_9 + 2z_8 + z_7))/8h_y \qquad (3.34)$$

where, as before, h_x is the grid spacing in x and h_y is the grid spacing in y in the same units as the elevation (188). This approach is illustrated here given a grid with the following values (in m):

$$47\ 48\ 44$$
$$42\ 40\ 38$$
$$43\ 37\ 35$$

for a grid spacing of 1 m we have:

$$\partial z/\partial x = ((44 + (2 \times 38) + 35) - (47 + (2 \times 42) + 43))/8 = -2.375$$

$$\partial z/\partial y = ((44 + (2 \times 48) + 47) - (35 + (2 \times 37) + 43))/8 = 4.375$$

and the gradient is then obtained with:

$$\tan G = [(\partial z/\partial x)^2 + (\partial z/\partial y)^2]^{0.5} = [(-2.375)^2 + (4.375)^2]^{0.5} = 4.978\text{m}$$
$$(3.35)$$

Gradient in degrees can be computed with:

$$\text{atan}\, G \times 57.29578 \qquad (3.36)$$

In the case of the example, this leads to:

$$1.37255 \times 57.29578 = 78.64130° \qquad (3.37)$$

In words, this is almost a shear face. Note that if the spatial resolution is changed to, for example, 10m (and the denominators become 8×10), the gradient is 0.498m and in degrees it is 26.464°.

The derivation of gradient and aspect as well as convexity and concavity (negative convexity) using the parameters of Equation 3.30 are illustrated by Zevenbergen and Thorne (403) and Burrough and McDonnell (70). Other approaches to estimating derivatives of altitude have been compared by Skidmore (334) and Hodgson (187). Gradient across Great Britain is shown in Figure 3.18 and aspect is given in Figure 3.19. Gradient was calculated in ArcGIS™ using the algorithm given by Horn (188), and illustrated above.

Derivatives of altitude are often obtained as a first stage in computing, for example, drainage networks or friction surfaces. Burrough and McDonnell (70) provide a good summary of such methods which are outside the focus of this book. Lloyd (245) also provides details of other relevant approaches.

FIGURE 3.18: Gradient in decimal degrees.

FIGURE 3.19: Aspect in decimal degrees (clockwise from north).

3.9 Overview

There is a huge range of additional tools for the analysis of gridded data. These include, for example, methods for pattern recognition and a variety of other spatial filters. Sonka et al. (338) provide details of some such methods. In this chapter, a range of methods for exploring spatial variation in single variables represented as grids were explored. These ranged from widely-used spatial filters for the analysis of gridded data through to derivatives of altitude. In the next chapter the focus is on the analysis of spatial variation in single variables represented using areas or as points with attached values and such approaches can also be applied to gridded data.

4

Spatial Patterning

In the previous chapter, the focus was on the analysis of spatial variation in single variables, with an emphasis on gridded data. In this chapter, the concern is with spatial variation in single variables represented using areas or points. In Section 4.1, the topic of locally estimated summary statistics is introduced, and in Section 4.2 geographically weighted summary statistics are outlined. These sections build on parts of the previous chapter which were concerned specifically with gridded data. Section 4.3 discusses the analysis of spatial autocorrelation, primarily with reference to areal data. In Section 4.4, some local measures of spatial autocorrelation are detailed. Section 4.5 briefly considers the analysis of spatial association and categorical data. The chapter goes on to review some other issues such as other analytical approaches with connections to the themes explored in the main part of the chapter.

The main focus of the latter sections is on variables available on areal units which are often of varying size and shape. A large proportion of the work conducted using areal data is in human geography and related disciplines. As such, many of the examples cited below are concerned with socioeconomic data, but some applications in physical geography are suggested. Selected approaches are illustrated using values in a remotely-sensed image that were aggregated over areas represented by vector polygons.

4.1   Local summary statistics

The most straightforward way of assessing local variations in single variables is simply to compute summary statistics in a moving window in the same way as is done with the focal operators described in Section 3.5. A more sophisticated approach is to weight observations as a function of their distance from the centre of the window. Such approaches are the subject of the following section.

4.2 Geographically weighted statistics

Any summary statistic can be computed locally using a geographical weighting scheme. Many weighting schemes, both binary and continuous, have been developed (see Section 2.4). A widely used weighting scheme is the Gaussian function (129). With this function, a weight at the observation i is obtained with:

$$w_{ij} = \exp[-0.5(d/\tau)^2] \tag{4.1}$$

where d is the Euclidean distance between the location of observation i, and the location j and τ is the bandwidth of the kernel. The Gaussian function (for a bandwidth of (i) 2500 and (ii) 5000 units) is illustrated in Figure 4.1.

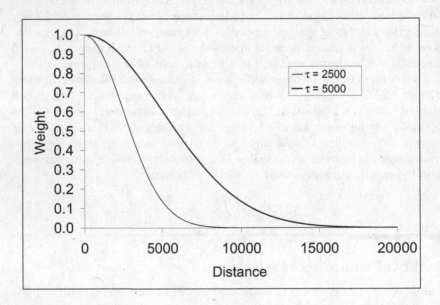

FIGURE 4.1: Gaussian curve for a bandwidth (i) 2500 and (ii) 5000 units.

The remainder of this section details geographically weighted variants of standard statistical summaries. The geographically weighted mean can be computed with (129):

$$\bar{z}_i = \frac{\sum_{j=1}^{n} z_j w_{ij}}{\sum_{j=1}^{n} w_{ij}} \tag{4.2}$$

Following Fotheringham et al. (129), if the weights are rescaled to sum to one, it is given by:

$$\bar{z}_i = \sum_{j=1}^{n} z_j w_{ij} \tag{4.3}$$

Table 4.1 shows the computation of the geographically weighted mean, with the weights given in unstandardised form, for a subset of the log-ratio data described in Section 1.8.6 for bandwidths of 1000, 2000, and 3000 m.

TABLE 4.1

Geographically weighted mean (GWmean) for log-ratio data. BW is bandwidth; d_{ij} is distance; w_{ij} is the weight.

ln(CathCB/ NonCathCB)	d_{ij} (m)	w_{ij}	$z_j w_{ij}$	w_{ij}	$z_j w_{ij}$	w_{ij}	$z_j w_{ij}$
−2.483	0.000	1.000	−2.483	1.000	−2.483	1.000	−2.483
−1.450	1000.000	0.607	−0.879	0.882	−1.280	0.946	−1.372
−1.413	1000.000	0.607	−0.857	0.882	−1.247	0.946	−1.337
−1.376	1000.000	0.607	−0.835	0.882	−1.214	0.946	−1.302
−1.673	1414.214	0.368	−0.615	0.779	−1.303	0.895	−1.497
−1.444	1414.214	0.368	−0.531	0.779	−1.125	0.895	−1.292
−1.474	2000.000	0.135	−0.199	0.607	−0.894	0.801	−1.180
−1.700	2000.000	0.135	−0.230	0.607	−1.031	0.801	−1.361
−1.728	2000.000	0.135	−0.234	0.607	−1.048	0.801	−1.384
−0.964	2236.068	0.082	−0.079	0.535	−0.516	0.757	−0.730
−1.187	2236.068	0.082	−0.097	0.535	−0.635	0.757	−0.899
−2.766	2236.068	0.082	−0.227	0.535	−1.481	0.757	−2.095
−1.923	2236.068	0.082	−0.158	0.535	−1.029	0.757	−1.457
−0.679	2236.068	0.082	−0.056	0.535	−0.363	0.757	−0.514
−2.572	2828.427	0.018	−0.047	0.368	−0.946	0.641	−1.649
−1.562	2828.427	0.018	−0.029	0.368	−0.575	0.641	−1.002
−1.195	3000.000	0.011	−0.013	0.325	−0.388	0.607	−0.725
−1.175	3000.000	0.011	−0.013	0.325	−0.381	0.607	−0.713
−1.259	3162.278	0.007	−0.008	0.287	−0.361	0.574	−0.722
−1.649	3162.278	0.007	−0.011	0.287	−0.472	0.574	−0.946
−1.035	3162.278	0.007	−0.007	0.287	−0.297	0.574	−0.594
−1.563	3605.551	0.002	−0.002	0.197	−0.308	0.486	−0.759
−1.439	3605.551	0.002	−0.002	0.197	−0.283	0.486	−0.699
−1.478	3605.551	0.002	−0.002	0.197	−0.291	0.486	−0.718
−1.624	4000.000	0.000	−0.001	0.135	−0.220	0.411	−0.668
−1.108	4123.106	0.000	0.000	0.119	−0.132	0.389	−0.431
−2.626	4123.106	0.000	−0.001	0.119	−0.314	0.389	−1.021
−1.499	4123.106	0.000	0.000	0.119	−0.179	0.389	−0.583
−1.673	4242.641	0.000	0.000	0.105	−0.176	0.368	−0.615
−1.355	4472.136	0.000	0.000	0.082	−0.111	0.329	−0.446
	Sum	4.456	−7.619	13.217	−21.084	19.766	−31.193
	GWmean		−1.710		−1.595		−1.578
	BW (m)		1000		2000		3000

The unweighted mean of the observations in the example is −1.569. Note
how the geographical weights at larger distances increase in proportion to
the geographical weights for smaller distances as the bandwidth is increased.
Lloyd (245) presents a further worked example of the geographically weighted
mean. The geographically weighted mean was computed using the Gaussian
function with a 2km bandwidth, given the log-ratio data described in Section
1.8.6, and it is illustrated in Figure 4.2. Comparison of Figure 1.10 with
Figure 4.2 demonstrates the smoothing effect of the local mean. Figure 4.3
shows the geographically weighted mean for a 5km bandwidth. Varying the
bandwidth in this way provides a means of assessing how far local patterns
persist across different spatial scales. The maps suggest that there is marked
variation in the local mean.

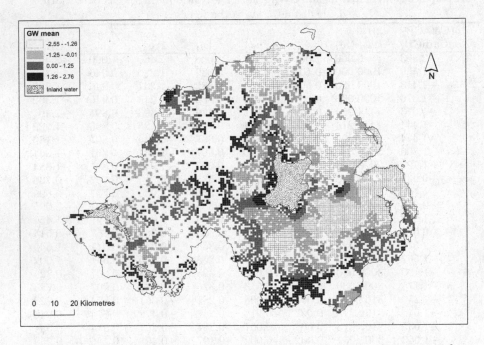

FIGURE 4.2: Geographically weighted mean for a Gaussian function with a
2km bandwidth: log-ratio of Catholics/Non-Catholics in Northern Ireland in
2001 by 1km grid squares. Northern Ireland Census of Population data — ©
Crown Copyright.

For standardised weights, the geographically-weighted standard deviation
is given by:

$$s_i = \left[\sum_{j=1}^{n} (z_j - \bar{z}_i)^2 \cdot w_{ij} \right]^{1/2} \tag{4.4}$$

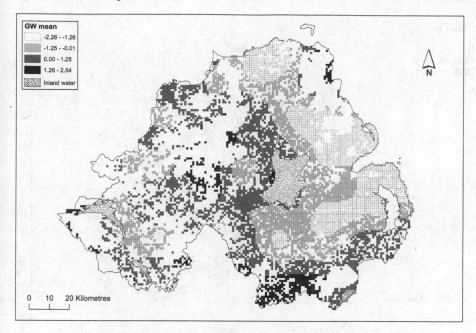

FIGURE 4.3: Geographically weighted mean for a Gaussian function with a 5km bandwidth: log-ratio of Catholics/Non-Catholics in Northern Ireland in 2001 by 1km grid squares. Northern Ireland Census of Population data — © Crown Copyright.

The geographically weighted standard deviation (for a 2km bandwidth) is given in Figure 4.4, and this highlights areas with mixed population characteristics by community background. For example, the large values in the Belfast region (i.e., the central eastern part of Northern Ireland), indicate neighbouring areas which are dominated by Catholics or Protestants. In contrast, areas with small standard deviations are more homogeneous in terms of the community background of their residents.

The geographically weighted standardised differences of the global and local means (termed the geographically weighted standard score by Fotheringham et al. (129)) provides a means of assessing locally marked deviations from the global mean:

$$zs_i = \frac{\bar{z}_i - \mu}{\sigma\sqrt{\sum_{j=1}^n w_{ij}^2}} \qquad (4.5)$$

where μ is the global mean, and σ is the global standard deviation. As before, the weights are standardised. An example of the geographically weighted standard score computed using the log-ratio data is given in Figure 4.5. The map indicates the high degree of variation in the geographically weighted

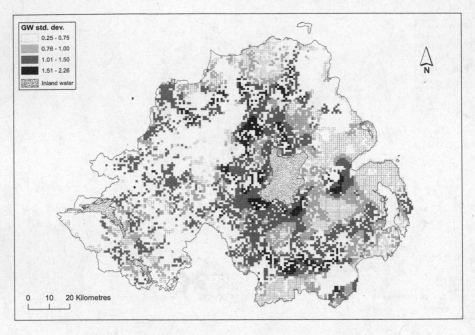

FIGURE 4.4: Geographically weighted standard deviation for a Gaussian function with a 2km bandwidth: log ratio of Catholics/Non-Catholics in Northern Ireland in 2001 by 1km grid squares. Northern Ireland Census of Population data — © Crown Copyright.

mean and the distinction between the west and the east of Northern Ireland is clear.

Following the previous definitions, the geographically weighted coefficient of variation is given by:

$$CV_i = s_i/\bar{z}_i \qquad (4.6)$$

When the mean is close to zero, as is the case with some of the local means (see Figure 4.2), the coefficient of variation is unstable. Therefore, the upper and lower ranges of the geographically weighted coefficient of variation in Figure 4.6 are simply given as being larger (for positive values) or smaller (for negative values) than the specified thresholds. There are several areas in the map with large (positive or negative) values. As one example, there is a group of contrasting values around the central Belfast area, illustrating very different characteristics of the population within parts of Belfast, and between the city and the suburbs.

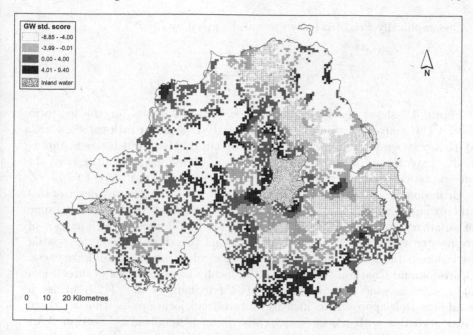

FIGURE 4.5: Geographically weighted standard score for a Gaussian function with a 2km bandwidth: log-ratio of Catholics/Non-Catholics in Northern Ireland in 2001 by 1km grid squares. Northern Ireland Census of Population data — © Crown Copyright.

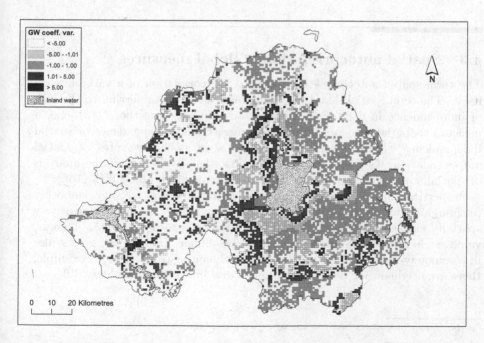

FIGURE 4.6: Geographically weighted coefficient of variation for a Gaussian function with a 2km bandwidth: log-ratio of Catholics/Non-Catholics in Northern Ireland in 2001 by 1km grid squares. Northern Ireland Census of Population data — © Crown Copyright.

Geographically weighted skewness can be given by (67):

$$b_i = \sqrt[3]{\frac{\sum_{j=1}^{n}(z_j - \bar{z}_i)^3 w_{ij}}{s_i^3}} \tag{4.7}$$

Figure 4.7 shows the geographically weighted skewness for the log-ratio data. The map suggests that there is marked variation in local skewness. With any geographically weighted statistic, altering bandwidth size is important. Exploration of maps generated using alternate bandwidths may aid interpretation of complex patterns such as those shown in Figure 4.7.

Brunsdon et al. (67) and Fotheringham et al. (129) present descriptions and example applications of geographically weighted versions of a range of summary statistics. Harris and Brunsdon (177) present an analysis of freshwater acidification critical load data, and they find evidence for spatial variation in the local mean, variance, coefficient of variation, and skewness. Brunsdon and Charlton (66) discuss the locally-based analysis of directional data such as wind direction data. Fotheringham et al. (129) utilise a randomisation approach to identify 'interesting' locations — that is, those local statistics with values which are deemed significant. Section 4.3.1 discusses a similar approach with respect to spatial autocorrelation. A range of geographically weighted statistics can be computed using the R package spgwr*.

4.3   Spatial autocorrelation: Global measures

The term spatial autocorrelation refers to the correlation of a variable with itself. The term spatial dependence has been defined as referring to the lack of independence in data at locations which are close together (173). So, a measure of spatial autocorrelation may suggest spatial dependence or spatial independence. In this section, the focus is on global measures of spatial autocorrelation. Cliff and Ord (87) discuss some key issues in the analysis of spatial autocorrelation; another introduction is given by Griffith (166).

Properties such as elevation and precipitation tend to vary smoothly (although there are, of course, exceptions), and they are usually positively spatially autocorrelated (at least at some scales of measurement). That is, values at locations close together tend to be similar. In contrast, grey scales in a remotely-sensed image may be negatively autocorrelated if, for example, there are neighbouring fields in an agricultural image that have very different

---

*http://cran.r-project.org/web/packages/spgwr/index.html

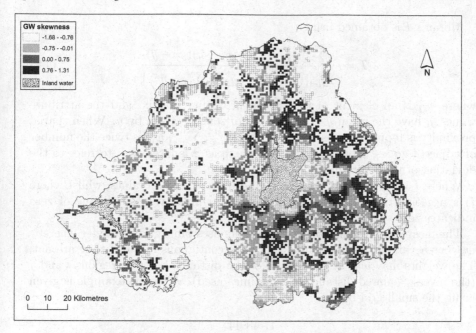

FIGURE 4.7: Geographically weighted skewness for a Gaussian function with a 2km bandwidth: log-ratio of Catholics/Non-Catholics in Northern Ireland in 2001 by 1km grid squares. Northern Ireland Census of Population data — © Crown Copyright.

characteristics. Getis (141) provides a discussion about the concept and measurement of spatial autocorrelation.

Various forms of local measures have been developed for use in cases where local variation in spatial autocorrelation is suspected. This section introduces global measures, and Section 4.4 discusses some local variants. In the case of areal data, the centroids of areas can be used to measure distances between areas. Alternatively, simple connectivity between areas could be used, as discussed below.

The joins-count statistic provides a simple measure of spatial autocorrelation. The method is used to assess structure in areas coded black (B) or white (W). Tests can be conducted based on the number each of the possible type of joins: BB, WW, BW, and WB (304). Comparisons between areas can be made using only neighbours that share edges (i.e., boundaries) (termed rook's case, as detailed further below) or neighbours that share edges or corners (or vertices) (termed queen's case and also detailed further below).

Two measures of spatial autocorrelation frequently encountered in the GISystems literature are Moran's I and Geary's c. Both global and local versions of these statistics are detailed below.

Moran's I is obtained using:

$$I = \frac{n \sum_{i=1}^{n} \sum_{j=1}^{n} w_{ij}(y_i - \overline{y})(y_j - \overline{y})}{\left(\sum_{i=1}^{n}(y_i - \overline{y})^2\right)\left(\sum_{i=1}^{n} \sum_{j=1}^{n} w_{ij}\right)} \qquad (4.8)$$

where w_{ij} is an element of the spatial proximity matrix, and the attribute values y_i have the mean $\overline{y}$. The number of zones is given by n. When spatial proximity is represented using contiguity, $\sum_{i=1}^{n} \sum_{j=1}^{n} w_{ij}$ is twice the number of adjacent zones. Note that y is used here as z is later used to refer to the deviation of y from its mean.

Where I is positive this indicates clustering of similar values, whilst where I is negative this indicates clustering of dissimilar values; a value of zero indicates zero spatial autocorrelation.

The most simple definition of the weights uses binary connectivity whereby w_{ij} has a value of 1 if regions i and j are contiguous and 0 if they are not. The weights may also be a function of the distance between regions i and j (the inverse squared distance is sometimes used). A simple example is given using the small grid below:

47 48 44
42 40 38
43 37 35

Approaches using various combinations of neighbours are common. When all common boundaries and vertices are used this is termed queen contiguity. As defined above, when only observations sharing an edge (rather than a corner or vertex) with a given observation are used this is termed rook contiguity. Queen contiguity is used in the example below — neighbours are pixels that are horizontally, vertically, or diagonally connected to another pixel. The appropriate values calculated are inserted into Equation 4.8 giving:

$$I = \frac{9 \times 167.012}{158.222 \times 40} = \frac{1503.111}{6328.889} = 0.237$$

Note that 40 is twice the number of adjacent zones. Where the weights are row-standardised (they sum to one with respect to the neighbourhood of each cell — e.g., where a cell has four neighbours, the weights are 0.25), $I = 0.286$. I is not forced to be within the $(-1,1)$ range, but the equation could be modified to ensure that this is the case. A further worked example of Moran's I is provided by O'Sullivan and Unwin (304).

Geary's contiguity ratio, c, is estimated with:

$$c = \frac{(n - 1) \sum_{i=1}^{n} \sum_{j=1}^{n} w_{ij}(y_i - y_j)^2}{2 \left(\sum_{i=1}^{n}(y_i - \overline{y})^2\right)\left(\sum_{i=1}^{n} \sum_{j=1}^{n} w_{ij}\right)} \qquad (4.9)$$

Small values of c indicate positive spatial autocorrelation, while large values of c indicate negative spatial autocorrelation. Both I and c can be generalised easily to enable measurement of spatial autocorrelation at several lags (that is, observations separated by a given distance or range of distances — directions could be considered too). Moran's I can be estimated at spatial lag h with:

$$I^{(h)} = \frac{n \sum_{i=1}^{n} \sum_{j=1}^{n} w_{ij}^{(h)} (y_i - \overline{y})(y_j - \overline{y})}{\left(\sum_{i=1}^{n}(y_i - \overline{y})^2\right)\left(\sum_{i=1}^{n} \sum_{j=1}^{n} w_{ij}^{(h)}\right)} \tag{4.10}$$

where $w_{ij}^{(h)}$ is an element of the spatial proximity matrix for lag h. The exploration of spatial autocorrelation is discussed further in Chapter 7 where the covariance function, correlogram (related to Moran's I), and variogram (related to Geary's c) are defined.

The Moran scatterplot provides one way of examining variation in spatial autocorrelation for a given dataset (11). The Moran scatterplot is a plot which relates the values of some property in one zone, i, to values in neighbouring zones. As for the Moran's I measure itself, neighbouring zones may be those that are adjacent, or a function of separation distance may be specified. The scatterplot may be particularly useful for identifying outliers that correspond to local anomalies (e.g., a value markedly different to its neighbours). This topic is explored further in Section 4.4.1.2.

Bivariate measures of spatial autocorrelation are used to assess spatial association between two different variables. The GeoDa™ software (12), (13), (14) provides functionality to compute bivariate local indicators of spatial association (LISAs; as defined in Section 4.4.1). Section 7.4.3 describes the cross-variogram which is used to characterise spatial covariation.

4.3.1 Testing for spatial autocorrelation

A test for spatial autocorrelation can be constructed where there are a sufficiently large number of observations. By assuming that the y_i are drawn independently from a normal distribution (they are observations on random variables Y_i) then if Y_i and Y_j are spatially independent (for $i \neq j$) I has a sampling distribution that is approximately normal with expected value of I given by (32):

$$\mathrm{E}(I) = -\frac{1}{(n-1)} \tag{4.11}$$

and the variance:

$$\mathrm{Var}(I) = \frac{n^2 S_1 - n S_2 + 3 S_0^2}{S_0^2 (n^2 - 1)} \tag{4.12}$$

where

$$S_0 = \sum_{i=1}^{n} \sum_{j=1}^{n} w_{ij} \tag{4.13}$$

and

$$S_1 = \frac{1}{2} \sum_{i=1}^{n} \sum_{j=1}^{n} (w_{ij} + w_{ji})^2 \tag{4.14}$$

and

$$S_2 = \sum_{k=1}^{n} \left(\sum_{j=1}^{n} w_{kj} + \sum_{i=1}^{n} w_{ik} \right)^2 \tag{4.15}$$

it is possible to test the observed value of I against the percentage points of the approximate sampling distribution. Where the value of I is "extreme", spatial autocorrelation is indicated (32).

An alternative approach is a random permutation test. The approach is based on the idea that, if there are n observations over a particular region, $n!$ permutations of the data are possible with different arrangements of the data over the region. The value of I may be obtained for any one of the permutations. An empirical distribution of possible values of I can then be constructed with random permutations of the data. Where the observed value of I is extreme with respect to the permutation distribution, this may be considered evidence of significant spatial autocorrelation. In practice, it is not usually possible to obtain $n!$ permutations, and a Monte Carlo approach may be used instead to approximate the permutation distribution. An appropriate number of values can then be drawn randomly from among the $n!$ permutations (32). If we assume that the process generating the observed data is random and the observed pattern is one of many possible permutations, then the variance of I is given by:

$$\mathrm{Var}(I) = \frac{nS_4 - S_3 S_5}{(n-1)(n-2)(n-3)S_0^2} \tag{4.16}$$

where

$$S_3 = \frac{n^{-1} \sum_{i=1}^{n} (y_i - \overline{y})^4}{(n^{-1} \sum_{i=1}^{n} (y_i - \overline{y})^2)^2} \tag{4.17}$$

and

$$S_4 = (n^2 - 3n + 3)S_1 - nS_2 + 3S_0^2 \tag{4.18}$$

and

$$S_5 = S_1 - 2nS_1 + 6S_0^2 \tag{4.19}$$

Further details about tests of spatial autocorrelation are provided by Fotheringham et al. (128). The topic is explored further in Section 4.4.1.2.

4.4   Spatial autocorrelation: Local measures

The measures outlined above provide global summaries. However, there have been various adaptations of the standard approaches to allow assessment of local variation in spatial autocorrelation and such approaches are the subject of this section. A review of the topic of spatial autocorrelation and local spatial statistics is provided by Fotheringham (126).

Getis and Ord (144) defined G, a global measure of association which enables the assessment of spatial clustering in the values of an attribute. The local form of G is then used to measure spatial patterning around each area. The ith observation (for example, area centroid) may or may not be included in the calculation. In the first case, the notation G_i^* is used. In the latter case (i may not equal j), the notation G_i is used. G_i is given as:

$$G_i = \frac{\sum_{j=1}^n w_{ij} y_j}{\sum_{j=1}^n y_j}, j \neq i \tag{4.20}$$

Getis and Ord (144) state that G_i is intended for use only with variables which have a natural origin, since transformations like logarithms will change the results. In the example which follows, log-ratio data are used and it should be remembered, of course, that the results relate to the log-ratios specifically (and the results would be different for alternative transforms). Local G is often expressed in standardised form, whereby the expectation, $\sum_{j=1}^n w_{ij}/n$, is subtracted from the statistic and then divided by the square root of its variance (144). G_i and G_i^* can be written as:

$$G_i = \frac{\sum_{j=1}^n w_{ij} y_j - w_i \bar{y}(i)}{s(i)\{[((n-1)w_{i(2)}) - w_i^2]/(n-2)\}^{1/2}}, j \neq i \tag{4.21}$$

$$G_i^* = \frac{\sum_{j=1}^n w_{ij} y_j - w_i^* \bar{y}^*}{s^*\{[(nw_{i(2)}^*) - w_i^{*2}]/(n-1)\}^{1/2}}, \text{all } j \tag{4.22}$$

The various terms are as follows (145):

$$w_i = \sum_{j=1}^n w_{ij}, j \neq i \tag{4.23}$$

$$w_i^* = \sum_{j=1}^n w_{ij} \tag{4.24}$$

and

$$w_{i(2)} = \sum_{j=1}^n w_{ij}^2, j \neq i \tag{4.25}$$

$$w_{i(2)}^* = \sum_{j=1}^{n} w_{ij}^2 \qquad (4.26)$$

the means are:

$$\bar{y}(i) = \frac{\sum_{j=1}^{n} y_j}{(n-1)}, j \neq i \qquad (4.27)$$

$$\bar{y}^* = \frac{\sum_{j=1}^{n} y_j}{n} \qquad (4.28)$$

and the variances:

$$s(i)^2 = \frac{\sum_{j=1}^{n} y_j^2}{(n-1)} - [\bar{y}(i)]^2, j \neq i \qquad (4.29)$$

$$s^{*2} = \frac{\sum_{j=1}^{n} y_j^2}{n} - \bar{y}^{*2} \qquad (4.30)$$

Figure 4.8 shows G_i^* (following Equation 4.22, and using adjacent cells) for the log-ratio data analysed in Section 4.2. ArcGIS™ was used to generate the values. Comparison of this map with Figure 1.10 shows that local G_i^* values are picking out areas dominated by Catholics or non-Catholics. Local forms of Moran's I and Geary's c have also been developed, and these are discussed in the next section.

4.4.1   Local indicators of spatial association

Anselin (10) defined a body of local indicators of spatial association (LISA). In that paper, local forms of Moran's I and Geary's c are defined. LISA statistics allow for decomposition of global indicators. A LISA, as defined by Anselin (10) is any statistic that fulfils two requirements:

1. The LISA for each observation indicates the extent of significant spatial clustering around the observation.

2. The sum of LISAs for all observations is proportional to a global indicator of spatial association (10). If the concern is to assess the degree to which a global statistic is representative of the average pattern of local association, then this requirement is important.

A LISA is given as a statistic $LISA_i$ for a variable y_i observed at the location i:

$$LISA_i = f(y_i, y_{L_i}) \qquad (4.31)$$

where f is a function, and y_{L_i} are the observed values (or deviations from the mean) in the neighbourhood L_i. The GeoDa™ package (12) includes the capacity to compute both univariate and bivariate LISAs.

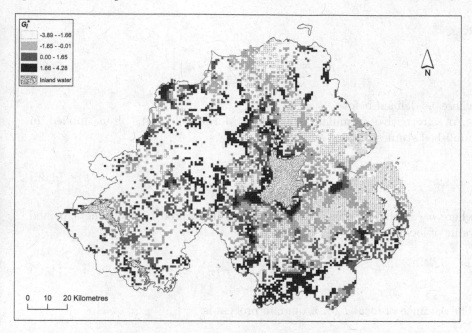

FIGURE 4.8: G_i^*: Log-ratio for Catholics/Non-Catholics in Northern Ireland in 2001 by 1km grid squares. Northern Ireland Census of Population data — © Crown Copyright.

4.4.1.1 Local Moran's I

The local form of Moran's I for observation i is given by (10):

$$I_i = z_i \sum_{j=1}^{n} w_{ij} z_j, j \neq i \qquad (4.32)$$

the observations z_i are deviations from the mean $(y_i - \bar{y})$. Usually, non-zero weights are assigned only to the nearest neighbours (in effect, the summation includes only the neighbouring zones). The spatial weights w_{ij} may be in row-standardised form (that is, the weights sum to one) to facilitate comparison of different sets of results.

The sum of local Moran's I is given as:

$$\sum_{i=1}^{n} I_i = \sum_{i=1}^{n} z_i \sum_{j=1}^{n} w_{ij} z_j \qquad (4.33)$$

and Moran's I is given as:

$$I = (n/S_0) \sum_{i=1}^{n} \sum_{j=1}^{n} w_{ij} z_i z_j / \sum_{i=1}^{n} z_i^2 \qquad (4.34)$$

or

$$I = \sum_{i=1}^{n} I_i / \left[S_0 \left(\sum_{i=1}^{n} z_i^2 / n \right) \right] \tag{4.35}$$

where, as defined before, $S_0 = \sum_{i=1}^{n} \sum_{j=1}^{n} w_{ij}$ (10).

An alternative formulation of local Moran's I (10), often applied in published studies, is:

$$I_i = (z_i/m_2) \sum_{j=1}^{n} w_{ij} z_j, j \neq i \tag{4.36}$$

where m_2 is the variance; the approach is illustrated below. The expected value of local I is:

$$E(I_i) = -\sum_{j=1}^{n} w_{ij}/(n-1) \tag{4.37}$$

The variance of local I for a random process is:

$$\text{Var}(I_i) = \frac{w_{i(2)}(n - b_2)}{(n-1)} + \frac{2w_{i(kh)}(2b_2 - n)}{(n-1)(n-2)} - \frac{w_i^2}{(n-1)^2} \tag{4.38}$$

where $b_2 = m_4/m_2^2$ and $m_r = \sum_{i=1}^{n} z_i^r/n$; $w_{i(2)} = \sum_{j=1}^{n} w_{ij}^2$, $j \neq i$, and $2w_{i(kh)} = \sum_{k=1}^{n} \sum_{h=1}^{n} w_{ik}w_{ih}$, $k \neq i$ and $h \neq i$. The distribution of I_i has not been explored but Getis and Ord (145) suggest a test based on the normal distribution :

$$Z(I_i) = (I_i - E(I_i))/\sqrt{\text{Var}(I_i)} \tag{4.39}$$

The probability distribution of I_i may, in practice, be poorly represented by a normal distribution. Bivand et al. (46) provide an illustrated example of the possible shortcomings of the normal approximation. An alternative approach to testing for significant spatial association is to use a random permutation test (such as the conditional randomisation approach detailed by Anselin (10)), as outlined in Section 4.3.1, and this is explored further in the following section in the context of the local I statistic. Alternative approaches are described by Tiefelsdorf (360) and Bivand et al. (46). Tiefelsdorf (360) provides a review of approximation methods and demonstrates that the saddlepoint approximation has benefits over other approaches in terms of its flexibility, accuracy, and computational costs.

Some research has suggested that the presence of global spatial autocorrelation has implications for tests based on permutations. This research indicates that, in such situations, the tests may be too liberal in that some coefficients may appear significant when they are not (337) (and see a review by Boots (52)). Tiefeldorf (359) offers a solution to testing significance of I_i in the presence of global spatial autocorrelation based on the global process as a Gaussian autoregressive or a Gaussian moving average spatial process.

It may be important to account for sample size when computing local I. The GeoDa™ software allows for consideration of sample size variation (e.g., difference in sizes of population from which percentages were computed) through Empirical Bayes (EB) adjustment using the method of Assunção and Reis (20).

4.4.1.2 Identifying local clusters

Anselin (10) describes an approach to testing for significant local autocorrelation based on a randomisation procedure (see Section 4.3.1). Limitations of such approaches are outlined by Waller and Gotway (375, pp. 238-239).

The GeoDa™ software offers the capacity to test the significance of local I using randomisation and to map significant clusters. The procedure entails random spatial relocation of data values given a specified number of permutations which are user-specified. Moran's I, whether global or local, is then computed given each permutation. The distribution of Moran's I values can then be compared to that derived using the observed values and the probability that the observed value comes from a random distribution can be determined.

One problem is that local probability values are not independent of one another. In short, the reference distributions for two areas are constructed by repeated drawings from the same population, and so they are not independent. In addition, I_i for two neighbouring areas is based on the values associated with one another and thus the test statistics are correlated (165). The Bonferroni adjustment has been suggested as a possible solution. For the overall significance level (that is, the probability of a type I error), ρ, with respect to the correlated tests and with k comparisons, the significance level ρ^* can be specified with (52):

$$\rho^* = \frac{\rho}{k} \tag{4.40}$$

However, if $k = n$ is specified, given the calculation of a local statistic for each data location, this is likely to be too conservative as not all n statistics are correlated (52). An alternative suggested by Getis and Ord (145) is given by:

$$\rho^* = 1 - (1 - \rho)^{\frac{1}{n}} \tag{4.41}$$

As noted by Boots (52), this test is only marginally less conservative than the correction in Equation 4.40 and both corrections may result in significant values being obtained only rarely. An alternative approach (not covered here) is based on controlling the false discovery rate (37).

The derivation of clusters is illustrated with reference to an example. In GeoDa™, the default is 99 permutations with a pseudo-significance level of

$\rho = 0.05$. In conducting the analysis on which the following example is based, larger numbers of permutations (e.g., 9999) were used to assess the sensitivity of the results. The problem of multiple testing can be taken into account to some degree by altering the pseudo-significance level and assessing changes in the clusters. Pseudo-significance levels of $\rho = 0.01$ and $\rho = 0.001$ were also applied in the following example for this reason.

Clusters are identified using the Moran scatterplot (11). As detailed in Section 4.3, the Moran scatterplot relates individual values to weighted averages of neighbouring values. The slope of a regression line fitted to the points in the scatterplot gives global Moran's I. The variables are given as deviations from the mean and the plot centres on 0, 0. The four quadrants of the plot represent different kinds of spatial association. The upper right and lower left quadrants represent positive spatial association while the upper left and lower right quadrants represent negative spatial association. Using GeoDa™, clusters are coded 'Low-low' (small values surrounded by small values; lower left quadrant of the Moran scatterplot), 'High-high' (large values surrounded by large values; upper right quadrant), 'Low-high' (small values surrounded by large values; upper left quadrant), 'High-low' (large values surrounded by small values; lower right), and 'Not significant' (at a default pseudo significance level of 0.05).

Figure 4.9 gives an example Moran scatterplot computed from the community background log-ratio data detailed in Section 1.8.6 and utilised in Section 4.2. The weights were computed using the Gaussian kernel defined in Equation 4.1 with a 1km bandwidth. In this case, global I is 0.739, whereas for queen contiguity (as used below) it is 0.752. Anselin at al. (15) describe software for brushing linked plots (such as the Moran scatterplot) and maps. Such tools allow, for example, points on the Moran scatterplot to be selected and their spatial position highlighted on the corresponding map. This kind of approach is a powerful means of identifying and assessing local heterogeneities. Haslett et al. (179), Brunsdon (63), Dykes (116), and Wilhelm and Steck (393) present approaches to visualising local statistics using linked maps and other graphical representations.

Local I for ln(CathCB/NonCathCB), for queen contiguity, is given in Figure 4.10 with clusters shown in Figure 4.11. The largest positive values are in the south of County Armagh and south west County Down, west County Antrim and north County Down, parts of the north coast, and parts of the mid-west of the province. The city of Derry/Londonderry also has fairly large positive values. The areas in the Belfast region are predominantly Protestant, as are those areas identified on the north coast, whereas those areas identified in the south and mid-west are predominantly Catholic (see Figure 1.10 for the log-ratios themselves). The map suggests that the areas most concentrated by community background are south Armagh, parts of the mid-west and west Belfast (Catholic dominated areas) and north west Down and south east Antrim (Protestant dominated areas).

Figure 4.11 shows clearly the distinction between the predominantly Catholic west and predominantly Protestant east. Although this general pattern is well-documented, the map of significant clusters offers a visually striking local perspective on high concentrations of Catholics and Protestants as well as on areas with locally large proportions of both (that is, high-low or low-high clusters). Altering the pseudo-significance level to 0.01 and 0.001 (from the default of 0.05) demonstrated that the clusters were persistent. It is worth noting that, in some areas, there are distinct boundaries between neighbourhoods dominated by each of the two groups. In such areas (perhaps 'interface areas' — where zones which are predominantly Catholic meet zones which are predominantly Protestant), mixing between different group members may be very limited, but this may be represented statistically by negative spatial autocorrelation which would (in this case wrongly) suggest diverse areas. Such regions are, however, the exception.

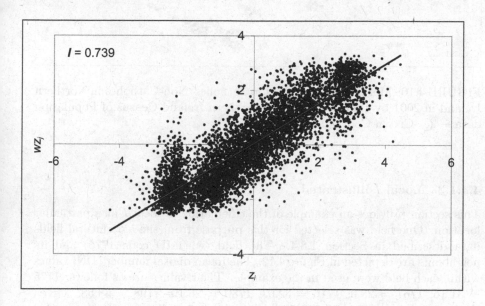

FIGURE 4.9: Moran scatterplot for community background log-ratio: Gaussian kernel with a 1km bandwidth.

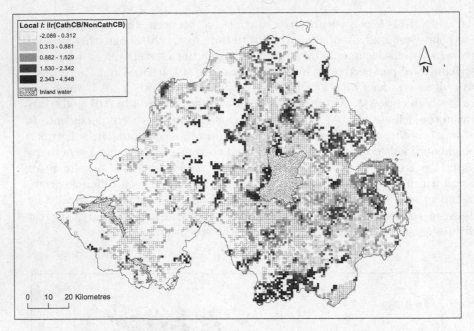

FIGURE 4.10: Moran's I: Log-ratio for Catholics/Non-Catholics in Northern Ireland in 2001 by 1km grid squares. Northern Ireland Census of Population data — © Crown Copyright.

4.4.1.3   Local I illustrated

This section provides an example of the calculation of local I at a particular location. One field was selected for this purpose from the agricultural fields data described in Section 1.8.4. The field (with ID code 1775) and its neighbours are depicted in Figure 4.12. The mean digital number (DN) values within each field were used in the example. Their values are as follows: 1775 = 31.16; 1767 = 25.0; 1716 = 32.63; 1780 = 26.19; 1798 = 38.58. Given the entire dataset, $\overline{x}$ is 52.793 and m_2 is 380.042 (following Getis and Ord (145), the value for the selected location, here field 1775, is excluded from the calculations), and there are 2361 fields.

Following Equation 4.36, and with the four neighbours included in the calculations ordered as above, this gives:

$$I_{1775} = (-21.633/380.042) \times ((0.25 \times -27.793) + (0.25 \times -20.163) + (0.25 \times -26.603) + (0.25 \times -14.213)) = 1.263$$

where the weights are row-standardised (there are four neighbours so the individual weights are 0.25). As an example, field 1767 has a mean DN value

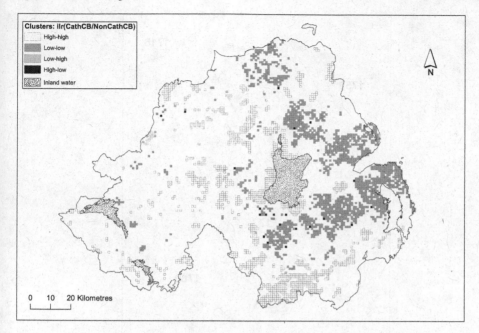

FIGURE 4.11: Clusters: Log-ratio for Catholics/Non-Catholics in Northern Ireland in 2001 by 1km grid squares. Northern Ireland Census of Population data — © Crown Copyright.

of 25.0, and the mean of all field mean DN values is 52.793. Thus, the deviation from the mean is $25.0 - 52.793 = -27.793$. The expected value of I_{1775} is given by:

$$E(I_{1775}) = -1/2360 = -0.000423$$

Getis and Ord (145) also give a worked example of Local I. Anselin (10) applies both local I and local Geary's C as well as the G_i^* statistic of Getis and Ord (144) to the analysis of the spatial distribution of conflict in African countries. A case study demonstrating the application of global and local Moran's I follows.

4.4.1.4 Case study

The application of selected summaries of spatial autocorrelation is illustrated in this section using the land use dataset described in Chapter 1. The average of DN within digitised boundaries was computed to enable this analysis (as shown in Figure 4.13). Local measures were calculated using the GeoDa™ software (12). In this case study, rook contiguity (common boundaries, but not vertices) rather than queen contiguity (all common boundaries and vertices) was used. The value of global Moran's I was 0.0527. Local Moran's I is

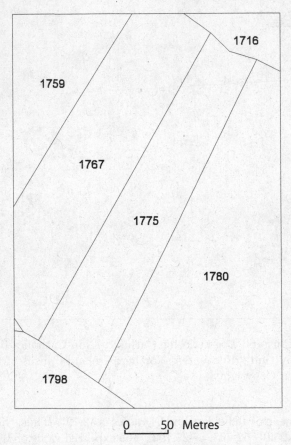

FIGURE 4.12: Subset of the land use data.

mapped in Figure 4.14. Values of local Moran's I range between -2.554 and 2.894. Most locations are negatively (or weakly positively) autocorrelated, but areas with large positive values of local I indicate that there are small clusters of neighbouring fields with similar values; it is clear that a global measure obscures such spatial variation. A possible next step would be to generate a map of significant clusters, as detailed in Section 4.4.1.2.

4.4.1.5   Local Geary's c

A local variant of Geary's c has also been defined by Anselin (10). The local Geary's c for location i, with the same notation as before, is given as:

$$c_i = \sum_{j=1}^{n} w_{ij}(z_i - z_j)^2 \qquad (4.42)$$

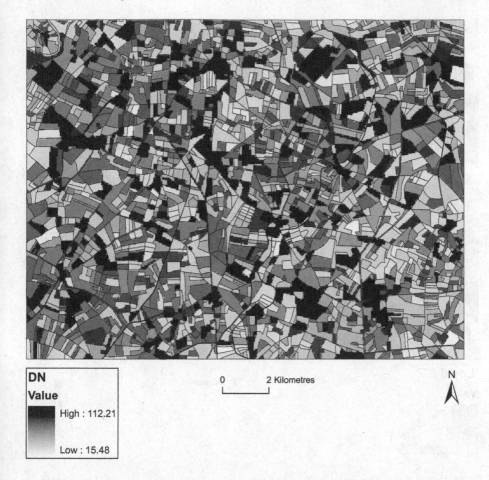

DN
Value

High : 112,21

Low : 15.48

0   2 Kilometres

N

FIGURE 4.13: Mean of DN values within fields.

This can be modified to:

$$c_i = (1/m_2) \sum_{j=1}^{n} w_{ij}(z_i - z_j)^2 \qquad (4.43)$$

Given Equation 4.43, the sum of local Geary's c is given as:

$$\sum_{i=1}^{n} c_i = n \left(\sum_{i=1}^{n}\sum_{j=1}^{n} w_{ij}(z_i - z_j)^2 / \sum_{i=1}^{n} z_i^2 \right) \qquad (4.44)$$

while Geary's c can be given by:

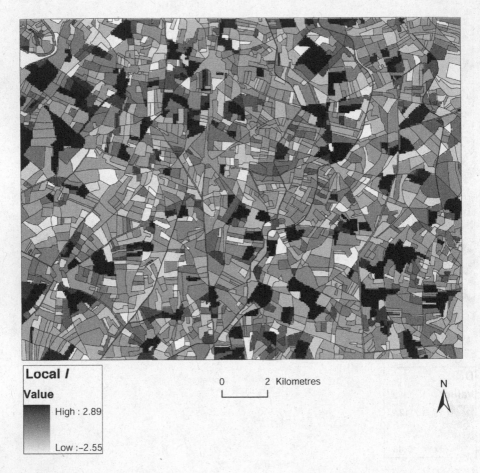

Local *I*

Value

High : 2.89

Low :-2.55

0   2 Kilometres

N

FIGURE 4.14: Moran's *I* for field means with vector field boundaries superimposed.

$$c = [(n-1)/2S_0] \left(\sum_{i=1}^{n} \sum_{j=1}^{n} w_{ij}(z_i - z_j)^2 / \sum_{i=1}^{n} z_i^2 \right) \qquad (4.45)$$

Clearly, small values of c_i indicate locations where differences between neighbouring values are small and large values indicate locations where differences are large. Figure 4.15 shows c_i for the log-ratio data used previously. In this case, queen contiguity is used. Areas which are dominated by members of one group (i.e., Catholics or non-Catholics), are indicated by small values of c_i. Large values pick out borders between areas dominated by different groups. Getis and Ord (145) define alternative versions of Geary-type statistics (termed K statistics) which are based on y rather than z. The

lack of standardisation by the global mean is beneficial in that the statistics
are not affected by nonstationarity in the mean (145).

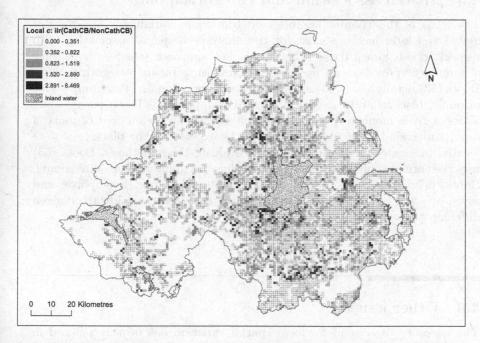

FIGURE 4.15: Geary's c: Log-ratio for Catholics/Non-Catholics in Northern
Ireland in 2001 by 1km grid squares. Northern Ireland Census of Population
data — © Crown Copyright.

4.4.2   Use of different approaches

The local measures of spatial association outlined capture different elements of
spatial structure in variables. For example, computing G_i and I_i will highlight
different characteristics of the data. Getis and Ord (144) present a comparison
using global G and global I while Anselin (10) compares local G^* and local
I in an analysis of spatial patterns of conflict in Africa. Since, as noted
above, the sum of LISA statistics is proportional to a global indicator, such
approaches can be used to explore local instability, and a LISA can be used
as a diagnostic for outliers with respect to the corresponding global measures
(10).

4.5   Spatial association and categorical data

The focus of this chapter has been on continuous variables. A variety of global and local methods exist for the analysis of spatial association with respect to categorical data. The joins-count approach (see Section 4.3 for a summary) provides one means of summarising binary categorical data. Boots (53) details a local measure of spatial association for (two group — for example, black/white) categorical data. His approach is based on assessing if, for a given number of black cells in a subregion, the number of joins of a particular kind differ from what would be expected if the black cells were located by chance in the subregion. To evaluate the hypothesis, Boots (53) uses the normal approximation of the sampling distribution of the joins counts. Boots, in working with gridded data, uses three-by-three, five-by-five, and seven-by-seven cell moving windows and different clusters are identified given different window sizes.

4.6   Other issues

The necessity to account for local spatial variation has been recognised in many contexts. This theme has connections to Section 3.5.4, which was concerned with spatial structure in gridded variables. Another context which has links with some of the topics explored in this chapter is the analysis of segregation — mixing between members of different population subgroups or exposure of members of one group to members of other groups (although there are further definitions relating to different domains of segregation). Local segregation indices are described and illustrated by Wong (394) and Feitosa et al. (120). On a similar theme, local measures of spatial association such as local I can be used to assess spatial concentrations in population subgroups (e.g., see Lloyd (242)).

4.7   Overview

This chapter introduces a variety of approaches for exploring spatial variation in single variables. The concept of geographical versions of univariate statistics was introduced as were approaches to measuring spatial autocorrelation both globally and locally. While global summaries such as the mean or standard deviation or global Moran's I may provide an overview, it is clear that the characteristics of many spatial properties vary geographically, and the approaches detailed in this chapter enable exploration of properties. The use of geographically weighted summary statistics is a sensible early step in the analyses of spatial data.

The focus in this chapter was on the analysis of single variables. The following chapter deals with the analysis of relations between spatially-referenced variables.

5

Spatial Relations

In this chapter, the concern is with spatial variation in the multivariate relations between variables. A range of approaches, including long-established methods and, for example, recent developments in the field of geographically weighted regression (GWR), are reviewed. This chapter has links to Chapter 6, in particular, where the use of regression for prediction purposes is discussed. Global regression is discussed in Section 5.1, while spatial and local regression approaches are the subject of Section 5.2 through Section 5.8. The remaining sections deal with spatially weighted classification and summarise the chapter.

The main focus of this chapter is, as would be expected given the book's focus, on explicitly local models, but global methods and methods which account for spatial structure, but provide only a single set of output coefficients, are also introduced for context. The techniques discussed allow exploration of variation at different spatial scales. For example, the spatial expansion method enables characterisation of large scale (global) trends, while GWR can be used to fit model parameters over very small scales (limited by the density and number of observations). The common link between the methods outlined in the latter part of this chapter is the ability they offer to account for variation in spatial relations. Several case studies are presented using different methods, and the benefits of applying the selected nonstationary models are considered.

5.1 Global regression

Relations between two or more variables are usually explored globally. For example, if the relation between precipitation and elevation is of interest, all precipitation observations may be plotted against all corresponding elevation observations, and the form of the relationship explored. Ordinary least squares (OLS) regression for k independent variables is given by:

$$z_i = \beta_0 + \beta_1 x_{1i} + \dots + \beta_k x_{ki} + \varepsilon \tag{5.1}$$

where i refers to a location, $x_{1i}, \dots, x_{ki}$ are the independent variables at location i, β_0, β_1, ..., β_k are the parameters that will be estimated, and ε is an error term. In matrix form this can be given by:

$$\mathbf{z} = \mathbf{X}\beta + \varepsilon \tag{5.2}$$

The parameters for the model can be estimated using OLS by solving:

$$\hat{\beta} = (\mathbf{X}^T\mathbf{X})^{-1}\mathbf{X}^T\mathbf{z} \tag{5.3}$$

and the standard errors are given by:

$$\mathrm{Var}(\hat{\beta}) = \sigma^2(\mathbf{X}^T\mathbf{X})^{-1} \tag{5.4}$$

As an example of the potential shortcomings of global regression, the elevation and precipitation data values described in Chapter 1 were plotted against one another, and OLS was used to fit a line to the plotted values (Figure 5.1).

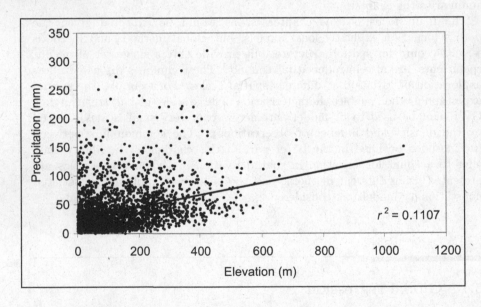

FIGURE 5.1: Elevation against precipitation: all data.

From this plot, elevation and precipitation appear to be weakly related. This does not correspond with prior expectations. For precipitation over periods of weeks or more, a strong positive relation with elevation would be expected, especially in areas with high elevation. In Section 5.8.1, two local regression approaches are used to assess spatial variation in the relations between elevation and precipitation. In that case study, it is demonstrated that the relations between the two variables vary markedly from region to region, and a global model masks this spatial variation.

Residuals from a global regression can be mapped to enable identification of areas where the global model fails to explain a large proportion of the variation. The residuals (i.e., prediction−observed) given the fitted line shown in Figure 5.1 were mapped and are shown in Figure 5.2. The residuals are clearly spatially structured (positively autocorrelated), with a marked tendency for the global model to under-predict (i.e., negative residuals) in the west of Scotland with, for example, a tendency to over-predict (positive residuals) in much of Southern England. Note that the ranges of values in Figure 5.2 were selected to best contrast values of residuals in particular areas.

Mapping residuals does not enable assessment of *how* modelled relations vary spatially — this is only possible using some form of local regression. It has been argued that local variation in parameter estimates may simply indicate that the model used is a misspecification of reality. In practice this is impossible to verify. Fotheringham et al. (129) argue that, if the form of the model we are currently employing results in marked local variations, examination of such spatial variation is important in helping to expand our understanding of the nature of any intrinsic spatial variation in behaviour or to suggest a model specification that is more appropriate. To examine spatial variation it is necessary to use some approach that, for example, estimates model parameters locally. The following sections detail various forms of regression that account for spatial structure and, more importantly given the focus of this book, spatial variation in relationships.

5.2 Spatial and local regression

A wide range of methods have been developed that account for spatially varying relationships, and some key approaches are outlined below. These approaches enable estimation of model parameters locally (the main focus in this chapter), or they allow model parameters to vary as a function of location. Alternatively, different scales of variation may be modelled as separate components. A review of such approaches has been provided by Fotheringham et al. (129).

The primary focus in the following sections is on methods for the local estimation of model parameters. However, for the purposes of providing context, some approaches to accounting for spatial dependence in global regression modelling are outlined. First, the benefits of generalised least squares are outlined (Section 5.3), then spatially autoregressive models are summarised (Section 5.4). Such approaches overcome the limitation of ordinary regression in that they account for spatial dependence. However,

FIGURE 5.2: Residuals for global regression of elevation against precipitation.

in their standard form they provide a single set of coefficients and, therefore, do not account for spatial variations in the relations between variables. The bulk of the remainder of the chapter focuses on methods which allow for geographical variation in coefficients and thus fit within the umbrella term of local models.

Multilevel modelling is described in Section 5.5. In the multilevel modelling framework, it is possible to consider variation in average relationships across hierarchical structures. Thus, variation within nested spatial structures (e.g., schools within local education authority areas) is explicitly accounted for. Methods that allow for spatial variation in model parameters are outlined in Section 5.6. The main focus of this chapter is on moving window regression (MWR; Section 5.7) and its more general form, geographically weighted regression (GWR; Section 5.8). As discussed below, MWR is a special case of GWR where the geographical weights are all equal to one. In Section 5.8.5, a case study applying MWR and GWR is presented.

5.3   Regression and spatial data

It is well known that use of OLS regression in the analysis of spatial data is problematic. If it is believed that spatial variation in a dataset is due to first-order effects (for example, there is a consistent trend in values of some property across the study region) it may be possible to identify an additional variable which accounts for this variation. However, as Bailey and Gatrell (32) argue, if after addition of such a covariate, the residuals from the regression are significantly spatially autocorrelated and a map of these residuals reveals no obvious spatial pattern which could be accounted for by additional covariates, then it is necessary to consider some alternative approach. In other words, if spatial dependence is due to more than simple first-order effects (spatial variation in the mean) then it may be considered worthwhile to adjust the model to allow for second-order effects (32). Generalised least squares (GLS) offers a way of relaxing the assumption of spatial independence. With GLS, the spatial structure in the data is represented by the variance-covariance matrix, $\mathbf{C}$. The GLS model, for k independent variables, can be given with:

$$z_i = \beta_0 + \beta_1 x_{1i} + ... + \beta_k x_{ki} + u \tag{5.5}$$

where u is a vector of errors with $\mathrm{E}(u) = 0$ and $\mathrm{E}(uu^T) = \mathbf{C}$. The β parameters can be estimated with (32):

$$\hat{\beta} = (\mathbf{X}^T \mathbf{C}^{-1} \mathbf{X})^{-1} \mathbf{X}^T \mathbf{C}^{-1} \mathbf{z} \tag{5.6}$$

and the standard errors are given by (32):

$$\text{Var}(\hat{\beta}) = (\mathbf{X}^T \mathbf{C}^{-1} \mathbf{X})^{-1} \qquad (5.7)$$

Since $\mathbf{C}$ is usually not known, some means for estimating it is required. One approach consists of estimating the covariances from the OLS residuals and using the derived information to conduct GLS. This is discussed further in Section 7.15.2. Fotheringham et al. (129) define models which utilise the covariance structure to account for local relationships, but which only provide global parameter estimates, as 'semi-local'.

5.4　Spatial autoregressive models

Building spatial structure into a GLS procedure could proceed by estimating the covariance matrix, $\mathbf{C}$, directly from the data, as summarised above. An alternative approach is to use an interaction scheme. Such a model utilises relationships between variables and their neighbours which specify indirectly particular forms of $\mathbf{C}$ (32). Standard linear regression can be expanded, such that z at a location $\mathbf{s}_i$ is a function of values of z at neighbouring locations. As an example, if $z(\mathbf{s}_i)$ represents mean commuting distance at location $\mathbf{s}_i$, then this may be a function of mean commuting distances at neighbouring locations (see Haining (173), for further examples). With autoregressive correlation structures, neighbouring correlation values are larger than those values that are not at neighbouring locations, and distant observations have the smallest correlations (284).

Two types of models which account for dependence in neighbouring values in regression are:

1. Spatially lagged dependent variable models.

2. Spatial error models.

With spatially lagged dependent variable models, a spatially lagged dependent variable is added to the right hand side of the regression equation (377). Such a model can be given, in matrix form, as:

$$\mathbf{z} = \mathbf{X}\beta + \rho \mathbf{W}\mathbf{z} + \varepsilon \qquad (5.8)$$

Ward and Gleditsch (377) devote a chapter to this kind of model and to ways in which they can be fitted. With this model, the errors are assumed to be uncorrelated. With a spatial error model, rather than assuming that nearby values are directly dependent on one another, the model errors are assumed to be spatially dependent. A simple spatial error model can be given in matrix form by (173), (375):

$$\mathbf{z} = \mathbf{X}\beta + \rho\mathbf{W}\mathbf{z} - \rho\mathbf{W}\mathbf{X}\beta + \varepsilon \tag{5.9}$$

The model can be given with (see Equation 5.12 for the expanded form as above):

$$z(\mathbf{s}_i) = \mu(\mathbf{s}_i) + \rho \sum_{j=1}^{n} w_{ij}[z(\mathbf{s}_j) - \mu(\mathbf{s}_j)] + \varepsilon(\mathbf{s}_i) \quad i = 1, ..., N \tag{5.10}$$

for n neighbours of the ith location and N regions in total and with $\mathbf{s}_i$ here denoting the ith spatial location. The mean can be given by:

$$\mu(\mathbf{s}_i) = \beta_0 + \beta_1 x_1(\mathbf{s}_i) + \beta_2 x_2(\mathbf{s}_i) + \cdots + \beta_k x_k(\mathbf{s}_i) \tag{5.11}$$

w_{ij} represents proximity between observations i and j, ρ is the interaction parameter, and ε is the error term with $\mathrm{E}(\varepsilon) = 0$ and $\mathrm{E}(\varepsilon\varepsilon^T) = \sigma^2 I$. Proximity could be represented using simple adjacency or some function of distance between observations, as discussed in Section 4.2. (375). When $\rho = 0$ the model reduces to the OLS model (325).

This type of model is termed a simultaneous autoregressive model (SAR), which in this case has one interaction parameter ρ. Equation 5.10 can be expanded:

$$z(\mathbf{s}_i) = \mu(\mathbf{s}_i) + \rho \sum_{j=1}^{n} w_{ij}z(\mathbf{s}_j) - \rho \sum_{j=1}^{n} w_{ij}\mu(\mathbf{s}_j) + \varepsilon(\mathbf{s}_i) \quad i = 1, ..., N \tag{5.12}$$

With this model, $z(\mathbf{s}_i)$ depends on the n surrounding values, $z(\mathbf{s}_j)$ through $\rho \sum_{j=1}^{n} w_{ij}z(\mathbf{s}_j)$, the general trend through $\mu(\mathbf{s}_i)$ and on neighbouring trend values through $\rho \sum_{j=1}^{n} w_{ij}\mu(\mathbf{s}_j)$ (32).

If ρ was known, $\mathbf{C}$ could be estimated from the data and GLS used to fit the model. However, ρ is not usually known, and it must be estimated from the data. Simultaneous estimation of β and ρ is problematic. In practice, the parameters of SAR models are most often estimated using maximum likelihood (ML) and more details are given by Schabenberger and Gotway (330). One simpler approach entails guessing a value for ρ and such an approach is outlined by Bailey and Gatrell (32).

An alternative class of models are conditional autoregressive models (CAR). Further descriptions of spatial autoregressive methods are given by Griffith (167), Bailey and Gatrell (32), Fotheringham et al. (128), Haining (173), Rogerson (325), and Waller and Gotway (375). Bivand et al. (48) present routines for spatial autoregressive modelling using the R language. The next section deals with a means of accounting for spatial hierarchies.

5.5   Multilevel modelling

Multilevel modelling is generally concerned with the separation of effects of personal characteristics and place characteristics (contextual effects) on behaviour. In essence, a multilevel model (MLM) represents an attempt to avoid problems that arise due to the atomistic fallacy (modelling the behaviour of individuals fails to account for the context in which behaviour occurs) and the ecological fallacy (modelling results at an aggregate level may not represent well individual behaviour). Multilevel modelling combines a model at the individual level (disaggregate behaviour) with a macrolevel model (contextual variation in behaviour) and a MLM is defined here. A microlevel model for individual i can be given by:

$$z_i = \beta_0 + \beta_1 x_{1i} + \varepsilon_i \tag{5.13}$$

The average value of a property can be varied using a macromodel of the form:

$$\beta_{0j} = \beta_0 + \mu_{0j} \tag{5.14}$$

The equation indicates that the average quantity, β_{0j}, is a function of the average across some region, β_0, plus a varying difference, μ_{0j}, for each subregion. A combination of the microlevel and macrolevel models leads to a model of the form:

$$z_{ij} = \beta_0 + \beta_1 x_{1ij} + (\mu_{0j} + \varepsilon_{ij}) \tag{5.15}$$

where z_{ij} represents the behaviour of individual i at location j, x is an explanatory variable, and the random component is in brackets. The objective is to estimate the fixed intercept, β_0, representing the average value plus the variance, σ_μ^2, representing variability around the average (207).

In addition to varying the intercept, the slope can also be varied, giving:

$$\beta_{ij} = \beta_1 + \mu_{1j} \tag{5.16}$$

The equation indicates that the subregion slope is the average region-wide slope plus variation from subregion to subregion (207). Equations 5.13, 5.14, and 5.16 are combined to give the random-effects model:

$$z_{ij} = \beta_0 + \beta_1 x_{1ij} + (\mu_{1j} x_{1ij} + \mu_{0j} + \varepsilon_{ij}) \tag{5.17}$$

Therefore, each location-specific parameter comprises an average value and a random component. Figure 5.3 illustrates (A) a simple regression model, (B) a random intercepts model, and (C) a random intercepts and slopes model. OLS cannot be used to estimate the parameters unless μ_{0j} and μ_{1j} are zero, so specialist multilevel modelling software has been developed, such as MLwiN*, for example.

A limitation of conventional MLMs is the use of arbitrary spatial units which correspond to different levels. In many cases, division into levels in this way may not be meaningful. Also, the assumption is that spatial variation is discontinuous — the model applies over a particular area but a different model applies in a neighbouring area (128). An alternative approach, which is not under-pinned by the assumption that spatial variation is discontinuous, is geographically weighted regression, discussed in Section 5.8.

MLMs (including extensions to the model outlined above) are described in more detail by Snijders and Bosker (336), Goldstein (147) and Subramanian (349). Introductions to MLM in a geographical context are given by Jones (206) and Fotheringham et al. (128). In addition to many applications in socioeconomic contexts, a multilevel approach has been employed in the study of various aspects of the physical environment. For example, Jayaraman and Lappi (202) estimated height-diameter curves in planted teak stands using a multilevel model, whereby plantations and plots within plantations provided different levels. Also, Hòshino (189) used a multilevel model to study land use in different regions, in that study a two-level model was employed with municipality and prefecture levels.

The remainder of the chapter focuses on a range of methods which allow for spatial variation in relations between variables. The methods outlined have received much attention from geographers in recent years.

5.6 Allowing for local variation in model parameters

The main focus in the preceding sections has been on regression methods which account for spatial structure in variables in some way. Such approaches do not allow assessment of spatial variation in the relations between variables in that only a single set of coefficients are generated. In cases where it is believed such spatial variation may exist, several existing solutions are available and these are here included in the umbrella term 'local regression'. Conceptually, the most simple approach is to conduct regression using spatial subsets (i.e., for different areas or for a moving window). An alternative solution is to make the coefficients a function of spatial location — an obvious approach is to model the spatial coefficient surface using a polynomial. More flexible alternatives include the use of (i) splines and (ii) geographical weighting schemes centred on each calibration location. Another class of approach which has been increasingly widely-used entails treating the

---

*multilevel.ioe.ac.uk/features/index.html

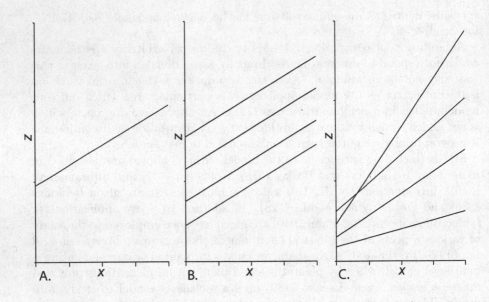

FIGURE 5.3: A) A simple regression model, B) a random intercepts model, and C) a random intercepts and slopes model.

spatially-varying coefficient surface as a realisation from a spatial process. The following sections provide more background on each of these classes of approach, although the main focus is on family of methods that are increasingly commonly encountered in GISystems contexts — geographically weighted regression and adaptations of this approach. Section 5.10 considers some pros and cons of the methods outlined.

There are various ways of allowing for trends in model parameters. These include the spatial expansion method (detailed below) and spatially adaptive filtering. Adaptive filtering is used in the analysis of time series data, and the method is used to account for trends in regression parameters through time. The approach works as a predictor-corrector. If the prediction is regarded as insufficiently accurate, then the model coefficients are adjusted such that the prediction accuracy is increased, which leads to new coefficients (129). This idea has been extended to exploring trends in parameters across space (124), (157). With temporal data, an observation may have a unique nearest neighbour in terms of time and the coefficients can be updated given this nearest neighbour. In the case of spatial data, the procedure is iterative to account for the fact that a location usually does not have a unique neighbour, and the coefficients will have to be updated more than once. With spatial data, the approach updates in two directions between paired neighbouring locations until convergence is, in some sense, achieved. If this is the case, a unique estimate of the β coefficients should be achieved for each case. These coefficients may then be mapped to enable exploration of spatial trends (129). The expansion method also allows for trends in model parameters and it is discussed next.

5.6.1 Spatial expansion method

With the expansion method (73), (74), the parameters of a global model may be made functions of other attributes such as location, and trends in parameter estimates may be explored. The spatial expansion method enables, therefore, the examination of trends across space (60). Firstly, a global model is given as:

$$z_i = \alpha + \beta x_{1i} + \cdots + \tau x_{ki} + \varepsilon_i \tag{5.18}$$

where the parameters $\alpha, \beta, ..., \tau$ are to be estimated, there are k independent variables, and ε_i is an error term. This global model may be expanded such that each of the parameters are made functions of other variables. The parameters can be varied across space and a linear expansion may be given by (128):

$$\alpha_i = \alpha_0 + \alpha_1 u_i + \alpha_2 v_i \tag{5.19}$$

$$\beta_i = \beta_0 + \beta_1 u_i + \beta_2 v_i \tag{5.20}$$

$$\tau_i = \tau_0 + \tau_1 u_i + \tau_2 v_i \tag{5.21}$$

where u_i, v_i refers to the ith set of spatial coordinates.

After the selection of a form of expansion, the parameters in the global model are replaced by their expansions. For the linear model outlined above this gives:

$$z_i = \alpha_0 + \alpha_1 u_i + \alpha_2 v_i + \beta_0 x_{1i} + \beta_1 u_i x_{1i} + \beta_2 v_i x_{1i} + \cdots \tau_0 x_{ki} + \tau_1 u_i x_{ki} + \tau_2 v_i x_{ki} + \varepsilon_i \tag{5.22}$$

The model may then be calibrated using OLS. The estimates of the parameters are then input into Equations 5.19 to 5.21 to derive parameter estimates that vary spatially. Since the estimates refer to location i, they can therefore be mapped. Rogerson (325) gives an example of the spatial expansion method in practice. Brown and Jones (60) use a second-order polynomial surface to model variation in the independent variable. That is, the β coefficients are made functions of geographic location (retaining the notation u, v to represent spatial coordinates), using the terms u, v, uv, u^2, v^2.

SEM is illustrated using data on residential and vehicle thefts, mean household income and mean housing value in Columbus, Ohio, in 1980 (summarised on Section 1.8.7). The zones within which the crimes and other

variables were reported are shown with ID numbers superimposed in Figure
5.4; note that the distance units are arbitrary and thus are not indicated on
the map. The dependent variable was residential and vehicle thefts per 1000
people and the explanatory variables were mean household income and mean
housing value.

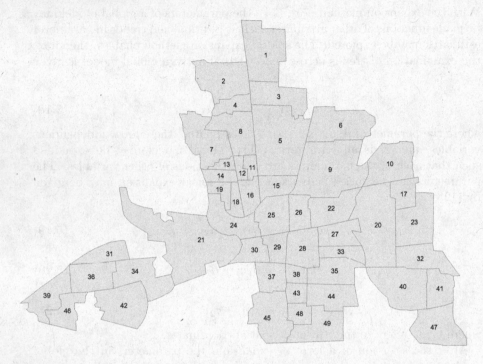

FIGURE 5.4: Zones with ID numbers for Columbus, Ohio, vehicle theft data.

For this example, the model, which includes a first-order trend, is given by:

$$z_i = \alpha + \beta_0 x_{1i} + \beta_1 u_i x_{1i} + \beta_2 v_i x_{1i} + \tau_0 x_{2i} + \tau_1 u_i x_{2i} + \tau_2 v_i x_{2i} + \varepsilon_i$$

where $x_{1i} =$ HH (household) income and $x_{2i} =$ HH value with:

$$\beta_i = \beta_0 + \beta_1 u_i + \beta_2 v_i$$
$$\tau_i = \tau_0 + \tau_1 u_i + \tau_2 v_i$$

In words, in this case, income and house value are made functions of the
coordinates u, v. Table 5.1 gives standard regression and SEM coefficients.
Anselin (9) also generated a SEM using these data.

Comparison of the results for the standard regression model and the SEM
indicate that inclusion of the functions of location increases R^2 (some methods

TABLE 5.1

Columbus crime data: standard
regression and SEM coefficients.

	Std.	SEM
Intercept	68.610	69.496
Income	−1.596	−4.086
House value	−0.274	0.404
$u \times$Income		−0.046
$v \times$Income		0.121
$u \times$House value		0.027
$v \times$House value		−0.049
R^2	0.552	0.633

for assessing model performance are outlined in Section 5.8.2). One obvious problem with the SEM results relates to the counter-intuitive sign of the house value coefficient, which suggests a positive association between house value and crime. In contrast, for the standard model, the coefficients for income and house value are both negative. A key problem with SEM is the collinearity (i.e., correlation between explanatory variables) induced by adding variables which are a function of location. In this case, the income variable is strongly correlated (i.e., $r > 0.9$) with $u \times$income and $v \times$income and the same is true for the house value variable, $u \times$house value and $v \times$house value. The issue of collinearity is discussed further in Section 5.8.4 and is dealt with briefly below.

Despite these problems, these coefficients were used to estimate the location specific coefficients. Figure 5.5 is the map of household income SEM coefficients while Figure 5.6 is the map of household value SEM coefficients. There are clear spatial trends in the two sets of coefficients, but the trends of larger (i.e., smaller negative) coefficients in the north than in the south for the income coefficients is reversed in the case of the household value coefficient which has positive values in the south and negative values in the north. Given the counterintuitive sign of the house value coefficient in Table 5.1 this is unsurprising.

Ridge regression can be used to overcome ill-conditioned design matrices (385). With ridge regression, the coefficients are shrinked through imposing a penalty on their size (180), (385). The parameters for ridge regression may be estimated by solving:

$$\hat{\beta}^R = (\mathbf{X}^T \mathbf{X} + \lambda \mathbf{I})^{-1} \mathbf{X}^T \mathbf{z} \tag{5.23}$$

where λ is the ridge regression parameter, and $\mathbf{I}$ is the identity matrix. For $\lambda = 0$, the standard regression and ridge regression results are the same.

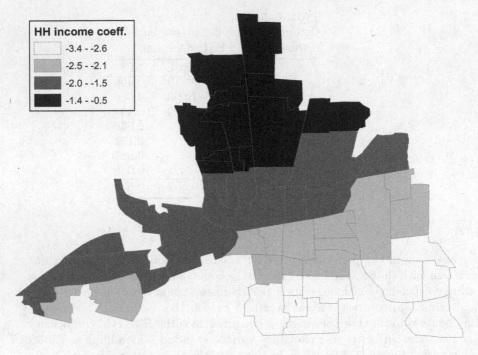

FIGURE 5.5: Household income SEM coefficient.

Ridge regression was applied as a way of overcoming collinearity effects in the SEM model. As is usual with ridge regression, the data were first scaled and centred. The data values, x and z, were transformed with:

$$zs_i = \frac{z_i - \mu_z}{\sigma_z}$$

$$xs_i = \frac{x_i - \mu_x}{\sigma_x}$$

Ridge regression is then conducted without a constant. Table 5.2 gives the coefficients, given the transformed data, for standard regression and values of the ridge regression parameter, λ, of 0.5 and 1.0. In the case of standard regression (λ=0.0) the counterintuitive sign of the house value coefficient is still apparent, whereas for ridge regression for both $\lambda = 0.5$ and $\lambda = 1.0$, the coefficient values are intuitively sensible.

The mapped coefficients given ridge regression follow broadly the same patterns as those shown in Figures 5.5 and 5.6. Therefore, the problem of counterintuitive local coefficients, as apparent in the two maps, remains. There are similar results, in terms of local coefficients, when second order polynomial expansion is used.

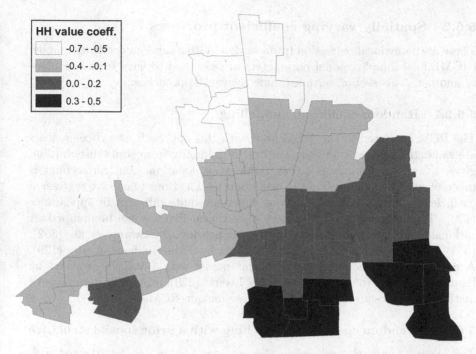

FIGURE 5.6: Household value SEM coefficient.

SEM maps allow exploration of general trends in the values of coefficients but, as stated by Fotheringham et al. (129), alternative approaches are likely to be better suited to assessing fine-scale spatial variation. Accordingly, these data are interrogated further in Section 5.8.4. Fotheringham et al. (128) include an example of the SEM in practice. In their example, the spatially-varying intercept is also obtained.

TABLE 5.2
Columbus crime data: SEM coefficients for centred data.

	$\lambda = 0.0$	$\lambda = 0.5$	$\lambda = 1.0$
Income	-1.393	-0.786	-0.678
House value	0.446	-0.258	-0.330
$u \times$ Income	-0.763	-0.334	-0.195
$v \times$ Income	1.305	0.441	0.243
$u \times$ House value	1.318	0.741	0.534
$v \times$ House value	-2.030	-0.738	-0.463

5.6.2 Spatially varying coefficient processes

There are many local regression frameworks. With a random coefficient model (RCM), for example, model parameters are assumed to vary from one place to another. This section outlines some relevant approaches.

5.6.2.1 Random coefficient modelling

The RCM is related to the MLM and, with this approach, the β coefficients are random variables, and they are drawn randomly from some distribution, given z_i and y_i. Estimates are available at all locations, but no account is taken of spatial location. That is, neighbouring locations may have regression coefficients drawn from distributions which are quite different in appearance (129). The approach is relevant here since the coefficients can be mapped. A rationale and description of the method is provided by Swamy et al. (352), (353), (354), and the approach is outlined by Fotheringham et al. (129). A comparison of the random coefficient model with GWR is provided by Brunsdon et al. (65). Kreft and De Leeuw (220) provide a summary of methods which relate to MLMs, and these include RCMs.

5.6.2.2 Random coefficient modelling with a prior spatial structure

The general approach outlined in the previous section can be adapted such that the random effects are spatially correlated. In this context, various Bayesian regression approaches exist whereby the coefficients are allowed to vary spatially. This section gives a very brief introduction to some key concepts in Bayesian statistical modelling and Bayesian regression as context to a summary of Bayesian spatially varying coefficient models.

In the Bayesian framework, the prior distribution reflects existing knowledge about some unobserved parameters $\boldsymbol{\theta}$. Given some observed data, knowledge about the parameter values is modified, and the posterior distribution is obtained. That is, posterior refers to the distribution *after* observing the data. The posterior can be defined as being proportional to the product of the likelihood and the prior distributions. For random variable Z with parameters $\boldsymbol{\theta}$, the likelihood can be defined as the joint distribution of the sample values (232):

$$L(\mathbf{z} \mid \boldsymbol{\theta}) = \prod_{i=1}^{n} L(z_i \mid \boldsymbol{\theta}) \tag{5.24}$$

For likelihood $L(\mathbf{z} \mid \boldsymbol{\theta})$ and the joint prior distribution of $\boldsymbol{\theta}$, $\mathbf{g}(\boldsymbol{\theta})$, the posterior distribution is given by (233):

$$p(\boldsymbol{\theta} \mid \mathbf{z}) \propto L(\mathbf{z} \mid \boldsymbol{\theta})\mathbf{g}(\boldsymbol{\theta}) \tag{5.25}$$

With $g_i(\theta_i)$ indicating the prior distributions for the $i = 1, ..., p$ components of the parameter $\boldsymbol{\theta}$, the posterior distribution is then given by (233):

$$p(\boldsymbol{\theta} \mid \mathbf{z}) \propto L(\mathbf{z} \mid \boldsymbol{\theta}) \prod_{i=1}^{p} g_i(\theta_i) \qquad (5.26)$$

Where information is lacking, a non-informative prior may be used — Brunsdon (64) gives the example of the case where θ takes a value between 0 and 1 and $f(\theta) = 1$, indicating a uniform distribution and no value of θ has a greater prior probability density than any other value. A short introduction to Bayesian inference is provided by Brunsdon (64).

A core idea in the Bayesian framework is that parameters arise from distributions, and this links to the concept of models with parameters which arise within hierarchies (233). Given the example of a bivariate linear regression, the parameters α and β are often estimated from the data. In a Bayesian framework, the parameters may be regarded as stochastic with these parameters having distributions. In this case, the distributions are termed hyperprior distributions and the parameters are termed hyperparamaters (233). Parameters may then exist in hierarchies. Lawson et al. (233) outline an example, with a Poisson data likelihood, that is a two level hierarchy. In their example, there is a relative risk parameter, θ, with a gamma prior denoted with Gamma(α, β) comprising the first level. The second level comprises hyperprior distributions for α and β.

Bayesian regression approaches have been applied widely. A good introduction to the topic is provided by Birkes and Dodge (44). Albert (4) provides another introduction along with guidance for conducting Bayesian regression using the R environment. In cases where a non-informative prior is used, the inferences from standard (frequentist) regression and Bayesian regression will be similar, but the basis behind the approaches and interpretations of confidence intervals and Bayesian intervals are quite different. One benefit of Bayesian regression is in variable selection whereby priors can be selected to screen predictors which are marginally important (89).

The focus here, of course, is on locally-adaptive models. Various approaches to Bayesian spatially varying coefficient modelling exist. Gelfand et al. (136) developed an approach whereby the spatially varying coefficient surface is modelled as a realisation from a spatial process. The degree of smoothness of the process realisation is modelled by the selected covariance function — unlike the RCM approach above, spatial location is accounted for. Different prior specifications for the β coefficients are defined in the literature. One approach is to assume that the coefficients follow an areal unit model such as the conditional autoregressive model (e.g., see Banerjee et al. (33)), while Gelfand et al. (136) and Wheeler and Calder (389) use a geostatistical approach.

Congdon (91) describes multivariate spatial priors and their application for modelling spatially varying predictor effects. The approach is illustrated using data on male and female suicides in England. Gelfand et al. (136) also use a Bayesian approach for modelling, and they provide an example application. Banerjee et al. (33) provide an introduction and detailed context to this approach. Other introductions to Bayesian varying coefficient models are given by Congdon (89), (91). Gelfand et al. (136) introduce a variety of models based on spatially- and temporally-referenced data. The authors describe models with locally varying (i) intercepts, (ii) slopes, and (iii) intercepts and slopes. The local β coefficients at a given location may be considered as a random spatial adjustment at that location to the relevant overall β coefficient, and the dependence between the coefficients is defined globally (389).

With models which have two or more levels to the hierarchy, the posterior distributions of the parameters may be complex, and it is often not possible to analytically derive these posterior distributions (233). A common solution is to apply simulation-based Markov chain Monte Carlo (MCMC) methods. With MCMC, Monte Carlo simulations of parameter values are generated from a Markov chain which has a stationary distribution. After convergence of the Markov chain, the simulated values will represent a sample from the posterior distribution (375). The Gibbs sampler is a simple MCMC algorithm. With Gibbs sampling, each parameter is taken in turn and the full conditional distribution (the distribution of the parameter conditional on all other model parameters and the data (255)) is sampled. Gilks (146) discusses the derivation of full conditional distributions as well as sampling from the full conditionals. The full conditional for a given parameter, for example β, consists only of the terms in the joint posterior distribution which involve β (146).

Schabenberger and Gotway (330) give a clear conceptual introduction to the Gibbs sampler, and they provide an example given a data vector $\mathbf{z}$ with two parameters ϕ_1 and ϕ_2 with the posterior distribution, $f(\phi_1, \phi_2 \mid \mathbf{z})$, which is intractable and thus an MCMC approach is required. The full conditional distributions are given by $f(\phi_1 \mid \phi_2, \mathbf{z})$ and $f(\phi_2 \mid \phi_1, \mathbf{z})$. Given starting values $\phi_2^{(0)}$, the iterations through the full conditionals are:

$$\phi_1^{(1)} \text{ is generated from } f(\phi_1 \mid \phi_2^{(0)}, \mathbf{z})$$
$$\phi_2^{(1)} \text{ is generated from } f(\phi_2 \mid \phi_1^{(1)}, \mathbf{z})$$
$$\phi_1^{(2)} \text{ is generated from } f(\phi_1 \mid \phi_2^{(1)}, \mathbf{z})$$
$$\phi_2^{(2)} \text{ is generated from } f(\phi_2 \mid \phi_1^{(2)}, \mathbf{z})$$
$$\vdots$$

Given a sufficiently large number of iterations, the values of ϕ_1 and ϕ_2 will become equivalent to samples from the joint posterior distribution.

Alternative approaches are necessary where the prior component and the likelihood component are not 'compatible' (255). In such a context, Gelfand et al. (136) outline a sliced (see Banerjee et al. (33)) Gibbs sampler which can be used to fit the Bayesian spatially varying coefficient process (SVCP) models which the authors detail. An alternative approach is outlined by Wheeler and

Calder (389). The paper by Wheeler and Calder (389) provides a detailed account of the derivation of the posterior distribution in the context of a SVCP model.

As an example, the spatially varying regression effect model detailed by Congdon (89) is summarised; Congdon (90) provides another account. Congdon builds on the work of Langford et al. (229) and Leyland et al. (237) and defines a model with respect to a count outcome with a Poisson likelihood given by $Z_i \sim \text{Poi}(E_i\theta_i)$ (with $\sim$ indicating 'is distributed as'), where E_i can be defined as the expected deaths in area i, and θ_i is the relative risk of mortality in area i. The Poisson distribution is given by (with $\mu = E_i \times \theta_i$):

$$\frac{e^{-\mu}\mu^z}{z!} \text{ for } z = 0, 1, 2... \tag{5.27}$$

Congdon (89) defines a convolution model (with subscript i here denoting location $\mathbf{s}_i$) as:

$$\ln(\theta_i) = \alpha + u_i + e_i + \beta_i x_{1i} + \gamma_i x_{2i} \tag{5.28}$$

With this model, unstructured errors v_{i1}, v_{i2}, v_{i3} underlie the spatially structured effects e_i, β_i, and γ_i. The unstructured errors are linked to the effects in Equation 5.28 using scaled weighing systems and, in the case detailed by Congdon (89), these are considered the same across the effects:

$$e_i = \sum_{j=1}^{n} w_{ij}v_{j1}$$

$$\beta_i = \sum_{j=1}^{n} w_{ij}v_{j2}$$

$$\gamma_i = \sum_{j=1}^{n} w_{ij}v_{j3}$$

with, in the following example, $w_{ij} = 1$ when i and j are neighbours and 0 otherwise. Congdon (89) notes that where in addition $u_i = v_{i4}$ then a multivariate prior of dimension 4 can be set on the v_{ij} with mean zero for v_{i1} and v_{i4} but with v_{i2} and v_{i3} having means β_μ and γ_μ. Given Equation 5.28, $\mu_v = (0, \beta_\mu, \gamma_\mu, 0)$ is the mean vector, and Σ_v is the dispersion matrix of dimension 4×4. For p covariates, Σ_v will be of dimension $(2 + p)$ (89). The model can be adapted to allow for interdependence of the v_{ij} — the spatial effects could be allowed to interact and to have different distance decays (89).

In an example making use of data on lip cancer in Scotland, Congdon (89) uses the WinBUGS[†] (255) software to implement the spatially varying coefficient model. Detailed introductions to WinBUGS are provided by

---

[†]http://www.mrc-bsu.cam.ac.uk/bugs/

Lawson et al. (233) and Ntzoufras (290). Given the likelihood and prior distributions, WinBUGS derives the conditional distributions (as detailed by Lunn et al. (255)).

In the example of Congdon (89), the Poisson likelihood is as given above: $Z_i \sim \text{Poi}(E_i\theta_i)$. In WinBUGS, this Poisson distribution is denoted by z $\sim$ dpois(mu). The data are for 56 regions and include a single independent variable which is 0.1 times the percentage of the labour force employed in agriculture and related occupations (defined by $x = \text{AFF}/10$). This model is given with:

$$\ln(\theta_i) = \alpha + e_i + \beta_i x_i$$

With priors defined as follows: $\alpha \sim \text{N}(0, 0.001)$ and $\mu_\beta \sim \text{N}(0, 1)$ where, in WinBUGS, the format is N(mean,precision), and this is denoted by z $\sim$ dnorm(mu,tau). The precision is $1/\sigma^2$ (denoted below by τ). For completeness, the normal distribution is given by:

$$\sqrt{\tfrac{\tau}{2\pi}}\exp\left(-\tfrac{\tau}{2}\left(z - \mu\right)^2\right) \; -\infty < z < \infty \qquad (5.29)$$

A bivariate Gaussian prior distribution is assigned to $V_i = (v_{i1}, v_{i2})$:

$$V_i \sim N_2\left(\begin{pmatrix} 0 \\ \mu_\beta \end{pmatrix}, \Sigma\right)$$

where Σ is a positive definite matrix which is the precision of (v_{i1}, v_{i2}). The multivariate normal distribution is denoted in WinBUGS by z $\sim$ dmnorm(mu[],Σ[]), and the distribution can be given by:

$$(2\pi)^{-d/2}\,|\Sigma|^{1/2}\exp\left(-\tfrac{1}{2}\left(z - \mu\right)^T \Sigma\left(z - \mu\right)\right) \; -\infty < z < \infty \qquad (5.30)$$

A Wishart prior is assigned to Σ (see Waller et al. (376) for a similar example). The Wishart distribution is denoted in WinBUGS with Σ[,] $\sim$ dwish(R[,], k) (here, with 1 for the diagonal elements of R and 0 for the off-diagonals), and it can be defined with:

$$|R|^{k/2}|\Sigma|^{(k-d-1)/2}\exp\left(-\frac{1}{2}\text{tr}\left(R\Sigma\right)\right) \qquad (5.31)$$

where the matrix Σ is $d \times d$, and k is the degrees of freedom of the matrix; tr indicates the trace of the matrix. In this example, $k = 2$. Spiegelhalter et al. (339) define the available distributions in WinBUGS.

WinBUGS code to implement the model is available through the website of Peter Congdon[‡]. This code was used to generate local coefficients for AFF/10, and these are shown in Figure 5.7. The mapped values are the posterior

---

[‡]http://webspace.qmul.ac.uk/pcongdon/

means. There is clear spatial variation in the derived local coefficients, with the most obvious areas with large values being in the north and parts of the south east of the country. The map was derived using WinBUGS and made use of boundaries provided as a part of one of the tutorial datasets distributed with the software. The analysis can, therefore, be replicated easily. WinBUGS allows monitoring of the convergence of the chains and the posterior distributions can be summarised in a variety of ways.

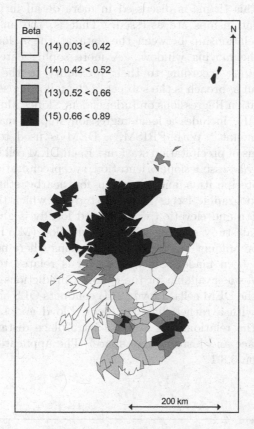

FIGURE 5.7: Beta coefficients for AFF/10. The numbers in parenthesis indicate the number of zones in each class.

One key distinction between the SVCP models and those detailed below is that the former entail a single likelihood whereas the moving window and geographically weighted approaches detailed below are based on *n* separate models (90). Some potential advantages of SVCP models relative to other frameworks, such as those detailed in the following sections, are outlined in Section 5.10.

5.7   Moving window regression (MWR)

A simple way of assessing spatial variations in relations between variables is to conduct regression in a moving window (e.g., (240), which uses regression and other techniques to predict values of precipitation at unsampled locations, given values of elevation at precipitation measurement locations and on a regular grid) and this theme is discussed in more detail in Chapter 6. In this case, local relationships are assessed. That is, the analysis is based on the modelled relationship between the data at the n locations closest to the centre of the moving window. A more sophisticated approach is to weight observations according to their proximity to the centre of the window, and such an approach is the subject of the following section. PRISM (Precipitation-elevation Regressions on Independent Slopes Model), described by Daly et al. (101), includes a local regression step to make predictions of precipitation amount. With PRISM, a DEM is used to estimate the orographic elevations of precipitation stations. Each DEM cell is assigned to a topographic facet by assessing slope orientation, and precipitation is estimated by regression of precipitation and elevation for nearby stations within a DEM grid cell's topographic facet. Another study in which the relationship between precipitation and elevation was assessed locally is given by Nicolau et al. (288). In that study, moving window regression (MWR) was used to predict precipitation amount in Portugal at locations where no precipitation measurements had been made. Since elevation is related to precipitation amount and a DEM was available for the country, predictions could be made at the locations of the DEM cells. Lloyd (244) conducts OLS and GLS MWR. In an application with a rather different focus, Lloyd and Lilley (252) use MWR to explore the relationships between intra-place distances measured in modern map space and historic map space. The application of MWR is illustrated in Section 5.8.1.

5.8   Geographically weighted regression (GWR)

In the last decade a major area of research in the analysis of spatial data has been focused in the development and use of geographically weighted regression (GWR). With GWR, the data are weighted according to their distance from the location **s**. That is, observations (often at points or centroids of areas) close to the centre of the window are weighted more heavily than observations farther away (a spatial kernel is used, as described in Section 2.4). For a full account of the principles of GWR see Fotheringham et al. (128), (129), (130).

When GWR is applied, key decisions concern the choice of (i) a weighting function (the shape of the kernel), and (ii) the bandwidth of the kernel. While (i) has been shown to usually have a minimal effect on results, (ii) may affect results markedly, so methods for the identification of an optimal bandwidth have been developed (129). These issues will be discussed below. Often, a kernel with a single bandwidth is used at all locations — that is, the same model parameters are used everywhere. But, the use of locally-adaptive kernel bandwidths is possible, as discussed below in this section.

The local estimation of the parameters with GWR is given by (for two independent variables):

$$z(\mathbf{s}_i) = \beta_0(\mathbf{s}_i) + \beta_1(\mathbf{s}_i)x_1 + \beta_2(\mathbf{s}_i)x_2 + \varepsilon \tag{5.32}$$

where $\mathbf{s}_i$ is the location at which the parameters are estimated — note that vector notation is used for locations in most of this section for clarity. Parameters can also be estimated at locations where there are no data. The parameters for GWR may be estimated by solving:

$$\hat{\beta}(\mathbf{s}_i) = (\mathbf{X}^T \mathbf{W}(\mathbf{s}_i)\mathbf{X})^{-1}\mathbf{X}^T \mathbf{W}(\mathbf{s}_i)\mathbf{z} \tag{5.33}$$

where $\mathbf{W}(\mathbf{s}_i)$ is an n by n matrix, the diagonal elements of which are the geographical weightings of observations around point i:

$$\mathbf{W}(\mathbf{s}_i) = \begin{bmatrix} w_{i1} & 0 & \cdots & 0 \\ 0 & w_{i2} & \cdots & 0 \\ \vdots & \vdots & \cdots & \vdots \\ 0 & 0 & \cdots & w_{in} \end{bmatrix} \tag{5.34}$$

where w_{in} is the weight assigned to the observation at location n.

Several weighting functions have been used. One of these, the Gaussian function, was defined in Equation 4.1. This function is applied in the GWR software (129). As noted above, GWR may also be applied with spatially adaptive kernels. A variety of approaches have been applied. One approach is termed the bi-square nearest neighbour scheme. In this case, where there are many observations in a given area the bandwidth is small, whereas where data are sparse a larger bandwidth is used. The weights are given by:

$$w_{ij} = \begin{cases} [1 - (d_i/\tau)^2]^2 & \text{if } d_i \leq \tau \\ 0 & \text{otherwise} \end{cases} \tag{5.35}$$

That is, when the distance d_i between regression location i and the location j is greater than the distance τ, the weight is zero. So, τ is the distance to the Nth nearest neighbour, and the value of τ varies locally such that each regression location has the same number of neighbouring points with non-zero weights. This method is applied in Section 5.8.5.

The goodness-of-fit of a GWR can be assessed using the geographically weighted coefficient of determination (129):

$$r_i^2 = (\text{TSS}^w - \text{RSS}^w)/\text{TSS}^w \tag{5.36}$$

where TSS^w is the geographically weighted total sum of squares:

$$\text{TSS}^w = \sum_{j=1}^{n} w_{ij}(z_j - \overline{z}_i)^2$$

and RSS^w is the geographically weighted residual sum of squares:

$$\text{RSS}^w = \sum_{j=1}^{n} w_{ij}(z_j - \hat{z}_j)^2$$

Standard linear regression is based on the assumption of a Gaussian distribution. The GWR software allows for the selection of Gaussian, logistic, and Poisson models (129). Poisson GWR is described and applied by Nakaya et al. (286). A worked example of standard GWR is given by Lloyd (245). Páez (306) explores anisotropy (i.e., directional variation) in the context of GWR and argues that anisotropy may be common in spatial processes. In his case study, utilising anisotropic variance functions resulted in significant performance gains. Several software environments exist for conducting GWR. These include the software written by the developers of GWR (129), routines written in the R language (48), and the GWR functionality of ArcGIS™ 9.3.

Wheeler and Tiefelsdorf (387) discuss the problem of multicollinearity with respect to GWR and its possible impact on the interpretation of the GWR coefficients. This is the subject of Section 5.8.4. LeSage (234) presents a Bayesian implementation of GWR, and he argues that his approach has several advantages over standard GWR including greater confidence in inference regarding spatial variation in the modelled relationships. Congdon (89) provides another description of Bayesian GWR along with an example.

5.8.1   Illustrated application of MWR and GWR

The application of MWR and GWR is illustrated using the elevation and precipitation dataset described in Chapter 1. First, regression was conducted using all of the data, as described in Section 5.1. A subset of the precipitation data was used to allow comparison of global regression, MWR and GWR with respect to one location. Table 5.3 gives, for the data subset, the elevation and precipitation observations, distances from the prediction location and weights assigned using a Gaussian kernel with a bandwidth of 11.074 km (identified using the GWR software; see Section 5.8.2 for more details). In Figure 5.8, the regression (with equal weights used to fit the model) using the data subset is shown. MWR is based on regression using subsets in this manner. The regression coefficients are written to the location at the centre of the window, and this procedure is then conducted at all other data locations. Here, regression is based on each observation and its 16 nearest neighbours. Where GWR is used with a fixed bandwidth, a discrete subset is not used in this way. That is, additional points outside of this subset would receive positive weights with GWR if the full dataset was used.

TABLE 5.3
Elevation and precipitation observations, distances from the prediction location, and weights assigned using a Gaussian kernel with a bandwidth of 11.074 km.

X (m)	Y (m)	Elevation (m)	Precipitation (mm)	Distance (m)	Weight
426100	420000	46	12.10	0	1.000
429300	423300	98	29.40	4596.738	0.917
422000	422500	50	18.20	4802.083	0.910
432000	419600	32	14.80	5913.544	0.867
434600	420400	24	12.70	8509.407	0.744
427300	408400	113	13.10	11661.904	0.574
424600	432400	123	18.50	12490.396	0.529
432300	431500	25	10.30	13064.838	0.499
439600	423900	16	28.80	14052.046	0.447
421300	405600	260	27.00	15178.933	0.391
427200	437500	100	16.75	17534.537	0.285
415300	405800	283	45.40	17840.404	0.273
409600	413000	250	25.70	17923.448	0.270
414100	405800	262	47.50	18591.396	0.244
444400	426500	18	22.65	19420.093	0.215
407200	425600	248	24.70	19712.179	0.205
410900	406800	254	50.50	20131.567	0.192

The regression equations for global regression, MWR and GWR were as follows:

Global: $z = 24.814 + 0.105x$ (based on all 3037 observations)
MWR: $z = 13.302 + 0.0872x$
GWR: $z = 12.824 + 0.0846x$ (using only the data in Table 5.3)
GWR: $z = 13.643 + 0.0741x$ (using all of the (geographically weighted) data)

Note that two sets of GWR results are referred to — those using (i) only the data detailed in Table 5.3 (to allow replication of the analysis by readers) and (ii) using all of the data. The second corresponds to the standard approach using a fixed bandwidth.

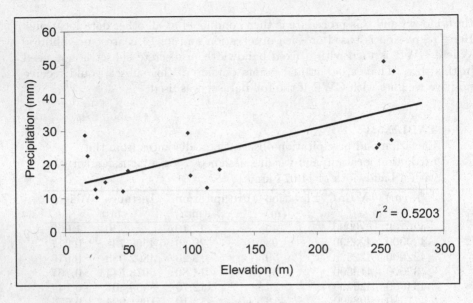

FIGURE 5.8: Elevation against precipitation: data subset.

The global coefficient of determination (r^2) was 0.111. For MWR r^2 was 0.520, while for GWR r^2 was 0.498 (using only the data in Table 5.3) and 0.599 (using all data). The predictions and residuals for the three methods are as follows: global model prediction = 29.644, error = 17.544, MWR prediction = 17.313, error = 5.213, GWR prediction (data subset) = 16.716, error = 4.616, GWR prediction (all data) = 17.047, error = 4.947. The figures suggest that there *is* spatial variation in the relation between elevation and precipitation. In addition, given that the GWR residual is smallest, the figures suggest that weighting the regression as a function of distance from the prediction location may increase the accuracy of predictions at locations where no observations are available.

5.8.2   Selecting a spatial bandwidth

Various procedures have been developed for selecting an appropriate bandwidth. One approach is to use a cross-validation (CV) procedure. That is, each point is extracted in turn and its value is predicted using the neighbouring observations. Specifically, the CV score is obtained with:

$$\mathrm{CV} = \sum_{i=1}^{n} [z_i - \hat{z}_{\neq i}(\tau)]^2 \tag{5.37}$$

where $\hat{z}_{\neq i}(\tau)$ is the fitted value of z_i with the model calibration conducted omitting the observation at location i. Therefore, the bandwidth, τ, with the smallest CV score is usually retained for use in analysis. An alternative to cross-validation is the Akaike Information Criterion (AIC) (129). The AIC can be given by:

$$\text{AIC}_c = 2n \ln(\hat{\sigma}) + n \ln(2\pi) + n \left\{ \frac{n + \text{tr}(\mathbf{A})}{n - 2 - \text{tr}(\mathbf{A})} \right\} \tag{5.38}$$

where $\hat{\sigma}$ denotes the estimated standard deviation of the error term, and $\text{tr}(\mathbf{A})$ is the trace of the hat matrix (see Section 6.3.7 for a discussion about the hat matrix in the context of thin plate splines). The hat matrix relates $\hat{\mathbf{z}}$ to $\mathbf{z}$:

$$\hat{\mathbf{z}} = \mathbf{A}\mathbf{z} \tag{5.39}$$

Each row of $\mathbf{A}$, $\mathbf{r}_i$, is given by (129):

$$\mathbf{r}_i = \mathbf{X}_i (\mathbf{X}^T \mathbf{W}(\mathbf{s}_i) \mathbf{X})^{-1} \mathbf{X}^T \mathbf{W}(\mathbf{s}_i) \tag{5.40}$$

Minimising the AIC allows for a trade-off between goodness of fit and degrees of freedom. The AIC has advantages over cross-validation — it is more generally applicable and can be applied in non-Gaussian GWR. The GWR software designed and described by Fotheringham et al. (129) allows use of the AIC for bandwidth selection, as the adjustment of the bandwidth alters the number of degrees of freedom in the model. A key concern in the application of local models is that they should be, in some sense, more appropriate than a global model, and the AIC can be used to inform selection of a global model or a GWR model.

Lloyd and Shuttleworth (253) apply GWR with several different bandwidths in the exploration of travel-to-work patterns in Northern Ireland. The need to explore changes in observed spatial variation with changes in bandwidths is stressed. In this case, rather than unmodified straight-line distances between area centroids, a modified interarea distance matrix was used which accounts for natural obstructions such as water bodies. The application entailed the use of different bandwidths to assess change in the output maps with change in the scale of the analysis. Clearly, when the bandwidth is small local features are visible, while for large bandwidths regional-scale trends are apparent. Such research demonstrates the need to match observed spatial variation, and thus interpretations, to the scale of analysis.

Applications of GWR in physical geography contexts include the work of Brunsdon et al. (68) and Lloyd (244). Research making use of GWR for exploring aspects of human populations is presented by Lloyd and Shuttleworth (253) and Catney (76).

5.8.3 Testing the significance of the GWR model

If a local approach is applied then it is, of course, useful to ascertain if the local model explains more variation than the global model. In the context of GWR there are various ways of assessing if parameter estimates are 'significantly nonstationary' (129). One approach is based on assessing individual parameter stationarity. The variance for parameter k can be given by:

$$V_k = \frac{1}{n} \sum_{i=1}^{n} \left(\hat{\beta}_k(\mathbf{s}_i) - \frac{1}{n} \sum_{i=1}^{n} \hat{\beta}_k(\mathbf{s}_i) \right)^2 \qquad (5.41)$$

Given the difficulty associated with deriving the null distribution of the variance under the hypothesis that the parameter is constant globally, a Monte Carlo approach is used. The Monte Carlo procedure reallocates randomly the observations across the spatial units (129). The variance can be obtained given each of the n permutations, and the actual variance of the coefficient is compared to this set of variances. In this way, an experimental significance level can be derived (see Section 4.3.1 for some related material). In addition to the Monte Carlo significance test, the GWR software offers a test based on the analytical approach of Leung et al. (235) which was proposed given the computationally intensive nature of the Monte Carlo test. However, Fotheringham et al. (129) note they have found that, in practice, the test has proven to be as computationally intensive as the Monte Carlo test.

5.8.4   Geographically weighted regression and collinearity

Wheeler and Tiefelsdorf (387) have demonstrated that geographical weighting can exacerbate the problem of collinearity in explanatory variables, and that moderate collinearity may lead to a strong dependence in local model coefficients with consequent difficulties in the interpretation of the coefficients. Wheeler (385), (388), (386) presents some diagnostic tools and suggests solutions to the problem. An overview is provided by Wheeler and Páez (390). Griffith (168), (169) offers another approach in a spatial filtering framework.

Several approaches to identifying explanatory variable collinearity have been developed and applied. These include variance decomposition using singular-value decomposition (SVD) and variance inflation factors (VIF). Wheeler (385) assesses the application of both approaches in a GWR context.

The VIF for the kth explanatory variable at location i is given by (385):

$$\text{VIF}_k(\mathbf{s}_i) = \frac{1}{1 - R_k^2(\mathbf{s}_i)} \qquad (5.42)$$

where R_k^2 is the coefficient of determination for x_k regressed on the other explanatory variables with respect to location $\mathbf{s}_i$. Where there are two explanatory variables, the geographically weighted correlation coefficient can be used to derive the VIF (which, in this case, will be the same for both variables).

Building on the discussion in Section 4.2, the geographically weighted covariance can be given by:

$$\text{cov}(x_1(\mathbf{s}_i), x_2(\mathbf{s}_i)) = \sum_{j=1}^{n} w(\mathbf{s}_j)(x_1(\mathbf{s}_j) - \bar{x}_1(\mathbf{s}_i))(x_2(\mathbf{s}_j) - \bar{x}_2(\mathbf{s}_i)) \quad (5.43)$$

The geographically weighted correlation coefficient for location $\mathbf{s}_i$ can then be given by:

$$r(\mathbf{s}_i) = \frac{\text{cov}(x_1(\mathbf{s}_i), x_2(\mathbf{s}_i))}{\sqrt{\sigma(x_1(\mathbf{s}_i))\sigma(x_2(\mathbf{s}_i))}} \quad (5.44)$$

For the calculation of r (and r^2) for the purposes of computing the VIF, the square root of the weights is used; the weights are then standardised to sum to one. The geographically weighted standard deviation can be estimated as detailed in Equation 4.4. In the case of more than two explanatory variables, the VIF can be obtained through a GWR of each explanatory variable on all of the others.

Variance-decomposition using singular-value decomposition (SVD) is another approach to assessing collinearity which is described by Belsley et al. (36) and detailed in a GWR context by Wheeler (385). The condition index and variance-decomposition as collinearity diagnostics derived from SVD are detailed below after a summary of SVD itself. For GWR, the SVD is given by:

$$\mathbf{W}^{1/2}(\mathbf{s}_i)\mathbf{X} = \mathbf{U}\mathbf{D}\mathbf{V}^T \quad (5.45)$$

where $\mathbf{W}^{1/2}(\mathbf{s}_i)$ is the square root of the geographical weights for location $\mathbf{s}_i$. For n observations and p explanatory variables, the (orthogonal) matrices $\mathbf{U}$ and $\mathbf{V}$ are $n \times (p+1)$ and $(p+1) \times (p+1)$ respectively, while $\mathbf{D}$ is a $(p+1) \times (p+1)$ diagonal matrix of singular values. The matrix of explanatory variables (including the constant) is given by $\mathbf{X}$. The product of $\mathbf{W}^{1/2}(\mathbf{s}_i)\mathbf{X}$ is scaled to have equal column lengths. In words, each case is divided by the column norm, $\|\mathbf{x}\|$, where each case has been multiplied by the appropriate geographical weight:

$$xs_i = \frac{x_i}{\|\mathbf{x}\|} \quad (5.46)$$

where

$$\|\mathbf{x}\| = \sqrt{\sum_{j=1}^{n} x_j^2} \quad (5.47)$$

Given the SVD, the variance-covariance matrix for the regression coefficients is given by:

$$\text{Var}(\hat{\beta}) = \sigma^2 \mathbf{V}\mathbf{D}^{-2}\mathbf{V}^T \quad (5.48)$$

The variance of the kth regression coefficient is given by:

$$\text{Var}(\hat{\beta}_k) = \sigma^2 \sum_{j=1}^{p} \frac{v_{kj}^2}{d_j^2} \tag{5.49}$$

where the v_{kj} are elements of the $\mathbf{V}$ matrix, and the d_j are the singular values. The proportion of the variance of the kth regression coefficient associated with the jth component of its decomposition is termed the variance-decomposition proportion (385) and these proportions are given by:

$$\pi_{jk} = \frac{\phi_{kj}}{\phi_k} \tag{5.50}$$

where

$$\phi_{kj} = \frac{v_{kj}^2}{d_j^2} \tag{5.51}$$

and

$$\phi_k = \sum_{j=1}^{p} \phi_{kj} \tag{5.52}$$

For column $j = 1, ..., p+1$ of $\mathbf{W}^{1/2}(\mathbf{s}_i)\mathbf{X}$, the condition index is given by:

$$\eta_j = \frac{d_{\max}}{d_j} \tag{5.53}$$

Belsley et al. (36) and Wheeler (385) detail some recommendations as to condition indexes and variance proportions which suggest collinearity. Variance proportions greater than 0.5 for two or more coefficients for each variance component are considered to indicate collinearity. For the conditional index, larger values indicate stronger collinearity amongst the columns of the matrix of explanatory variables $\mathbf{X}$, and Belsley et al. (36) suggest that moderate to strong relations are indicated by condition indexes of above 30. The remainder of this section focuses on potential solutions when collinearity is indicated.

The preceding text focuses on ways of identifying collinearity. As noted in Section 5.6.1 ridge regression offers a way of reducing the effect of this problem. Wheeler (385) applies ridge regression in the GWR framework (GW ridge regression — GWRR). As noted by Wheeler (385), cross-validation or generalised cross-validation (see Section 6.3.7) can be used to estimate the ridge regression parameter.

Application of ridge regression in the GWR context necessitates the removal or isolation of the intercept term (385). Wheeler (385) outlines two approaches to removing the intercept — global centring and local centring. Global centring entails the following steps:

1. Centre the x variables, removing the portion of the intercept when $x = 0$ which leaves the global mean of z.

2. Centre the dependent variable, removing the global mean of z.

3. Scale the x and z variables.

4. Remove the local x and z mean deviations from the global means resulting in an intercept of zero for each of the local models.

Wheeler (385) notes that one approach is to fit a GWR model to the globally centred data and subtract the fitted intercept from the local z. This has the effect of removing the intercept from the ridge regression constraint and allows the penalised coefficients to be estimated. The scaling of the y and z variables by their standard deviations is justified by the fact that the ridge regression results are scale dependent. Failure to scale the variables would result in the ridge regression solution being more influenced by variables with a large variance (385). With global centring, the GWRR coefficients can be estimated with:

$$\hat{\beta}(\mathbf{s}_i) = (\mathbf{X}^{*T}\mathbf{W}(\mathbf{s}_i)\mathbf{X}^* + \lambda\mathbf{I})^{-1}\mathbf{X}^{*T}\mathbf{W}(\mathbf{s}_i)\mathbf{z}^* \tag{5.54}$$

where $\mathbf{X}^*$ is the matrix of standardised explanatory variables, and $\mathbf{z}^*$ is the standardised dependent variable. As for standard ridge regression (see Equation 5.23), the ridge parameter must be estimated before the regression coefficients. Following estimation of the regression coefficients, the dependent variable predictions are obtained by adding the local mean deviation from the global z mean to the prediction based on global centring (i.e., $\hat{z}^*(\mathbf{s}_i) = \mathbf{X}^*(\mathbf{s}_i)\hat{\beta}(\mathbf{s}_i)$. The estimated z is transformed to the original units with $\hat{\mathbf{z}} = \hat{\mathbf{z}}^*\text{std}(z) + \bar{z}$, where $\text{std}(z)$ is the standard deviation of z.

The local centring procedure can be summarised as follows:

1. Globally scale the x and z variables.

2. For each model locally centre the x and z variables by calculating the weighted mean using the square root of the geographical weights ($\mathbf{W}^{1/2}(\mathbf{s}_i)$) and subtract the weighted mean from each variable.

3. Apply the weights ($\mathbf{W}^{1/2}(\mathbf{s}_i)$) to the centred values.

Given this procedure, the GWRR coefficients can be estimated with:

$$\hat{\beta}(\mathbf{s}_i) = (\mathbf{X}_w^T\mathbf{X}_w + \lambda\mathbf{I})^{-1}\mathbf{X}_w^T\mathbf{z}_w \tag{5.55}$$

where $\mathbf{X}_w$ is the matrix of weighted locally centred explanatory variables, and $\mathbf{z}_w$ are the weighted dependent variable cases. Given the estimated regression coefficients, the dependent variable predictions are obtained by adding the local mean $\bar{z}_w$ to $\hat{\mathbf{z}}_w(\mathbf{s}_i) = \mathbf{X}_w(\mathbf{s}_i)\hat{\beta}(\mathbf{s}_i)$. The estimated z is transformed to the original units through scaling by the standard deviation of z (that is, $\hat{\mathbf{z}} = \hat{\mathbf{z}}_w \text{std}(z)$).

Wheeler (385) recommends the use of global centring only if the desire is to compare results to global standardised regression and GWR models.

The use of CV and GCV to estimate the ridge parameter was mentioned above. Wheeler (385) outlines several possible approaches to estimating the kernel bandwidth and the ridge parameters with CV:

1. Estimate the kernel bandwidth first followed by the ridge parameter.

2. Estimate the kernel bandwidth, then the ridge parameter, and repeat using previous values until the parameters converge.

3. Conduct a search for the kernel bandwidth and conduct a search for the ridge parameter for each value of the kernel bandwidth considered.

4. Estimate simultaneously the kernel bandwidth and the ridge parameter using some constrained optimisation technique.

In a study making use of the crime data applied in Section 5.6.1 (see Figure 5.4), Wheeler (385) evaluates the use of variance-decomposition proportions and VIFs for assessing mulicollinearity. As noted above, the dependent variable was residential and vehicle thefts per 1000 people while the explanatory variables were mean household income and mean housing value. Wheeler (385) applies GWR and GWRR, with kernel bandwidth and ridge parameter estimation conducted using a range of approaches.

Wheeler concludes that GWRR produces more intuitive regression coefficients than GWR, for the particular case study presented, and also gives smaller prediction errors than does GWR. That analysis was replicated using global regression (using raw data and centred ('C') data), GWR and GWRR (with local centring — indicated by 'LC'). The global regression, GWR and GWRR coefficients are summarised in Table 5.4. For GWR and GWRR the bi-square nearest neighbour weighting function was used (see Equation 5.35) with 11 nearest neighbours. In the global (unstandardised) regression case, the VIF was 1.333. This indicates, as argued by Wheeler (385), no evidence for collinearity in the global model case.

TABLE 5.4

Columbus crime data: Global regression, GWR and GWRR coefficient summary.

Method	Global Raw	Global C			GWR Raw	GWR LC	GWRR LC
Intercept	68.610		Min.		37.285		
			Median		63.669		
			Max		89.492		
			Mean		62.271		
Income	−1.596	−0.544	Min.		−5.746	−1.816	−0.572
			Median		−1.038	−0.355	−0.235
			Max		1.091	0.373	0.153
			Mean		−1.380	−0.476	−0.255
House value	−0.274	−0.302	Min.		−1.957	−1.940	−0.971
			Median		−0.115	−0.113	−0.147
			Max		1.321	1.308	0.117
			Mean		−0.164	−0.169	−0.206

For GWR, condition indexes (see Equation 5.53), variance-decomposition proportions and VIFs were computed using purpose-written Fortran 77 code calling the svdcmp routine published by Press et al. (319). The condition index results were also checked using the GWR functionality of ArcGIS™. The condition indexes have values of greater than 30 at three locations (49, 44, and 47 on Figure 5.4), thus suggesting that the application of GWRR may be useful in this case. In addition, the condition indexes have a value of greater than 20 at a further seven locations. Table 5.5 summarises these cases (for the largest variance component) as well as those locations where the VIF was greater than five. Figure 5.9 is a map of condition index values, and this shows that the largest values are in the south central and south eastern regions.

Using approaches like variance decomposition or VIFs, it is possible to ascertain if multicollinearity is likely to have an impact on parameter estimates and if, therefore, the application of an approach like GWRR or the geographically weighted lasso (388), rather than standard GWR, is likely to be beneficial. In this example, the large condition index and VIF values at some locations suggest that the application of GWRR is justified.

TABLE 5.5
Columbus crime data: selected condition indexes, variance-decomposition proportions and VIFs.

ID	η_j	π_{j1}	π_{j2}	π_{j3}	VIF
22	25.999	0.053	0.988	0.981	17.815
27	21.961	0.000	0.984	0.984	21.117
33	17.106	0.021	0.975	0.964	10.881
40	25.358	0.100	0.981	0.955	8.306
41	23.979	0.053	0.965	0.919	4.157
43	21.917	0.007	0.882	0.919	2.189
44	41.748	0.493	0.930	0.994	4.525
45	20.542	0.167	0.974	0.761	2.001
47	31.024	0.038	0.975	0.956	8.382
48	29.065	0.067	0.911	0.967	2.871
49	46.693	0.390	0.948	0.992	5.706

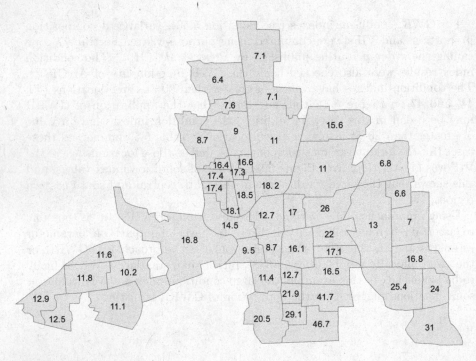

FIGURE 5.9: Condition indexes for Columbus, Ohio, vehicle theft GWR analysis.

Wheeler (388) presents maps of GWR and GWRR coefficients given the Columbus crime data. To allow comparison with the SEM-derived maps shown in Figures 5.5 (household income coefficient) and 5.6 (household value coefficient), maps of coefficients derived using GWRR (with $\lambda = 0.8$) are given in Figures 5.10 and 5.11. The maps have very different characteristics to those generated using the SEM and they show, as expected, a much greater degree of local variability.

FIGURE 5.10: Household income coefficient for GWRR with $\lambda = 0.8$.

5.8.5   Case study: MWR and GWR

The case study is concerned with the analysis of the relations between precipitation in Great Britain (for the month of July 1999) and elevation using the data described in Chapter 1. It is well known that precipitation and altitude tend to be related (at least for periods of weeks or more) (21), and this is discussed further by Lloyd (240). Brunsdon et al. (68) used GWR to explore spatial variations in relations between elevation and precipitation in Britain. A related approach is presented by Kelly and Atkinson (216), who were concerned with predicting snow depth at unsampled locations in the UK.

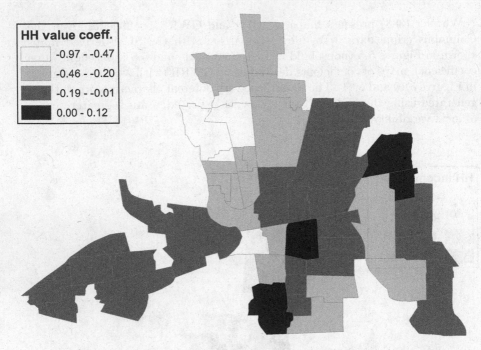

FIGURE 5.11: Household value coefficient for GWRR with $\lambda = 0.8$.

In that study, regression between elevation and snow depth was conducted in four mountainous regions of the UK.

As the first stage of the case study, global regression was conducted. The scatter plot and fitted line is given in Figure 5.1, and the mapped residuals are shown in Figure 5.2. MWR was carried out using purpose-written Fortran computer code. Some results of analysis of the data using MWR were presented by Lloyd (240). GWR was conducted using the GWR software (129). The GWR bandwidth was selected using the AIC while the MWR neighbourhoods were selected arbitrarily. Using the AIC, a bandwidth of 29.922 km was selected. GWR was also applied with a variable bandwidth. In this case, a neighbourhood of 155 nearest neighbours was selected (thus, the bandwidth varies according to the density of observations). The range of the MWR and GWR model parameters and r^2 values are given in Table 5.6. The GWR 3.0 software returns r^2 as the squared geographically weighted correlation coefficient between the observed and predicted values. For the purposes of this example, r^2 was computed using Equation 5.36 with purpose-written software.

TABLE 5.6

Precipitation vs. elevation: model parameters. BW is bandwidth.

Method	N. data	BW / window	Min. intercept	Max.	Min. slope	Max.	Min. r^2	Max.
MWR	8	n/a	−57.183	192.522	−5.187	4.600	0.000	0.963
MWR	16	n/a	−27.933	161.733	−1.622	0.978	0.000	0.969
MWR	32	n/a	−14.293	148.534	−0.303	0.686	0.000	0.859
MWR	64	n/a	3.974	145.174	−0.179	0.470	0.000	0.761
MWR	128	n/a	4.879	130.980	−0.135	0.347	0.000	0.717
GWR	n/a	29922.516	4.159	142.032	−0.517	0.498	0.000	0.777
GWR	155	n/a	5.173	139.020	−0.121	0.398	0.000	0.635

Comparison of the global r^2 and the local values in Table 5.6 indicates clearly that the global model explains much less of the variation in the relations between elevation and precipitation. The median r^2 for GWR with a fixed bandwidth was 0.104, while that for the adaptive bandwidth was 0.103.

With GWR, parameter estimates can be made at any location (that is, at both sample and non sample locations). · In this study, the coefficients (given a fixed bandwidth) were initially derived at the observation locations (as summarised in Table 5.6). For visualisation purposes the coefficients were also derived on nodes of a grid with a 2km spacing. The corresponding GWR intercept values are mapped in Figure 5.12, and GWR slopes are mapped in Figure 5.13 (page 140). Note that the differences in the ranges of values in Table 5.6 and in these figures is due to the latter being derived on a grid and the former at the observation locations only.

Intercept values tend to be larger in the west of Scotland than elsewhere. The largest negative slope values are in parts of the Western Isles of Scotland, the mid west of Scotland, and the Shetland Islands in the far north, as well as central southern England. There are distinct patterns elsewhere — for example, there are large positive values in the Essex region and parts of Norfolk. Where intercept values are large, this suggests that precipitation amounts are large, irrespective of the elevation. Where slope values are large, then this suggests that a small increase in elevation corresponds to a large increase in precipitation amount.

Values of r^2 are mapped in Figure 5.14 (page 141). The largest values are in the Orkney Islands, southwest Scotland, northwest Wales, southwest Wales, the northern midlands of England, and the Cornwall/Devon area. These areas correspond with relatively large elevation values. That is, precipitation appears to be related to elevation in areas where elevation values are large. This is particularly clear, for example, in mountainous areas of northwest Wales. This study indicates that use of a nonstationary model such as GWR (with fixed or adaptive kernel bandwidths) provides much more information than use of a global model. This has implications for methods that utilise the elevation-precipitation relation for spatial prediction, and this theme is explored further in Section 7.16.4.

5.8.6   Other geographically weighted statistics

The principle of geographical weighting can be expanded to include other univariate and multivariate techniques. These include summary statistics such as the mean, standard deviation and skewness (see Section 4.2 for summaries of some such statistics). Nakaya (285) presents a geographically weighted variant of a spatial interaction model — that is, a model of flows (or moves) between locations of which migration or commuting are common examples.

Fotheringham et al. (129) describe an approach for geographically weighted principal components analysis (GWPCA). With GWPCA, geographically weighted means and GW variances and covariances around the means are obtained (129), with the result that there is a set of GW means, variances and covariances for each of the n data locations. Once the geographically weighted variance-covariance matrix is obtained with respect to a given location, conducting GWPCA is straightforward. The outputs from GWPCA are extensive — with component scores and loadings at all data locations, summarising the information is problematic (129). Fotheringham et al. (129) present an example application of GWPCA to the exploration of educational attainment in the southern USA. In that application, the first PC was computed for the nodes of a grid overlaid on the area. The variable with the largest loading at each location was then indicated on a map. In addition, where the largest loading was more than 5% or more larger than the second largest loading the grid cell underlying the relevant node was indicated. Another application of GWPCA is presented by Lloyd (241) who was concerned with characterising population subgroups in Northern Ireland.

5.9   Spatially weighted classification

There is a range of classification procedures that take into account spatial locations. This topic is outside of the focus of the book, but some approaches are outlined briefly. In their review of image segmentation techniques, Haralick and Shapiro (175) discuss spatial clustering. Theiler and Gisler (357) present a contiguity enhanced k-means clustering algorithm. Oliver and Webster (296) presented an approach whereby the dissimilarity matrix is modified using the variogram (variograms are discussed in Section 7.4.2). The modified dissimilarity matrix is then used in a k-means classification procedure. Atkinson (24) includes a review of spatially-weighted supervised classification.

FIGURE 5.12: GWR intercept: elevation against precipitation.

FIGURE 5.13: GWR slope: elevation against precipitation.

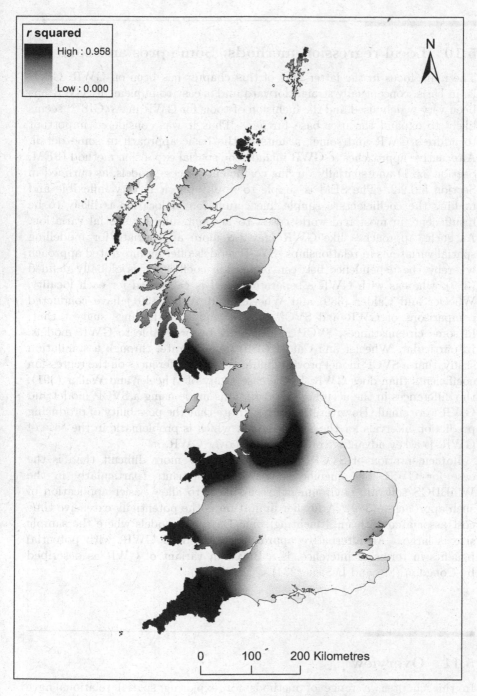

FIGURE 5.14: GWR r^2: elevation against precipitation.

5.10   Local regression methods: Some pros and cons

The main focus in the latter part of this chapter has been on GWR. GWR is, in basis, conceptually straightforward and is easy to implement. GWR has been very widely used and the inclusion of tools for GWR in ArcGIS™ seems likely to expand the user base further. Thus it was considered important to address GWR and some variants of the basic approach in some detail. Alternative approaches to GWR include the spatial expansion method (SEM; Section 5.6.1) and spatially varying coefficient process models, as outlined in Section 5.6.2.2. The SEM is simple to apply, but it is very inflexible and making the coefficients a simple linear function of location is likely to be insufficient, in most real world cases, to account fully for spatial variation. As such, approaches like GWR may be more appropriate for modelling spatial variations in relationships. SVCP models offer an integrated approach whereby the dependence between regression coefficients is globally defined (389), whereas with GWR a separate model is calibrated for each locality. Wheeler and Calder (389) and Wheeler and Waller (391) have conducted comparisons of GWR and SVCP models and their findings suggest that, in some circumstances, SVCP models may be preferable to GWR models. In particular, Wheeler and Calder (389) demonstrate, through a simulation study, that a SVCP model provides more accurate inferences on the regression coefficients than does GWR. In the case study of Wheeler and Waller (391), the differences in the accuracy of predictions made using a SVCP model and GWR were small. However, the authors note that the possibility of producing prediction intervals for the SVCP model, which is problematic in the case of GWR, is a key advantage of such models over GWR.

Implementation of SVCP models is generally more difficult than is the case for GWR, but ongoing software developments (particularly in the WinBUGS software environment) seem likely to allow easier application of such approaches (389). A further limitation is the potentially extensive time cost associated with implementing some Bayesian models where the sample size is large. An alternative approach to standard GWR, with potential benefits in terms of inference, is a Bayesian variant of GWR as described by Congdon (89) and LeSage (234).

5.11   Overview

In this chapter, a range of methods for exploring spatial relations have been outlined. The analysis of spatial relationships has been a particular growth area in the last decade. Techniques like GWR are conceptually straightforward, but a powerful means of exploring spatial relations, and the use of such approaches seems likely to expand markedly in the near future.

In terms of multivariate relationships, selection of a spatial regression procedure may not be straightforward. In cases where a hierarchy of administrative units are available, a MLM approach may be appropriate. In contrast, where the data are continuous, an approach like GWR may be more suitable. This chapter provides a summary of the main features of some approaches and some applications are outlined. There is a wide range of published applications to provide further guidance and this will hopefully help inform selection of an appropriate technique.

Some of the approaches described in this section have been used for spatial prediction and in the next chapter some univariate and multivariate methods for spatial prediction are detailed.

6

Spatial Prediction 1: Deterministic Methods, Curve Fitting, and Smoothing

Interpolation methods, like other methods discussed in this book, can be divided into two broad groups: global and local methods. A widely used distinction between interpolation methods is that global methods, such as trend surface analysis, use all data for prediction, while local methods, including inverse distance weighting, usually use only some subset of the data to make each prediction (that is, a moving window approach is employed). One benefit of local methods is that computational time is reduced by using only some subset of the data to make a prediction. Some methods may make use of all available data but may take into account distance from the prediction location. Such methods may still be considered local; therefore, many widely used interpolation techniques are local methods. A global approach may be used to remove large scale variation, to allow focus on local variation and spatial prediction can proceed with the residuals from a fitted trend model.

Lam (227) and Mitás and Mitásová (275) provide overviews of interpolation methods. In this chapter, some of the most widely used approaches are discussed and some key locally-based techniques are illustrated through a case study. Interpolation methods may be divided into two groups, point interpolation and areal interpolation, each with two subdivisions as outlined below.

Point interpolation is based on samples available at specific locations that can be approximated as points. A further division of methods comprises exact methods and approximate methods. Exact interpolators honour the data, while approximate interpolators do not. In other words, exact interpolators retain the data values in the output.

Areal interpolation is used where the data comprise measurements made over areas, and the desire is to convert between the existing zonal system and another zonal system. Alternatively, various procedures have been developed for interpolating from a zonal system to a surface. Areal interpolation methods can be divided into nonvolume preserving methods and volume preserving methods. Volume preservation is defined in Section 6.6.3.

This chapter discusses some techniques used widely in GISystems as well as recent developments. In Sections 6.1 through 6.3, several widely used local point interpolation procedures are outlined, while the subject of Sections 6.4 through 6.6 is areal interpolation. In both cases, a short discussion of global approaches is given for context prior to more in-depth discussion about local methods. Case studies illustrating the application of point and areal interpolation methods are given in the text. Some of the advantages and disadvantages of these methods are outlined in Section 6.7. The application of selected point interpolation methods is illustrated using the monthly precipitation dataset described in Section 1.8.1. (the application of kriging to the same dataset is explored in Chapter 7). An approach to areal interpolation is illustrated using the Northern Ireland 2001 Census data described in Section 1.8.6.

6.1 Point interpolation

This section outlines some widely used classes of local point interpolation procedures. Such methods use the available observations, $z(\mathbf{s}_i)$, $i = 1, 2, ..., n$, or some subset of these observations, to make a prediction at the location $\mathbf{s}_0$. This is illustrated in Figure 6.1.

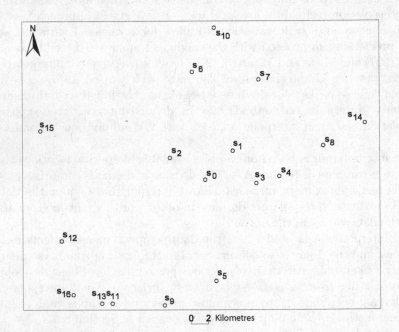

FIGURE 6.1: A neighbourhood of 16 observations.

As noted above, point interpolation methods may be divided into exact and approximate interpolators. Exact interpolators make a prediction at a data location that is the same as the data value, whereas approximate interpolators do not honour the data in this way. There are exact and approximate variants of some techniques (including thin plate splines). Examples of both kinds of interpolators are outlined below.

The performance of a predictor may be assessed in several ways. Jackknifing, where the values at some locations are predicted using a secondary set of data is one approach. Cross-validation is an approach that requires only a single set of observations. Cross-validation (also outlined in the previous chapter in the context of geographically weighted regression) comprises several steps: (i) a data point is removed, (ii) its value is predicted using the remaining data (or neighbouring data), and (iii) the data point is returned. This process continues until all data points have been predicted in this manner and prediction error statistics can be obtained.

6.2   Global methods

In the absence of spatial structure, the mean value provides a prediction of all other values. However, in practice spatially-referenced variables tend to have some spatial structure — that is, they tend to be positively spatially autocorrelated in at least some regions. Assuming some spatial structure, various purely global approaches can be employed. The simplest approach to generating a surface from sparse data is to fit a trend surface (that is, a polynomial which is typically of low order) to the data using least squares. Polynomials of order 1, 2, and 3 are given as:

Order 1:

$$m(\mathbf{s}) = \beta_0 + \beta_1 x + \beta_2 y \tag{6.1}$$

Order 2:

$$m(\mathbf{s}) = \beta_0 + \beta_1 x + \beta_2 y + \beta_3 x^2 + \beta_4 y^2 + \beta_5 xy \tag{6.2}$$

Order 3:

$$m(\mathbf{s}) = \beta_0 + \beta_1 x + \beta_2 y + \beta_3 x^2 + \beta_4 y^2 + \beta_5 xy + \beta_6 x^3 + \beta_7 y^3 + \beta_8 x^2 y + \beta_9 y^2 x \tag{6.3}$$

where the value z is derived at location $\mathbf{s}$, the β_i are coefficients to be estimated, and x and y are the spatial coordinates of observations. Figure

6.2 shows the values of monthly precipitation generated when a first-order polynomial was fitted to the precipitation data described in Section 1.8.1.

A global regression model may be used for spatial prediction where measurements of one or more secondary variables, which are related to the sparsely sampled primary variable, are available. The secondary variable(s) must be present at all locations where the value of the primary variable will be predicted. But, in cases such as that illustrated in Section 5.1 a local regression approach is likely to be necessary for prediction.

6.3   Local methods

The first two methods outlined in this Section are local only in that they are based, in some way, on the nearest neighbours to the prediction location. Section 6.3.1 discusses Thiessen polygons, the simplest form of point interpolation, while Section 6.3.2 is concerned with triangulation (generation of vector-based surfaces). The following sections detail local point interpolation methods as follows: Section 6.3.3: natural neighbour interpolation; Section 6.3.4: local fitting of polynomials to spatial data; Section 6.3.5: linear regression for spatial prediction (the principles of linear regression were outlined in Chapter 5); Section 6.3.6: inverse distance weighting; Section 6.3.7: thin plate splines; Section 6.3.8: thin plate splines case study; Section 6.3.9: generalised additive models; Section 6.3.10: finite difference methods; Section 6.3.11: locally adaptive approaches for constructing digital elevation models.

6.3.1   Thiessen polygons: Nearest neighbours

Thiessen (also called Dirichlet or Voronoi) polygons assign values to unsampled locations that are the same as the value of the nearest observation. Thiessen polygons are used widely to provide a simple overview of the spatial distribution of values. Intuitively, the method is more suitable for mapping values of variables like population counts which may not vary smoothly, whereas variation in properties such as precipitation is not well modelled by Thiessen polygons. However, for illustrative purposes, Thiessen polygons were generated from the precipitation data described in Section 1.8.1. The same data were analysed by Lloyd (239), (240) using several approaches. Thiessen Polygons were generated using ArcGIS™ and the map generated is shown in Figure 6.3 (page 150). There is a clear contrast between large values, most notably in Western Scotland, and small values in, for example, southeast England.

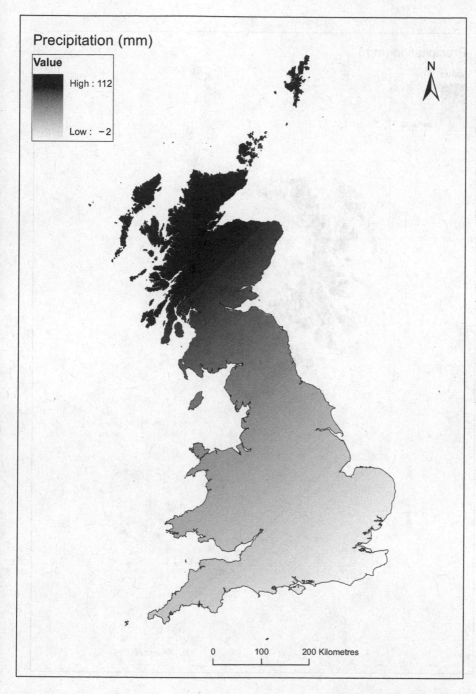

FIGURE 6.2: Monthly precipitation map generated with a polynomial of order 1.

FIGURE 6.3: Monthly precipitation map generated with Thiessen polygons.

6.3.2 Triangulation

A widely used vector representation of a (topographic) surface is the triangulated irregular network (TIN). A TIN is constructed by joining known point values to form a series of triangles based on Delaunay triangulation (70). Peuker et al. (316) provide a detailed summary of TINs and their construction. A key advantage of TINs over raster-based representations is that the sampling density can be varied as a function of the nature of the topography. Marked breaks of slope are usually sampled and there tends to be a larger number of observations in areas with more variable elevations and fewer samples in relatively flat areas. Observations can be selected in a variety of ways where the objective is to represent the surface as precisely as possible while minimising redundancy. The Very Important Points (VIP) algorithm has been used widely to select points from a regular grid. Using the VIP algorithm, points are assigned a significance which is a function of the difference between each observation point and its neighbours. Then a specific number of the most significant points can be retained such that the loss of accuracy is minimised given some criterion (see Li et al. (238) for more details).

TINs are based on neighbouring data points only and this is, therefore, a local approach. Two vertices are connected by Delaunay triangulation if their Thiessen polygons share an edge and Delaunay triangulations can be derived from Thiessen polygons. Figure 6.4 shows a TIN and Thiessen polygons derived (using ArcGIS™) from a subset of the Walker Lake data described by Isaaks and Srivastava (200). Lloyd (245) provides a further example.

Delaunay triangulation forms triangles such that the circumcircle of each triangle (which runs through the three vertices that belong to a given triangle) does not contain any vertices other than those that make up that triangle. Figure 6.5 gives an example.

6.3.3 Natural neighbours

Two observations are natural neighbours if there is a location or region equally close to both observations and no other observation is closer (379). So, any three observations are natural neighbours if no other observation is located within their circumcircle (see Section 6.3.2). The natural neighbour relation gives the basis for the Delaunay triangulation. The Voronoi tessellation shows that each observation has a unique natural neighbour region (Voronoi or Thiessen polygon).

Natural neighbour interpolation (332) is based on local coordinates (with values ranging from zero to one) that relate the position of the interpolation point to each observation in the neighbourhood subset. A neighbourbood co-ordinate is formed from the overlapping area of the Thiessen polygon surrounding the interpolation location and the Thiessen polygon surrounding a neighbouring observation where the interpolation location is not included.

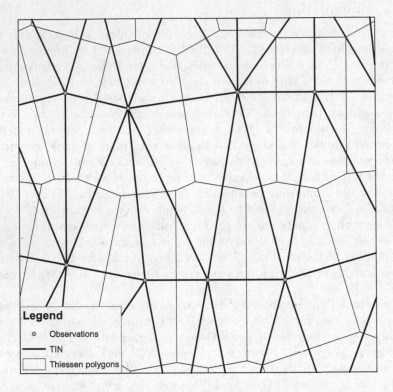

FIGURE 6.4: TIN and Thiessen polygon subset.

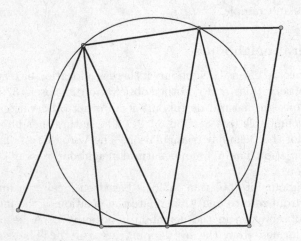

FIGURE 6.5: Circumcircle for selected triangle.

In other words, (i) Thiessen polygons are computed given the locations of the observations, (ii) Thiessen polygons are computed with the observation locations *and* the prediction location. The weights are then determined given the area of overlap between the prediction location Thiessen polygon generated in step (ii) and the Thiessen polygons generated in step (i).

The area of the overlapping region is the coordinate of the interpolation location with respect to a given observation after normalisation — division by the sum of all overlaps with natural neighbours of the interpolation location. Watson and Philip (379) identify as a key advantage of natural neighbour prediction over distance based approaches that, when the data are not regularly distributed, the neighbourhood coordinates vary anisotropically between two adjacent interpolation locations.

The natural neighbour coordinates of an interpolation location can be applied as proportions to the observed values of its natural neighbours. In the two-dimensional (2D) case, surfaces generated in this way have cone-like peaks and pits at the data locations. This can be corrected by adding blended gradient information to generate a smooth surface (379). A summary of natural neighbour interpolation is provided by Mitás and Mitásová (275).

6.3.4 Trend surface analysis and local polynomials

Fitting of polynomials may be conducted locally. That is, a low-order polynomial can be fitted to the n data $z(\mathbf{s}_i)$, $i = 1, 2, ..., n$, in the neighbourhood of the prediction location, and the order of the polynomial may be determined by selecting the lowest order polynomial that passes through all data points. One problem with such an approach is the potential for discontinuities at the edge of the piecewise polynomials (227).

Polynomials may be fitted to the data as a prior stage to interpolation, rather than being used for interpolation itself. In some applications, polynomials are fitted to the data globally or locally and analysis proceeds with the residuals from the polynomial fit. In this case, global variation may be removed, enabling a focus on local spatial variation. Such an approach is outlined in Sections 7.15 and 7.16, where the spatial structure of the residuals is explored, interpolation is conducted using the residuals, and finally the trend is added back to the predicted values.

6.3.5 Linear regression

To assess how relations between properties vary spatially, it is necessary to use some form of local regression. Local regression procedures were outlined in Chapter 5. Such approaches can be applied to prediction in cases where the value of the independent variable is known at locations in addition to those at which measurements of the dependent variable are available. In some cases, the independent variable(s) may be known at all locations in the area of interest. An example of this is the prediction of precipitation using elevation data as a secondary variable. Quite high quality digital elevation models (DEMs) are available for most parts of the world, so it is possible to use the elevation-precipitation relation (since the two are related over periods of weeks or more) to inform predictions of precipitation on a regular grid. Lloyd (240) has applied moving window regression (MWR) in this way. GWR (using a spatial kernel to weight observations) is a natural extension of MWR. The application of MWR for spatial prediction is illustrated below.

MWR was used to generate a regular grid from the monthly (July 1999) precipitation data (see Section 1.8.1). In this case, the relation between elevation (data on which are available in various formats at all locations in Britain) and precipitation was used to make predictions of the precipitation amount. The gridded elevation values detailed in Section 1.8.2 were used for this purpose. MWR was conducted using purpose-written computer code. Cross-validation prediction error summary statistics are given in Table 6.1. It is noticeable that the root mean square error (RMSE) becomes larger as the number of observations increase.

TABLE 6.1
Cross-validation prediction errors for MWR: precipitation in July 1999.

N. data	Mean error (mm)	Std. Dev. (mm)	Max. neg. error (mm)	Max. pos. error (mm)	RMSE (mm)
8	0.115	15.118	−159.538	98.231	15.118
16	0.384	15.751	−155.912	107.313	15.756
32	0.389	16.638	−181.154	85.910	16.643
64	0.185	17.941	−195.500	77.429	17.942

The map generated using MWR (using 32 nearest neighbours) is given in Figure 6.6. The map is very smooth in appearance — there is a gradual transition between small and large precipitation amounts. It is clear that predicted precipitation amounts are large where there are large elevations. Lloyd (240), (244) details the use of MWR for exploring and mapping precipitation amounts.

6.3.6　Inverse distance weighting (IDW)

Weighted moving averaging is a widely used approach to interpolation. A variety of different weighting functions have been used, but inverse distance weighting (IDW) is the most common form in GISystems. IDW is an exact interpolator, so the data values are honoured (as defined in Section 6.1). The IDW predictor, defined in Section 2.2, can be given as:

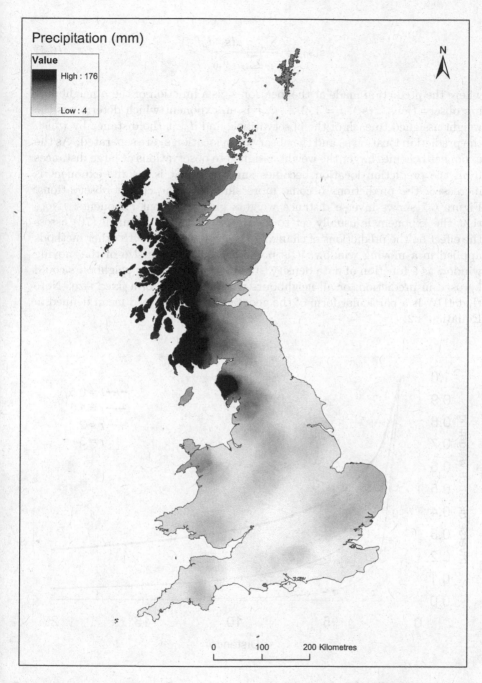

FIGURE 6.6: Monthly precipitation map generated with MWR, 32 nearest neighbours.

$$\hat{z}(\mathbf{s}_0) = \frac{\sum_{i=1}^{n} z(\mathbf{s}_i) d_{i0}^{-r}}{\sum_{i=1}^{n} d_{i0}^{-r}} \qquad (6.4)$$

where the prediction made at the location $\mathbf{s}_0$ is a function of the n neighbouring observations, $z(\mathbf{s}_i)$, $i = 1, 2, ..., n$, r is an exponent which determines the weight assigned to each of the observations, and d_{i0} is the distance by which the prediction location $\mathbf{s}_0$ and the observation location $\mathbf{s}_i$ are separated. As the exponent becomes larger the weight assigned to observations at large distances from the prediction location becomes smaller. That is, as the exponent is increased, the predictions become more similar to the closest observations. Figure 6.7 shows inverse distance weights given different exponents. Note that the exponent is usually set to 2. Burrough and McDonnell (70) assess the effect on the predictions of changing the exponent. As with other methods applied in a moving window, it is possible to adapt the size of the moving window as a function of data density. That is, the n nearest neighbours could be used in prediction, or all neighbours within a radius of a fixed size. Note that IDW is a particular form of the geographically weighted mean defined in Equation 4.2.

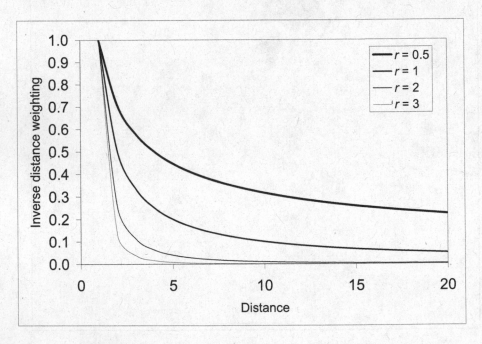

FIGURE 6.7: Inverse distance weights against distance.

IDW only predicts within the range of input data values. Therefore, minima or maxima that have not been sampled will not be predicted by IDW (378). As with other moving window based methods, as the neighbourhood becomes larger, the interpolated map becomes smoother. A suitable neighbourhood size could be identified through cross-validation or jackknifing.

When the sample is more dense in some areas than in others, use of IDW may be problematic. This is frequently the case when IDW is used to derive DEMs from the vertices of vector contour lines. In such cases, predictions near to the contour line values will be very similar to those contour values, resulting in steps around the contour lines. In any case where there are clusters of similar values, IDW may generate predictions that are a function of sampling configuration rather than spatial variation in the property of interest. A common feature of IDW-generated maps is the presence of spikes or pits around the data locations, since isolated data have a marked influence on predictions in their vicinity.

IDW has been modified to account for local anisotropy (directional variation). Tomczak (363) provides an application of anisotropic IDW, where the anisotropy ratio is selected using cross-validation or a jackknife approach.

IDW has been used widely in GISystems contexts. Lloyd and Atkinson (247) use IDW to make predictions from elevation measurements made using an airborne LiDAR (Light Detection And Ranging) sensor. In that application, IDW is shown to provide cross-validation absolute errors which are only a little larger than those obtained using more complex methods, due to the small sample spacing and the similarity in elevation values over small areas.

6.3.6.1 Illustrating IDW

In this section, a worked example of IDW is given using four precipitation observations, with the objective of predicting at another location. This is an example of local prediction, and the procedure would be followed at all prediction locations if a four observation neighbourhood were used. However, this is a very small sample size and, in practice, the minimum number of nearest neighbours used is likely to be larger.

Since an observation is available at the prediction location, but it has been removed for the present purpose, it is possible to assess the accuracy of the predictions. The data are given in Table 6.2, where the first observation is treated as unknown, and prediction is conducted using the other four observations. IDW was implemented with an exponent of 2.

In Table 6.3, inverse squared distances and the precipitation values multiplied by the inverse squared distances are given. The IDW prediction is obtained by 0.0000286 / 0.000000593 = 48.174.

TABLE 6.2
Precipitation: s_0 is treated as unknown and is
predicted using observations s_1 to s_4.

ID	x (m)	y (m)	Precipitation (mm)	Distance from x_0
s_0	369000	414300	51.6	0
s_1	370400	415700	52.6	1979.899
s_2	370900	412400	40.8	2687.006
s_3	366100	414700	48.3	2927.456
s_4	367100	411400	46.7	3466.987

TABLE 6.3
Precipitation: IDW prediction using observations s_1 to s_4. ISD is inverse squared distance.

ID	x (m)	y (m)	Precip. (mm)	Distance from x_0	ISD	Obvs. $\times$ ISD
s_1	370400	415700	52.6	1979.899	0.000000255	0.0000134
s_2	370900	412400	40.8	2687.006	0.000000139	0.00000565
s_3	366100	414700	48.3	2927.456	0.000000117	0.00000564
s_4	367100	411400	46.7	3466.987	0.0000000832	0.00000389
				Sum	0.000000593	0.0000286

6.3.6.2 Inverse distance weighting case study

The monthly (July 1999) precipitation data described in Section 1.8.1 were used to generate regular grids using IDW. IDW was carried out using the Geostatistical Analyst (205) extension of ArcGIS™. Before generating maps, cross-validation was used to help assess the performance of IDW given different data neighbourhoods. Global cross-validation prediction error summary statistics are given in Table 6.4. The smallest RMSE is for the smallest prediction neighbourhood.

For the map, predictions were made to a grid with a spacing of 661 m using a 32 observation neighbourhood (the same neighbourhood is used for other approaches later in this chapter). The map generated using IDW is given in Figure 6.8. In the IDW map, spikes are apparent at the data locations. As noted above, this is a common feature of IDW-derived maps.

6.3.7 Thin plate splines

Thin plate splines (TPS) constitute one of the most widely used approaches to spatial interpolation. Splines are piecewise functions; they may be used to fit a small number of data points exactly. Splines are used to smooth lines (B-splines, which are the sums of splines which have a value of zero

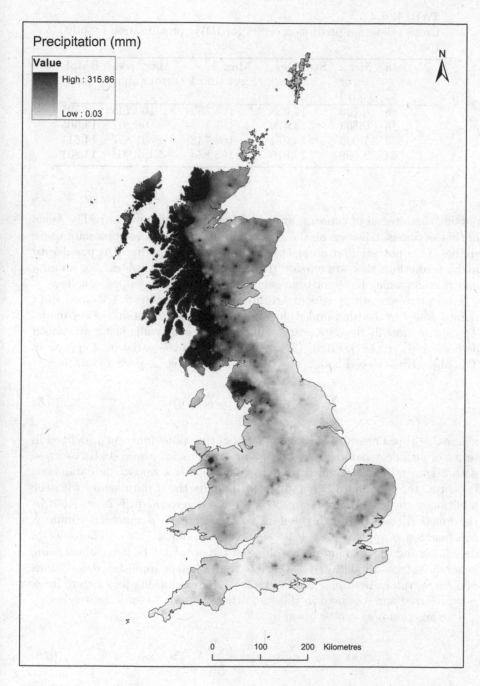

FIGURE 6.8: Monthly precipitation map generated with IDW, 32 nearest neighbours.

TABLE 6.4

Cross-validation prediction errors for IDW: precipitation in July 1999.

N. data	Mean error (mm)	Std. Dev. (mm)	Max. neg. error (mm)	Max. pos. error (mm)	RMSE (mm)
8	0.708	14.508	-169.355	109.295	14.525
16	0.863	14.479	-167.763	98.807	14.505
32	1.033	14.574	-164.748	91.857	14.611
64	1.099	14.810	-165.834	87.903	14.851

outside the interval of concern, are widely used for this purpose). The focus in this section is, however, on the generation of surfaces from point data using splines. One potential advantage of splines over IDW is that it is possible to make predictions that are outside the range of the data values. So, minima and maxima which have not been sampled may be predicted using splines.

Thin plate smoothing splines are fitted to the $z(\mathbf{s}_i)$, $i = 1, 2, ..., n$, data points closest to the (unsampled) location at which a prediction is to be made. The spline must fit the data points sufficiently closely, while being as smooth as possible (43), (161), (198), (371); this is termed the variational approach. The objective is to estimate $z(\mathbf{s})$ with a smooth function g by minimizing:

$$\sum_{i=1}^{n} (z(\mathbf{s}_i) - g(\mathbf{s}_i))^2 + \rho J_m(g) \qquad (6.5)$$

where $J_m(g)$ is a measure of the roughness of the spline function g (defined in terms of mth degree derivatives of g), λ are the unknown weights (in contrast with a known weight signified by w), while $\rho > 0$ is a smoothing parameter. In words, the function g is found which best fits the n data values with the addition of the smoothing parameter. When ρ is zero the fit is exact (that is, the function passes through the data points) while as ρ approaches infinity, the function g approaches a least squares polynomial (196). The order of derivative and the amount of smoothing of the data may be determined using generalized cross-validation (GCV). Note that, in many published descriptions of TPS, λ, rather than ρ, is used to indicate the smoothing parameter. Here, ρ is preferred and λ is used in this and other chapters to indicate weights.

The function $g(\mathbf{s})$ can be given by:

$$g(\mathbf{s}) = a_0 + a_1 x + a_2 y + \sum_{i=1}^{n} \lambda_i R(\mathbf{s} - \mathbf{s}_i) \qquad (6.6)$$

where a are trend model coefficients and $R(\mathbf{s} - \mathbf{s}_i)$ is a basis function (a scalar function of the Euclidean distance between $\mathbf{s}$ and $\mathbf{s}_i$). Billings et al. (42) provide a useful discussion about basis functions.

Where the dimension $d = 2$ and $m = 2$ (where $m - 1$ gives the order of the polynomial trend), $J_m(g)$ can be given by:

$$J_2(g) = \int \int_{\Re^2} \left[\left(\frac{\partial^2 g}{\partial x^2} \right)^2 + 2 \left(\frac{\partial^2 g}{\partial x \partial y} \right)^2 + \left(\frac{\partial^2 g}{\partial y^2} \right)^2 \right] dx dy \qquad (6.7)$$

This is termed a bending energy function; $\Re^2$ indicates a two-dimensional space. The corresponding basis function, $R(\mathbf{s} - \mathbf{s}_i)$, as shown by Duchon (113), is the TPS basis function and this can be given by:

$$d_i^2 \ln d_i \qquad (6.8)$$

where $d_i^2 = (x - x_i)^2 + (y - y_i)^2$. More generally, the basis function can be given by:

$$d_i^{2m-d} \ln d_i \qquad (6.9)$$

Thus, for the case of two-dimensional data (d, in superscript, is 2) and, as noted above, where the order of the base polynomial trend fitted to the data is given by $m - 1$, then the basis function is as given in Equation 6.8.

In matrix form, the coefficients a_k and λ_i, as indicated in Equation 6.6, are the solution of:

$$\begin{bmatrix} \mathbf{R} + \rho \mathbf{I} & \mathbf{F} \\ \mathbf{F}^T & \mathbf{0} \end{bmatrix} \begin{bmatrix} \lambda \\ \mathbf{a} \end{bmatrix} = \begin{bmatrix} \mathbf{z} \\ \mathbf{0} \end{bmatrix} \qquad (6.10)$$

where $\rho \mathbf{I}$ ($\mathbf{I}$ indicating the $n \times n$ identity matrix) can be interpreted as white noise added to the variances at the data locations, but not to the variance at the prediction location (371). So, the smoothing parameter is added to the diagonals (this is illustrated in Section 6.3.7.1). That is, if no smoothing parameter is used then TPS is an exact predictor, but if a smoothing parameter is used then the requirement to fit the spline to the observations is removed. Green and Silverman (161) show that the TPS system is non-singular and that, therefore, it has a unique solution.

The TPS system of Equation 6.10 can also be given by:

$$\mathbf{R}\lambda = \mathbf{z} \qquad (6.11)$$

With the full matrices given by:

$$\mathbf{R} = \begin{bmatrix} R(\mathbf{s}_1 - \mathbf{s}_1) & \cdots & R(\mathbf{s}_1 - \mathbf{s}_n) & 1 & x_1 & y_1 \\ \vdots & \vdots & \vdots & \vdots & \vdots & \vdots \\ R(\mathbf{s}_n - \mathbf{s}_1) & \cdots & R(\mathbf{s}_n - \mathbf{s}_n) & 1 & x_n & y_n \\ 1 & \cdots & 1 & 0 & 0 & 0 \\ x_1 & \cdots & x_n & 0 & 0 & 0 \\ y_1 & \cdots & y_n & 0 & 0 & 0 \end{bmatrix}$$

with the possible addition of a smoothing parameter to the diagonals. The TPS weights (λ) and the vector of observations $\mathbf{z}$ are given by:

$$\lambda = \begin{bmatrix} \lambda_1 \\ \vdots \\ \lambda_n \\ a_0 \\ a_1 \\ a_2 \end{bmatrix} \quad \mathbf{z} = \begin{bmatrix} z(\mathbf{s}_1) \\ \vdots \\ z(\mathbf{s}_n) \\ 0 \\ 0 \\ 0 \end{bmatrix}$$

The weights are obtained with:

$$\lambda = \mathbf{R}^{-1}\mathbf{z} \tag{6.12}$$

Hutchinson and Gessler (198) provide a statistical interpretation of the thin plate spline. The influence (or 'hat') matrix $\mathbf{A}$ is the matrix that expresses the fitted values (given by Equation 6.6) as a linear function of the data values, $z(\mathbf{s}_i)$ (198), (372); the hat matrix is outlined by Myers et al. (284) in the context of regression analysis. The matrix $\mathbf{A}$ satisfies (372), (373):

$$\begin{bmatrix} g_\rho(\mathbf{s}_1) \\ \vdots \\ g_\rho(\mathbf{s}_n) \end{bmatrix} = \mathbf{A}(\rho)\mathbf{z} \tag{6.13}$$

The trace (that is, the sum of the diagonal elements) of $\mathbf{I} - \mathbf{A}$, where, as above, $\mathbf{I}$ is the identity matrix, can be interpreted as the degrees of freedom of the residual sum of squares (noise), as given in the first term of Equation 6.5. The variance can then be estimated with:

$$\hat{\sigma}_\in^2 = \frac{(\mathbf{z} - \mathbf{A}(\rho)\mathbf{z})^T(\mathbf{z} - \mathbf{A}(\rho)\mathbf{z})}{\text{tr}(\mathbf{I} - \mathbf{A}(\rho))} \tag{6.14}$$

where $\mathbf{z}$ is the n dimensional vector of observations $z(\mathbf{s}_i)$ (198).

The value of $\hat{\sigma}_\in^2$ can be used to obtain an estimate of the true mean squared error of the fitted surface over all the data points, which is given by:

$$T(m, \rho) = (\mathbf{z} - \mathbf{A}\mathbf{z})^T(\mathbf{z} - \mathbf{A}\mathbf{z})/n - 2\sigma_\in^2\text{tr}(\mathbf{I} - \mathbf{A})/n + \sigma_\in^2 \tag{6.15}$$

If the variance $\sigma_\in^2$ is unknown then this estimate cannot be minimised to obtain m (where the degree of the polynomial in the model is $m - 1$) and ρ, but even when $\sigma_\in^2$ is known, determination of m and ρ by GCV (which does not require an estimate of $\sigma_\in^2$) is usually preferred (198). Hutchinson and Gessler (198) provide a summary account of GCV for ascertaining the order and the amount of data smoothing. The GCV is given by:

$$\text{GCV}(m, \rho) = \frac{(\mathbf{z} - \mathbf{A}\mathbf{z})^T(\mathbf{z} - \mathbf{A}\mathbf{z})/n}{[\text{tr}(\mathbf{I} - \mathbf{A})/n]^2} \tag{6.16}$$

It has been demonstrated that minimising Equation 6.16 is asymptotically equivalent to minimising Equation 6.15 (198). GCV is illustrated in the example below. Routines and further details about GCV are given by Bates et al. (34). The R package 'Fields' [§] includes a function called 'tps' which fits a thin plate spline surface to irregularly spaced data. The function allows the selection of a smoothing parameter using GCV.

The application of the TPS function can be problematic in areas where there are large gradients producing overshoots. A second problem is that the second-order derivatives diverge in the data points (although this would not be true if a higher order spline was used). This affects analysis of surface geometry. Some variants of the standard TPS method have been developed to overcome these limitations, and some such approaches are outlined in Section 6.3.7.3.

6.3.7.1 Illustrating TPS

Here, a worked example of TPS is given using the four precipitation observations detailed in Table 6.2, and to which IDW was applied in Section 6.3.6.1. This is an example of local TPS based prediction which would be conducted at all locations for which predictions are required. As indicated in Section 6.3.6.1, this small number of observations is used purely for ease of illustration of the method.

In this case, the coordinates were converted from metres to kilometres to prevent the need for exponentiation, since the coordinates are six figure grid references. The solution of the TPS coefficients was given in matrix form in Equation 6.10. The distances separating the observations used in prediction are input into the TPS basis function Equation (6.8). The TPS system was solved with **R** matrix entries to five decimal places, but for presentation purposes figures are rounded to three decimal places. For this example, the TPS system is given by:

$$
\begin{bmatrix}
0 & 13.427 & 28.942 & 49.657 & 1 & 370.400 & 415.700 \\
13.427 & 0 & 47.367 & 21.129 & 1 & 370.900 & 412.400 \\
28.942 & 47.367 & 0 & 14.718 & 1 & 366.100 & 414.700 \\
49.657 & 21.129 & 14.718 & 0 & 1 & 367.100 & 411.400 \\
1 & 1 & 1 & 1 & 0 & 0 & 0 \\
370.400 & 370.900 & 366.100 & 367.100 & 0 & 0 & 0 \\
415.700 & 412.400 & 414.700 & 411.400 & 0 & 0 & 0
\end{bmatrix}
\begin{bmatrix}
\lambda_1 \\ \lambda_2 \\ \lambda_3 \\ \lambda_4 \\ a_0 \\ a_1 \\ a_2
\end{bmatrix}
=
\begin{bmatrix}
52.6 \\ 40.8 \\ 48.3 \\ 46.7 \\ 0 \\ 0 \\ 0
\end{bmatrix}
$$

where the smoothing parameter, ρ, is zero (nothing is added to the diagonals in the **R** matrix).

The solutions to the matrix system are: $\lambda_1 = 0.261$, $\lambda_2 = -0.293$, $\lambda_3 = -0.252$, $\lambda_4 = 0.283$, $a_0 = -510.876$, $a_1 = -0.603$, $a_2 = 1.886$.

[§]http://cran.r-project.org/web/packages/fields/fields.pdf

To obtain the prediction, distances separating the prediction location and the observations used in prediction must be input into the basis function equation. This gives: $s_1 = 2.678$, $s_2 = 7.136$, $s_3 = 9.205$, $s_4 = 14.944$.

The prediction is obtained with: $-510.876 + (-0.603 \times 369.000) + (1.886 \times 414.300) + (0.261 \times 2.678) + (-0.293 \times 7.136) + (-0.252 \times 9.205) + (0.283 \times 14.944) = 48.475$.

For a smoothing parameter of 10 (10 is added to the diagonals in the **R** matrix) the solutions to the matrix are: $\lambda_1 = 0.128$, $\lambda_2 = -0.144$, $\lambda_3 = -0.123$, $\lambda_4 = 0.139$, $a_0 = -510.929$, $a_1 = -0.593$, $a_2 = 1.877$. The prediction is: $-510.929 + (-0.593 \times 369.000) + (1.877 \times 414.300) + (0.128 \times 2.678) + (-0.144 \times 7.136) + (-0.123 \times 9.205) + (0.139 \times 14.944) = 48.377$.

In this case, for $\rho = 0$ the predicted value (48.475) is closer to the observed value (51.6) than when ρ is set to 10 (48.377). Thus, increasing the value of ρ to 10 decreases the accuracy of the prediction in this case. As noted above, in practice GCV is often used to ascertain an appropriate order of polynomial and value of the smoothing parameter, ρ.

Equation 6.13 shows the hat (or influence) matrix, **A**, required for GCV, relates the fitted values to observations. The hat matrix can be obtained as follows:

$$\mathbf{g} = \begin{bmatrix} \mathbf{R} & \mathbf{F} \end{bmatrix} \begin{bmatrix} \mathbf{R} + \rho \mathbf{I} & \mathbf{F} \\ \mathbf{F}^T & \mathbf{0} \end{bmatrix}^{-1} \begin{bmatrix} \mathbf{z} \\ \mathbf{0} \end{bmatrix} = \mathbf{A}^* \begin{bmatrix} \mathbf{z} \\ \mathbf{0} \end{bmatrix} \qquad (6.17)$$

And the hat matrix **A** is the first n columns of $\mathbf{A}^*$. The GCV is computed given the four observations plus the fifth observation which is treated as unknown above. With $\rho=10$ this gives:

$$
\begin{bmatrix}
10 & 2.678 & 7.136 & 9.205 & 14.944 & 1 & 369 & 414.3 \\
2.678 & 10 & 13.427 & 28.942 & 49.657 & 1 & 370.4 & 415.7 \\
7.136 & 13.427 & 10 & 47.367 & 21.129 & 1 & 370.9 & 412.4 \\
9.205 & 28.942 & 47.367 & 10 & 14.718 & 1 & 366.1 & 414.7 \\
14.944 & 49.657 & 21.129 & 14.718 & 10 & 1 & 367.1 & 411.4 \\
1 & 1 & 1 & 1 & 1 & 0 & 0 & 0 \\
369 & 370.4 & 370.9 & 366.1 & 367.1 & 0 & 0 & 0 \\
414.3 & 415.7 & 412.4 & 414.7 & 411.4 & 0 & 0 & 0
\end{bmatrix}
\begin{bmatrix}
\lambda_1 \\ \lambda_2 \\ \lambda_3 \\ \lambda_4 \\ \lambda_5 \\ a_0 \\ a_1 \\ a_2
\end{bmatrix}
=
\begin{bmatrix}
51.6 \\ 52.6 \\ 40.8 \\ 48.3 \\ 46.7 \\ 0 \\ 0 \\ 0
\end{bmatrix}
$$

then:

$$
\mathbf{A}^* =
\begin{bmatrix}
0.4995 & 0.2095 & 0.1039 & 0.1353 & 0.0517 & 87.3244 & 0.0261 & -0.2541 \\
0.2095 & 0.7956 & 0.0874 & 0.0558 & -0.1483 & 986.7845 & -0.8884 & -1.5917 \\
0.1039 & 0.0874 & 0.8315 & -0.1542 & 0.1313 & 32.3692 & -1.4582 & 1.2191 \\
0.1353 & 0.0558 & -0.1542 & 0.8551 & 0.1080 & -125.6919 & 1.5766 & -1.1025 \\
0.0517 & -0.1483 & 0.1313 & 0.1080 & 0.8573 & -990.7862 & 0.7440 & 1.7291
\end{bmatrix}
$$

and this is verified with:

$$\mathbf{g} = \mathbf{Az} = \begin{bmatrix} 49.9871 \\ 51.9937 \\ 42.5725 \\ 49.9702 \\ 45.4764 \end{bmatrix}$$

These are the predicted values at the five data locations. GCV can then be computed following Equation 6.16. The various terms are, with $n = 5$:

$$\mathbf{z} - \mathbf{Az} = \begin{bmatrix} 1.6129 \\ 0.6063 \\ -1.7725 \\ -1.6702 \\ 1.2236 \end{bmatrix}$$

these are the prediction residuals.

$(\mathbf{z} - \mathbf{Az})^T(\mathbf{z} - \mathbf{Az}) = 10.3974$; $(\mathbf{z} - \mathbf{Az})^T(\mathbf{z} - \mathbf{Az})/n = 2.0795$

$$\mathbf{I} - \mathbf{A} = \begin{bmatrix} 1\ 0\ 0\ 0\ 0 \\ 0\ 1\ 0\ 0\ 0 \\ 0\ 0\ 1\ 0\ 0 \\ 0\ 0\ 0\ 1\ 0 \\ 0\ 0\ 0\ 0\ 1 \end{bmatrix} - \begin{bmatrix} 0.4995 & 0.2095 & 0.1039 & 0.1353 & 0.0517 \\ 0.2095 & 0.7956 & 0.0874 & 0.0558 & -0.1483 \\ 0.1039 & 0.0874 & 0.8315 & -0.1542 & 0.1313 \\ 0.1353 & 0.0558 & -0.1542 & 0.8551 & 0.1080 \\ 0.0517 & -0.1483 & 0.1313 & 0.1080 & 0.8573 \end{bmatrix}$$

$$= \begin{bmatrix} 0.5005 & -0.2095 & -0.1039 & -0.1353 & -0.0517 \\ -0.2095 & 0.2044 & -0.0874 & -0.0558 & 0.1483 \\ -0.1039 & -0.0874 & 0.1685 & 0.1542 & -0.1313 \\ -0.1353 & -0.0558 & 0.1542 & 0.1449 & -0.1080 \\ -0.0517 & 0.1483 & -0.1313 & -0.1080 & 0.1427 \end{bmatrix}$$

$\mathrm{tr}(\mathbf{I} - \mathbf{A}) = 1.1610$; $[\mathrm{tr}(\mathbf{I} - \mathbf{A})/n]^2 = 0.0539$

This leads to: $\mathrm{GCV}(2,10) = 2.0795 / 0.0539 = 38.5674$

GCV could be calculated using a variety of values of m and ρ and those which result in the smallest GCV retained. Bates et al. (34) offer approaches for efficiently identifying optimal values of m and ρ using GCV.

6.3.7.2 Partial thin plate splines

The partial thin plate spline (372) is a means of taking account of multiple variables to inform spatial prediction. Partial TPS are a version of splines which incorporate linear submodels (198). They may be solved using the same equation structure as ordinary TPS. Hutchinson and colleagues have published several applications of partial TPS models (see, for example, Hutchinson (193)).

6.3.7.3 Variants of thin plate splines

Equation 6.6 defined the function $g(\mathbf{s})$ given TPS. The general form is given by:

$$g(\mathbf{s}) = \sum_{k=0}^{K} a_k f_k(\mathbf{s}) + \sum_{i=1}^{n} \lambda_i R(\mathbf{s} - \mathbf{s}_i) \qquad (6.18)$$

where $f_k(\mathbf{s})$ are a set of K low order monomials.

Several different choices for the basis function $R(\mathbf{s} - \mathbf{s}_i)$ have been developed which overcome problems that may be encountered using standard TPS (as discussed briefly below). TPS with tension (TPST) is one such approach. With TPST, the tension can be changed so that the surface may resemble a stiff plate or an elastic membrane (275). TPS with tension can be adapted such that second-order (and possibly higher order) derivatives of the function are regular. For thin plate splines with tension (132) the basis function for two dimensions is:

$$-\frac{1}{2\pi\varphi^2} \left[\ln\left(\frac{d\varphi}{2}\right) + C_E + K_0(d\varphi) \right] \qquad (6.19)$$

where φ is the tension parameter, $C_E = 0.577215...$ (the Euler constant), and $K_0(d\varphi)$ is an approximation of the modified zeroth-order Bessel function (274). Polynomial-like approximations of Bessel functions are given by Abramowitz and Stegun (2), and see Press et al. (319). For TPST, the trend term (see Equation 6.18) comprises a constant (i.e. a_0). Chang (78) provides a worked example of TPST. Mitásová et al. (276) use ordinary cross-validation to ascertain optimal tension and smoothing parameters, although GCV could be applied.

Another variant of TPS uses the regularised spline basis function. Regularised splines for two dimensions are approximated with the following basis function (274):

$$\frac{1}{2\pi} \left\{ \frac{d^2}{4} \left[\ln\left(\frac{d}{2\tau}\right) + C_E - 1 \right] + \tau^2 \left[K_0\left(\frac{d}{\tau}\right) + C_E + \ln\left(\frac{d}{2\pi}\right) \right] \right\} \qquad (6.20)$$

where τ is a weight. In this case, the trend is specified as in Equation 6.6 (i.e., the monomials are a_0, a_1, a_2).

Mitásová et al. (278) developed the regularised spline with tension, also known as the completely regularised spline (CRS), which combines the desirable properties of the TPS approach. The basis function for a CRS (277) of two dimensions is:

$$-\left\{ \ln\left[\left(\frac{\varphi d}{2}\right)^2 \right] + E_1\left[\left(\frac{\varphi d}{2}\right)^2 \right] + C_E \right\} \qquad (6.21)$$

where E_1 is the exponential integral function (2), and φ can be regarded as a generalised tension parameter. Mitásová and Mitás (277) show how the tension parameter may be varied for different directions. For CRS, the trend term comprises a constant (a_0). In the specific context of DEM generation, Mitasova et al. (276) argue that tuning of the smoothing and tension parameters helps minimise overshoots, artificial pits, and the clumping of values around contours, as sometimes observed using standard TPS. Lloyd (243) applies univariate and multivariate (making use of a secondary variable for prediction purposes) CRS for generating maps of monthly precipitation in the UK.

6.3.8   Thin plate splines case study

As for other methods, TPS and CRS were used to generate grids from the monthly precipitation data detailed in Section 1.8.1. Both methods were applied using the Geostatistical Analyst (205) extension of ArcGIS™. Generalised tension parameters (determining degree of smoothness) for CRS are given in Table 6.5. In ArcGIS™, the optimal parameter value is selected by minimising the RMSE of the cross-validation predictions. Note that ArcGIS™ (version 9.3) allows for only exact TPS.

TABLE 6.5
CRS optimal generalised tension parameters by
data neighbourhood.

$n=8$	$n=16$	$n=32$	$n=64$
0.0017165	0.00085931	0.0017159	0.0017158

Global cross-validation prediction error summary statistics are given in Table 6.6. The mean error closest to zero is for TPS using 8 observations (0.045 mm), while the smallest RMSE is for CRS using 64 observations (14.005 mm). For TPS, the RMSE becomes smaller as the number of observations increase. For CRS, there is no consistent trend in the RMSE as the number of nearest neighbours changes.

The maps generated using TPS and CRS are given in Figures 6.9 (page 169) and 6.10 (page 170). Predictions were made using a 32 observation neighbourhood. Note that, in this case, the spline methods interpolated outside of the data minima and maxima and, for presentation purposes, the minima (which were predicted to be negative for both spline methods) have been set to zero as negative precipitation values are clearly nonsensical. Problems such as this must be addressed on a case-by-case basis when interpolation is employed and the spline model parameters may have to

TABLE 6.6

Cross-validation prediction errors for TPS and CRS: precipitation in July 1999.

Method	N. data	Mean error (mm)	Std. Dev. (mm)	Max. neg. error (mm)	Max. pos. error (mm)	RMSE (mm)
TPS	8	0.045	19.818	−351.833	191.491	19.814
TPS	16	0.079	17.986	−335.535	156.888	17.983
TPS	32	0.055	17.788	−323.736	153.391	17.786
TPS	64	0.088	17.519	−297.082	153.100	17.516
CRS	8	0.365	14.179	−175.250	92.210	14.182
CRS	16	0.255	16.037	−296.249	142.979	16.036
CRS	32	0.262	14.020	−170.942	92.143	14.020
CRS	64	0.231	14.005	−171.091	92.143	14.005

be adjusted, or a different method employed, if nonsensical values (such as negative precipitation amounts) are obtained. All of the maps show large precipitation amounts in the north west of Britain. The two spline-derived maps are similar in appearance, although the predicted maxima are quite different.

6.3.8.1   Other issues

The equivalence of splines and kriging (a focus of Chapter 7) has been demonstrated by several authors, and this theme is discussed in Section 7.10. Multiquadratic interpolation is not conducted through a variational approach, but there are similarities between the two approaches (275).

6.3.9   Generalised additive models and penalised regression splines

One limitation of the standard TPS framework, as outlined above, is that such models are computationally demanding, and there are as many unknown parameters as there are data (395), (396). A solution to this is to construct a generalised additive model (GAM) using penalised regression splines, as outlined by Wood and Augustin (397). With a GAM, the predictor depends partly on a sum of smooth functions of the predictor variables (396). Wood (396) provides a detailed account of GAMs and outlines approaches for representing the smooth functions, varying the smoothness of these functions, and selecting an appropriate degree of smoothness. Wood (395) applies regression and penalised regression models and seeks to identify optimal approximations to TPS which, firstly, overcome computational limitations to the use of TPS and, secondly, minimise deterioration in model performance associated with the approximation. The theme of thin plate regression splines

FIGURE 6.9: Monthly precipitation map generated with TPS, 32 nearest neighbours.

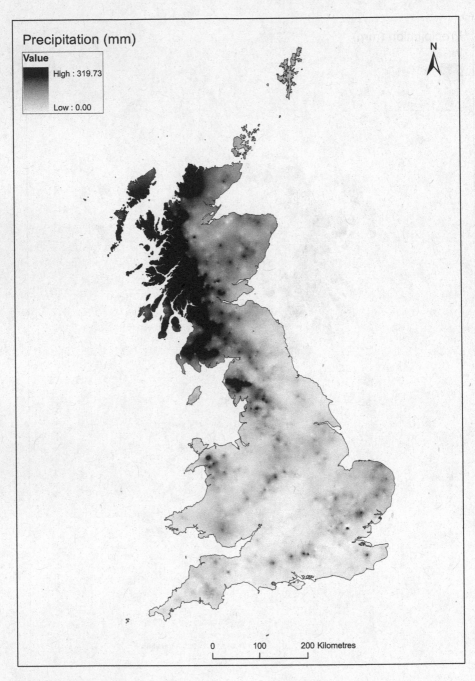

FIGURE 6.10: Monthly precipitation map generated with CRS, 32 nearest neighbours.

is addressed further by Wood (396) who provides illustrative examples of such an approach.

6.3.10 Finite difference methods

Finite difference methods work on the principle that the surface required conforms to a set of ordinary and partial differential equations. These equations are approximated using finite differences, and the equations are solved iteratively (227). Lam (227) gives the example of the Laplace equation: finding a function, z, whereby:

$$\frac{\partial^2 z}{\partial x^2} + \frac{\partial^2 z}{\partial y^2} = 0 \qquad (6.22)$$

inside a region while $z(x_i, y_i) = 0$ on the boundary. A finite difference approximation of the equation is given by:

$$z_{i-1,j} + z_{i+1,j} + z_{i,j-1} + z_{i,j+1} - 4z_{i,j} = 0 \qquad (6.23)$$

So, the value z at location i, j is the average of the four neighbouring cell values.

$$z_{i,j} = (z_{i-1,j} + z_{i+1,j} + z_{i,j-1} + z_{i,j+1})/4 \qquad (6.24)$$

Different differential equations may be used to derive a smoother surface. As Lam (227) states, this approach has similarities with the pycnophylactic areal interpolation method of Tobler (362) (see Section 6.6.3). In the following section, an approach to interpolation based on finite differences is discussed.

6.3.11 Locally adaptive approaches for constructing digital elevation models

Hutchinson (192) describes a morphological approach to generating DEMs through interpolation using irregularly distributed elevation point data and stream line data while contour data can also be utilised. The procedure combines minimisation of a terrain-specific roughness penalty with a drainage enforcement algorithm. The drainage enforcement algorithm removes spurious sinks or pits from the fitted grid on the grounds that such sinks are uncommon in nature. The interpolation problem is solved using an iterative finite difference interpolation technique based on minimising a discretised rotation invariant roughness penalty. The approach is based on the method of Briggs (58), and Swain (350) provides computer code based on the method of Briggs. The method of Hutchinson was designed to have the efficiency of local methods, while having the continuity and rotation invariance of global methods. Using this approach, the fitted DEM follows the sharp changes in terrain associated with streams and ridges. The roughness penalty is defined in terms of first- and second-order partial derivatives of the function g:

$$J_1(g) = \int \int_{\Re^2} \left[\left(\frac{\partial g}{\partial x} \right)^2 + \left(\frac{\partial g}{\partial y} \right)^2 \right] \mathrm{d}x \mathrm{d}y \qquad (6.25)$$

and

$$J_2(g) = \int \int_{\Re^2} \left[\left(\frac{\partial^2 g}{\partial x^2} \right)^2 + 2 \left(\frac{\partial^2 g}{\partial x \partial y} \right)^2 + \left(\frac{\partial^2 g}{\partial y^2} \right)^2 \right] \mathrm{d}x \mathrm{d}y \qquad (6.26)$$

The range of integration is the region covered by the fitted grid (192). Minimisation of $J_1(g)$ corresponds to discretised minimum potential interpolation, while minimising $J_2(g)$ corresponds to the discrete form of minimum curvature interpolation of thin plate splines (see Section 6.3.7 and Equation 6.7) (192). A useful account of the application of the roughness penalties in an analysis of wind fields is given by Testud and Chong (356). Smith and Wessel (335) outline an application of discretised splines with tension. In this context, they outline the solution by iteration of finite difference equations.

The grid is calculated at successively finer spatial resolutions (starting with an initial coarse grid), halving the grid spacing at each step, until the user specified resolution is obtained. For each grid resolution the data points are allocated to the nearest grid location. Values at grid locations which do not correspond with data points are calculated using Gauss-Seidel iteration with overrelaxation conditional on the roughness penalty specified and ordered chain constraints. The starting values of the first (coarse) grid are calculated from the average height of all data values. Starting values for each successive fine grid are linearly interpolated from the preceding coarser grid. Since the fitted values are available at every stage during the iterations, the drainage properties of the DEM can be monitored and morphological constraints can be imposed via ordered chain constraints on the grid points. In practice, the grid is typically examined every five Gauss-Seidel iterations for sinks (which have an elevation no higher than their eight immediate neighbours) and the accompanying saddle point. The saddle point has at least two neighbours higher than itself interleaved by neighbours no higher than itself when moving in a clockwise direction or an anticlockwise direction through the eight neighbours of the grid location. A detailed account of the approach is provided by Hutchinson (192).

When a DEM is generated for hydrological modelling, it is sensible to use an interpolation procedure that imposes directional constraints relating to the direction of surface water flow (192), (194), and this is one objective of Hutchinson's method. It is in this adaptation to local topography that the method has benefits over general interpolation methods, such as the thin plate spline variants discussed in Section 6.3.7. Modifications to the approach are detailed by Hutchinson (194). The procedure is available in a program called ANUDEM (195), and it has been applied for mapping topography in a range of areas including Australia (192), (197), and Africa (199) and for mapping river depths in Idaho and Oregon, USA (204). A version of the ANUDEM algorithm is provided in ArcGIS™, where the relevant tool is called TOPOGRID.

6.4 Areal interpolation

In many contexts, spatial data are available over areal units such as pixels (e.g., remotely-sensed images) or irregularly shaped zones (e.g., zones used to output population counts in a census), and often there is a desire to compare data recorded using different areal units. There are many possible sets of aggregations for individual data, and relationships observed at one scale may not exist at another scale. The problem of inferring characteristics of individuals from aggregations is referred to as the ecological fallacy (also discussed in Section 1.5). There are many areal units that could be used to display information about, for example, a population. It may be argued that none of these has intrinsic meaning about the underlying populations — the units are 'modifiable.' This issue, often called the modifiable areal unit problem (MAUP; (303), (300)), consists of the scale or aggregation problem (how many zones — what is the level of aggregation?) and the zoning problem (which zoning scheme should be used at a given level of aggregation?). It is, therefore, possible that any pattern apparent in mapped areal data may be due as much to the zoning system used as to underlying distribution of the variable (260). Figure 6.11 gives a simple synthetic illustration of scale (variation in the number of zones) and zonation (variation in the shape of zones) differences.

In the geostatistical terminology, the term support is used to refer to the geometrical size, shape, and orientation of the units associated with the measurements (159). The change of support problem thus concerns transferring between different supports. Gotway Crawford and Young (159) outline some methods that may be used to address this issue. Most applications of the kinds of approaches detailed in the following sections are in the social sciences. There are also some applications in agricultural contexts. For example, Geddes et al. (135) are concerned with disaggregation of land use statistics from large source zones to smaller target zones.

Martin et al. (263) discuss various issues concerning linkage of population censuses from different years. Particular problems include that the geography of different censuses may vary (that is, different zones are used) or that the questions asked in a census may vary (so, the variables available to researchers change). The concern here is with converting variables from one set of areal units to another set of areal units, or to a surface. Conversion of data from one set of areal units to another requires some form of interpolation where

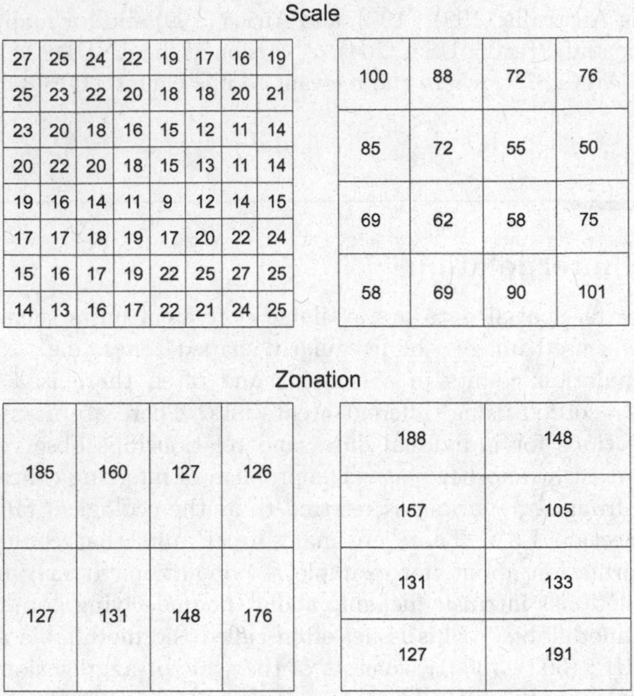

Scale

27	25	24	22	19	17	16	19
25	23	22	20	18	18	20	21
23	20	18	16	15	12	11	14
20	22	20	18	15	13	11	14
19	16	14	11	9	12	14	15
17	17	18	19	17	20	22	24
15	16	17	19	22	25	27	25
14	13	16	17	22	21	24	25

100	88	72	76
85	72	55	50
69	62	58	75
58	69	90	101

Zonation

185	160	127	126
127	131	148	176

188	148
157	105
131	133
127	191

FIGURE 6.11: Simplified representation of scale and zonation differences.

the boundaries cross one another (that is, there are not common boundaries). In other words, areal interpolation entails the reallocation of values from the source zones to the target zones.

Not all of the methods discussed below fall in the umbrella term 'local,' but applications which are global are discussed to provide a context for the methods which can be considered local. Summaries of areal interpolation methods are provided by Flowerdew and Green (123) and Langford et al. (230). Most applications are in human geography and so, by necessity, cited published applications are drawn from human geography. A geostatistical solution to predicting from point values to areas is discussed briefly in Section 7.12.

Martin et al. (263) identify three main approaches that can be used to link data on incompatible areal units:

1. Remodel the data by generating a surface-based representation (261). This entails redistributing values from area counts to the nodes of a regular grid.

2. Use areal interpolation to transfer data from one area to another (150).

3. Use look-up tables to transform one set of areal units to another on a best fit basis (176). That is, areas in dataset one are matched with the areas in dataset two within which they best fit.

These approaches may be used in combination. For example, a look-up table may be used with weights that assign parts of a source area to a target area. Examples of types 1 and 2 are discussed below, but type 3 is outside the remit of this book.

6.5   General approaches: Overlay

Overlay provides the simplest approach to converting between one set of zones and another. That is, one set of zones may be split into smaller zones using a second set of zones which are overlaid. However, using a standard overlay procedure such as the union operator, if a value attached to the original is, say, 1200 and the zone is then split into five sections, each of these new smaller zones will also have values of 1200. Clearly this is nonsensical if the variable was, for example, population (the extensive case, as defined later in this section). By converting from population to population density, meaningful results would be obtained. It may be assumed that the value of the variable can be divided based on the areas of each of the new zones. So, if one of the new zones covers one-third of the area of the original zone (with a population of 1200), then the value attached to it would become 400. This is termed the areal weighting method (151); (123) and an illustrative example is given in Figure 6.12. Lloyd (245) gives a further example.

In practical terms, the source and target zones often do not share common boundaries. For count data or absolute figures, once the areas common to the source and target zones (A_{ts}) are obtained, an estimate of the value (count) z at the target zone t can be given by:

$$\hat{z}_t = \sum_{s=1}^{n} \frac{A_{ts}}{A_s} z_s \tag{6.27}$$

where z_s indicates the value of the $s = 1, ..., n$ source zone which overlaps that target zone, and A_s is the area of source zone s. The first listed zone in Figure 6.12 is given as an example. It is assumed that area of the source zone (i.e., the left hand side of Figure 6.12), A_s, is 1000m^2. To three decimal places the area of the target zone, A_{ts}, is 293.618m^2 and $z_s = 633$. In this example, the target zone intersects only one source zone and this leads to:

$$\frac{293.618}{1000} \times 633 = 185.86$$

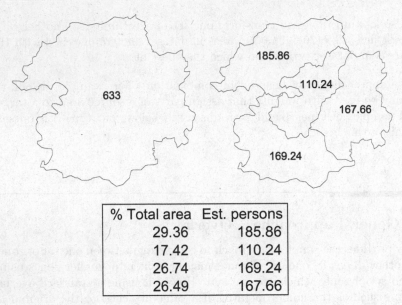

% Total area	Est. persons
29.36	185.86
17.42	110.24
26.74	169.24
26.49	167.66

FIGURE 6.12: The areal weighting method.

For ratios (for example, percentages), Equation 6.27 is modified to become:

$$\hat{z}_t = \sum_{s=1}^{n} \frac{A_{ts}}{A_t} z_s \qquad (6.28)$$

and $\hat{z}_t$ is a proportion.

The areal weighting approach is based on the assumption that the source zone data are distributed homogeneously across their areas. This assumption is clearly unrealistic in many cases. For example, the population may be concentrated in a very small part of a zone. Some more sophisticated approaches to areal interpolation are outlined below and some of these offer solutions to the problem of heterogeneity in variables across zones.

Goodchild and Lam (151) use the term 'intensive' to describes cases where weighted averages are appropriate. The term 'extensive' is used in cases where the value for a large zone is equal to the sum of the values of its component parts. The latter is the case with population counts, for example, and is the main concern here.

In Section 6.6.1, the focus is on generation of surfaces from areal data, Section 6.6.2 presents an application of population surface generation, and Section 6.6.3 is concerned with local volume preserving methods. Section 6.6.4 represents a change of focus — in which methods that make use of prior information about the spatial structure of the primary variable are used. As such, all of these approaches are local in some respect, but vary in whether the model is local in some sense or simply in that local information is used as an input to the interpolation (or reallocation) procedure. Finally, Section 6.6.5 discusses briefly the issue of uncertainty in areal interpolation outputs.

6.6 Local models and local data

In the remainder of this chapter, the focus is on areal interpolation methods that are either local in approach (based, in this case, on moving windows) or make use of additional data in recognition of the fact that the variable of interest (for example, human population count) varies from place to place.

6.6.1 Generating surface models from areal data

This class of methods relates to the point based interpolation procedures discussed in Section 6.1. With nonvolume preserving methods, whereby the output count over a given area is not necessarily the same as the input count over the same area, the usual approach is to select a point location to represent each area (for example, the centroid of an area) and to interpolate to a regular grid. The predictions may then be averaged within the target area.

Martin (258) outlines a method for mapping population from zone centroids. With this approach, each zone centroid is visited in turn and the mean intercentroid distance is calculated within a predefined search radius. This measure indicates the unknown areal extent of zones in the region, and it is used to calibrate a distance-decay function that assigns weights to cells in the output grid. The cells that are closest to the zone centroid receive the largest weights, while those cells estimated to be located in the maximum areal extent of the zone receive the smallest weights. The population is then redistributed in the surrounding region using these weights. A given cell in the output grid may receive population values from one or more centroids, or may remain unpopulated as it is beyond the area of influence of any of the centroid locations. Assignment of values to each cell in a grid may be conducted with:

$$\hat{z}_i = \sum_{j=1}^{n} z_j w_{ij} \qquad (6.29)$$

where $\hat{z}_i$ is the estimated population of cell i, z_j is the population recorded at point j, n is the total number of data points (locally), and w_{ij} is the weight of cell i with respect to point j (262). An appropriate distance function (based on work by Cressman (97)) is given by:

$$w_{ij} = \begin{cases} \left(\frac{\tau_j^2 - d_{ij}^2}{\tau_j^2 + d_{ij}^2}\right)^r & \text{for } d_{ij} < \tau_j \\ 0, & \text{for } d_{ij} \geq \tau_j \end{cases} \qquad (6.30)$$

where w_{ij} is the weight for distance d_{ij}, τ_j is the adjusted width of the window centred on point j, and r is an exponent. A modified version of this method is outlined below. To preserve the total population, it is necessary that the weights are constrained to sum to one:

$$\sum_{i=1}^{n} w_{ij} = 1.0 \qquad (6.31)$$

Population is preserved globally (the sum of populations in the zones is the same as the sum of populations in the population surface), but the sum of the number of people in a given zone does not necessarily correspond to the overlapping area in the population surface. Gotway Crawford and Young (158), (159) classify this approach as a spatial smoothing method.

Bracken and Martin (56) apply the method for linking 1981 and 1991 censuses of Britain. An illustrated example of the method is given in Section 6.6.1.1 and a real world case study in Section 6.6.2. The method of Bracken and Martin is a form of adaptive kernel estimation (32), as discussed in Section 8.10.2. Approaches such as this are considered problematic by some researchers on the grounds that population centroids used as the basis of the reallocation of population counts are usually not objectively defined.

6.6.1.1   Illustrating generation of a population surface

The application of the approach to redistributing population from zone centroids to grid cells described in Martin (258) is demonstrated here in a simple example. Table 6.7 gives population counts at four zone centroids for an artificial case. In this example, the population counts are redistributed to a grid of five-by-five cells whose centres have a minimum x and y coordinate of zero and a maximum x and y coordinate of four. The mean intercentroid distance is calculated within a predefined search radius, which in this case was 2.5 units. This gives the mean intercentroid distances specified in Table 6.8.

TABLE 6.7
Population counts.

ID	X	Y	Population
1	2	1	22
2	1.5	2.5	24
3	3.5	2.5	27
4	4	4	25

TABLE 6.8

Mean intercentroid distances.

No	X	Y	Mean distance
1	2	1	1.851
2	1.5	2.5	1.791
3	3.5	2.5	1.901
4	4	4	1.581

Then the distance between each centroid and each cell is calculated. The weights are then calculated using Equation 6.30. The weights are then scaled to sum to one. Finally, the weights are multiplied by the population values in each of the centroids. For example, the cell located at 2, 2 receives population from zone centroids 1 (1 unit in distance from the cell), 2 (0.71 units from the cell), and 3 (1.58 units from the cell).

Following this example, the weight for zone centroid 1 is calculated:

$$\left(\frac{1.851^2 - 1^2}{1.851^2 + 1^2}\right)^1 = \left(\frac{3.426 - 1}{3.426 + 1}\right)^1 = \left(\frac{2.426}{4.426}\right)^1 = 0.548$$

the weight for zone centroid 2:

$$\left(\frac{1.791^2 - 0.71^2}{1.791^2 + 0.71^2}\right)^1 = \left(\frac{3.208 - 0.504}{3.208 + 0.504}\right)^1 = \left(\frac{2.703}{3.713}\right)^1 = 0.728$$

and the weight for zone centroid 3:

$$\left(\frac{1.901^2 - 1.58^2}{1.901^2 + 1.58^2}\right)^1 = \left(\frac{3.614 - 2.496}{3.614 + 2.496}\right)^1 = \left(\frac{1.118}{6.110}\right)^1 = 0.183$$

The weights referring to each centroid are then scaled to sum to one. For example, the weights for centroid 1 sum to 4.245, with population assigned to nine cells, each weight is then calculated as a proportion of 4.245. These proportions are used to reassign population thereby ensuring the total output population is the same as the total input population (in this example, 98 people in total). The first of the three weights above becomes 0.129 (0.548 is 12.9% of 4.245), the second weight is 0.187, and the third weight is 0.044. The population assigned to cell 2,2 is then calculated using Equation 6.29, giving:

$$(22 \times 0.129) + (24 \times 0.187) + (27 \times 0.044) = 2.838 + 4.488 + 1.188 = 8.516 \text{ people}$$

The output grid is shown in Figure 6.13. Note that the sum of values in the population surface is the same as the sum of input data values. In the following section, a short real world case study is presented.

0.000	0.759	0.759	6.637	13.895
0.759	4.481	5.674	7.131	10.404
0.759	5.844	8.516	7.082	4.960
0.000	3.601	5.942	4.035	1.193
0.000	1.363	2.841	1.363	0.000

FIGURE 6.13: Population distributed to grid cells.

6.6.2   Population surface case study

In this case study, population data from the 2001 Census of Northern Ireland
are used to illustrate areal interpolation from centroids of 'Output Areas' (one
of the sets of zones for which population counts are provided) to a regular
grid. The Output Areas and the population counts were given in Figure
1.8. A population surface was derived from these data using the method of
Martin (258), and it is shown in Figure 6.14. The population values were
redistributed to a grid with a 200 m cell spacing, and the Cressman decay
function (defined in Equation 6.30) was used with a value of one for the decay
parameter. The urban area around Belfast (in the east of the province) and
other urban areas is clear, as is the fact that population is sparse across most
of the region. Approaches which distribute values from irregularly-shaped
zones to surfaces are useful in allowing comparison of counts from different
time periods for which the original zonal systems are incompatible. So, the
procedure applied here to 2001 population counts could be applied to counts
from earlier censuses, and the surfaces compared to allow assessment of spatial
variation in population change.

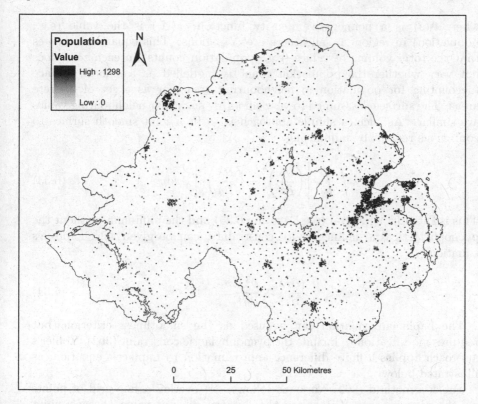

FIGURE 6.14: Population distributed to grid cells in Northern Ireland.

6.6.3 Local volume preservation

The term 'pycnophylactic' refers to the property of areal interpolators whereby the population of the target zones should be the same as the population of the constituent source zones — counts from a source zone should be allocated to target zones with which they intersect. Thus, volume is preserved locally.

Tobler's pycnophylactic method was designed to generate continuous surfaces from data represented as areas (for example, census district boundaries) (362). The method reallocates data — it is mass-preserving in that the volume of the attribute in a given area is the same in the original data (that is, in discrete polygons) and in the continuous surface derived from it. Summaries of the method are provided by Burrough and McDonnell (70) and Waller and Gotway (375).

The key condition for preserving mass (for example, maintaining the population count in the input data) is the invertibility condition:

$$\int_{A_i} \lambda(\mathbf{s}) \, d\mathbf{s} = z(A_i) \text{ for all } i \tag{6.32}$$

where $\lambda(\mathbf{s})$ is a nonnegative density function, $z(A_i)$ is the value (e.g., population) in region A_i, and there are N regions. This equation indicates that the total volume of values (e.g., population counts) in each zone does not vary whether the population count is modelled as a smooth surface (accounting for population in neighbouring areas) or as a set of discrete areas. The surface is assumed to vary smoothly such that neighbouring values are similar. As such, a simple approach is to fit a joint smooth surface to contiguous regions by minimising:

$$\int\int_{\Re^2} \left[\left(\frac{\partial \lambda}{\partial x}\right)^2 + \left(\frac{\partial \lambda}{\partial y}\right)^2 \right] \mathrm{d}x \mathrm{d}y \qquad (6.33)$$

This is Dirichlet's integral (see Equation 6.25) and the minimum, without the pycnophylactic and nonnegativity constraints, can be given with Laplace's equation:

$$\frac{\partial^2 \lambda}{\partial x^2} + \frac{\partial^2 \lambda}{\partial y^2} = 0 \qquad (6.34)$$

The Laplacian equation can be used as the smoothness criterion, but requires modification to include the pycnophylactic constraint (362). Tobler's approach applies a finite difference approximation to Laplace's equation, as illustrated below.

Tobler's method (362) works as follows: (i) overlay a fine grid of points on the source zones, (ii) assign the average density value to each point falling within a given zone, and (iii) then adjust these values iteratively with a smoothing function and a volume-preserving constraint until there is no significant change in the grid values between iterations. The approach is conceptually similar to the method detailed in Section 6.3.11.

Tobler (362) gives a small example of the application of the principles of his pycnophylactic method and part of that material is illustrated here. In Tobler's illustration, he uses the example of a histogram and calculating differences between adjacent bars as a measure of smoothness. Following Tobler (362), a 2D example can be illustrated with a grid representing cells belonging to two different regions:

$$1\ 1\ 1$$
$$1\ 2\ 2$$
$$2\ 2\ 2$$

Subtraction and squaring of adjacent cell values (that is, heights using the histogram parallel) can be defined with:

$$T_0 = \sum_{i=1}^{I} \sum_{j=1}^{J-1} (\lambda_{i,j+1} - \lambda_{i,j})^2 + \sum_{i=1}^{I-1} \sum_{j=1}^{J} (\lambda_{i+1,j} - \lambda_{i,j})^2 \qquad (6.35)$$

where I is the number of rows in the array, and J is the number of columns. Adding the pycnophylactic constraint leads to:

$$T = T_0 + \sum_{i=1}^{N} \psi_i(z_i - \sum_{k \in A_i} \lambda_k) \qquad (6.36)$$

where ψ is the Lagrangian multiplier. Setting the partial derivatives of T with respect to each λ (cell location) and each ψ to zero gives a system of $n + N$ linear equations where N are the number of regions, and n are the number of lattice points. The system is then, for a three-by-three matrix, $C\lambda = z$:

$$
\begin{bmatrix}
-2 & 1 & 0 & 1 & 0 & 0 & 0 & 0 & 0.5 & 0 \\
1 & -3 & 1 & 0 & 1 & 0 & 0 & 0 & 0.5 & 0 \\
0 & 1 & -2 & 0 & 0 & 1 & 0 & 0 & 0.5 & 0 \\
1 & 0 & 0 & -3 & 1 & 0 & 1 & 0 & 0.5 & 0 \\
0 & 1 & 0 & 1 & -4 & 1 & 0 & 1 & 0 & 0.5 \\
0 & 0 & 1 & 0 & 1 & -3 & 0 & 1 & 0 & 0.5 \\
0 & 0 & 0 & 1 & 0 & 0 & -2 & 1 & 0 & 0.5 \\
0 & 0 & 0 & 0 & 1 & 0 & 1 & -3 & 1 & 0.5 \\
0 & 0 & 0 & 0 & 0 & 1 & 0 & 1 & -2 & 0.5 \\
1 & 1 & 1 & 1 & 0 & 0 & 0 & 0 & 0 & 0 \\
0 & 0 & 0 & 0 & 1 & 1 & 1 & 1 & 1 & 0 & 0
\end{bmatrix}
\times
\begin{bmatrix}
\lambda_1 \\ \lambda_2 \\ \lambda_3 \\ \lambda_4 \\ \lambda_5 \\ \lambda_6 \\ \lambda_7 \\ \lambda_8 \\ \lambda_9 \\ \psi_1 \\ \psi_2
\end{bmatrix}
=
\begin{bmatrix}
0 \\ 0 \\ 0 \\ 0 \\ 0 \\ 0 \\ 0 \\ 0 \\ 0 \\ z_1 \\ z_2
\end{bmatrix}
$$

and the unique solution is $\lambda = C^{-1}z$. In Tobler's example, $z_1 = 8$ and $z_2 = 5$ (that is, if the counts represent population, there are 8 individuals in one region and 5 in the other). This gives:

$$
\begin{array}{ccc}
2.240 & 2.060 & 1.956 \\
1.740 & 1.306 & 1.170 \\
0.996 & 0.816 & 0.711
\end{array}
$$

with $T = 2.734$. The module surrounding -4 is the finite difference approximation to the two-dimensional Laplacian. Taking the first (top) row with respect to the λ value positions we have:

$$
\begin{array}{ccc}
-2 & 1 & 0 \\
1 & 0 & 0 \\
0 & 0 & 0
\end{array}
$$

The fifth row corresponds to the case with a full set of neighbours:

$$
\begin{array}{ccc}
0 & 1 & 0 \\
1 & -4 & 1 \\
0 & 1 & 0
\end{array}
$$

In the computer program written by Tobler a nonnegativity constraint has been added (362).

Tobler's method is intuitively sensible where the property in concern varies continuously (for example, rainfall). However, population does not tend to vary smoothly. For this reason, a variety of other approaches have been developed that make use of additional information about how these populations are likely to vary from place to place and some such methods are outlined in Section 6.6.4.

With the approach presented by Martin and Bracken (262), no account is taken of zone boundary location. That is, population may be gained from or lost into neighbouring zones through applying the method. For this reason, Martin (259) has presented an adapted version of the method. In this modified version, a rasterised zone map with the same cell size as the required output surface is acquired first. As population is redistributed, the weights for cells located outside the current zone are automatically set to zero — it is locally mass preserving.

6.6.4   Making use of prior knowledge

Another class of areal interpolators have been developed which may be termed 'intelligent' interpolators. Such methods take into account prior knowledge about the area of study. For example, if a zone comprises 50 per cent high lying land and 50 per cent low lying land, population is likely to be concentrated in the latter (123). As such, these methods make use of local information to inform areal interpolation. Gregory (162), (163) gives a summary of some such approaches and outlines two classes of approaches: dasymetric techniques and the use of control zones.

Dasymetric mapping (324) accounts for 'limiting variables' which provide information on the distribution of a variable within the source zone. For example, if part of a source zone is covered by water then the population of that area would be zero. Fisher and Langford (121) discuss the use of dasymetric mapping. With approaches based on control zones, ancillary information from another set of zones (target zones or another set of zones) is used as a part of the reallocation procedure (150), (230).

Gregory (163) and Gregory and Ell (164) apply a range of methods for reallocating counts from different historical censuses in Great Britain. Gregory aggregated counts from 1991 census enumeration districts (EDs) into districts and parishes (the latter providing information at the sub-district scale) based on the districts and parish in which the ED centroid was located. The data were then interpolated onto synthetic target districts generated in the same way. One set of target zones (synthetic 1911 Registration Districts) and three different sets of source zones (synthetic 1881 Registration Districts, synthetic Local Government Districts from 1931 and 1971) were used. Methods employed were (i) areal weighting, (ii) a dasymetric approach (with source parish zones as limiting variables), (iii) the expectations maximum likelihood (EM) algorithm (defined below; using the district-level variable of interest (e.g., people in a certain age range) for the target zones to construct control zones), (iv) the EM algorithm (using the total populations of the target parishes to create control zones), (v) a combined dasymetric and EM algorithm approach (using the district-level variable of interest for the target zones to create control zones), and (vi) a combined dasymetric and EM algorithm approach (using the total population of target parishes to create control zones). The different approaches are described below.

Gregory (163) outlines necessary conditions to apply the dasymetric method. In his study, population data over one set of zones (parishes) were used to provide information about intrasource zone distribution of the variable of concern. Two assumptions must be fulfilled: the distribution of a subpopulation variable z at parish level, z_p, can be estimated using the source zone (district) level value of z and the proportion of the total population of the district that is contained by the parish and, secondly, z should be distributed homogeneously across each parish. The values of a variable z are estimated for a set of source zones s onto a set of target zones t using the populations of one set of source zones (in Gregory's case, districts), p_s, and another set of source zones (in Gregory's case, parishes), p_p which nest within the districts. The estimate of the variable of interest for each source zone (parish) is given by:

$$\hat{z}_p = \frac{p_p}{p_s} z_s \qquad (6.37)$$

where $\hat{z}_p$ is the estimate of the value of the subpopulation variable for a parish, and z_s is the published value of the variable for the source zone. Following this step, the value for each target zone is estimated using a modified form of the areal weighting formula:

$$\hat{z}_t = \sum_{p=1}^{n} \frac{A_{pt}}{A_p} \hat{z}_p \qquad (6.38)$$

where A_{pt} is the zone of intersection between the parish and the target zone, and A_p is the area of the source parish. Gregory (163) notes that, for this approach to improve on standard areal weighting, the source zones (districts) should be subdivided well by the component secondary source zones (parishes). A theoretically superior approach is to lessen the assumption of intra-source zone homogeneity (163) and some relevant approaches are outlined below.

Approaches that utilise control zones are presented by Goodchild et al. (150) and Langford et al. (230); neither of the methods are limited by the pycnophylactic criterion defined in Section 6.6.3. Langford et al. (230) presented a technique where target zones behave as control zones. With this approach, satellite sensor imagery was used to estimate population over an arbitrarily defined area. With this procedure, landcover was classified using the imagery. Each pixel (of 30 m by 30 m) was classified as dense residential, residential, industrial, agricultural, or unpopulated. The number of pixels of each landcover type within each census area (ward) was then ascertained through overlay. Several regression procedures were applied and assessed to express the ward populations as a function of pixel counts of each land cover type. Poisson regression was selected, following·the work of Flowerdew and colleagues (see, for example Flowerdew and Green (123), and see below), on the grounds that the response variable (population) is a count variable. In the final model, the two landcover types dense residential and residential were summed to create a new variable which was regressed against population. No intercept term was used so that areas with no residential cover had zero population. The resulting model was then used to estimate population over a kilometre square grid.

Goodchild et al. (150) used external control zones (indexed by c) that could be assumed to have constant densities. The (unknown) population density of these control zones, (λ_c), is then estimated and used to obtain target-zone estimates. The areas of overlap between the source zone s and the control zone c are given by A_{cs}, while that between the control zone c and the target zone t is given by A_{ct}. If there are fewer control zones than source zones then the control zone densities can be estimated from the equation:

$$z_s = \sum_{c=1}^{n} \lambda_c A_{cs} \qquad (6.39)$$

where z_s is the observed source zone population. For counts, a suitable approach is to estimate the densities as coefficients using Poisson regression, for the reasons outlined above. This estimated population density is used to allocate data to the target zones using areal weighting based on the population density of the control zones and the areas of zones of intersection between the control zones and target zones:

$$\hat{z}_t = \sum_{c=1}^{n} \lambda_c A_{ct} \qquad (6.40)$$

As Goodchild et al. (150) note, a key advantage of this approach is that the control zones do not need to be congruent with either the source zones or the target zones.

Flowerdew and Green (123) outline a method that uses information about a binary variable in the target zone (used as control zones) to make estimates of a count variable. With this method, data about the target zones are assumed to provide information about the intrasource zone distribution of the variable of interest. The count is considered to be the outcome of two separate Poisson processes, the parameters of which are dependent on the value of the binary variable. For example, where the count variable is population, and the binary variable distinguished farmland and woodland, the population of a farmland target zone is expected to be the outcome of a Poisson process with one parameter, while the population of a woodland target zone is expected to be the outcome of a Poisson process with another parameter. Where a target zone includes both farmland and woodland, the expected population is then the average of the two parameters which are weighted according to the area of each of the two land use types. The Poisson parameters are estimated by Poisson regression conducted on the source zones. For example, the population is regressed (as in the cases mentioned above, through the origin) on area under farmland and area under cropland (123). This approach has been generalised using the EM algorithm. In this case, the EM algorithm is used to determine how much of the population should be distributed to different target zone types.

The EM algorithm is summarised by Flowerdew and Green (123). In basis, the algorithm comprises a set of methods designed to overcome problems of imperfect observations. In the case of missing values, if we wish to apply an estimation procedure that requires a complete dataset but some values are missing, the EM solution is to impute missing values and apply the estimation procedure to the dataset, including the imputed values. If we have counts over source zones, z_s, the desire may be to infer values of z for target zones z_t. In this case, the problem is to infer values of z for the zones of intersection between the source zones s and the target zones t. These estimates, z_{st}, are made using the source zone values, z_s, and ancillary information. The estimates, z_{st}, are then summed over the source zones to give estimates of z_t.

The EM algorithm entails iterations of a two step procedure. These are the E-step and the M-step. With the E-step, the conditional expectation of the missing data is computed given the model and the observed data. With the M-step, the model is fitted using maximum likelihood to the data, including the values imputed in the E-step. Values in the zones of intersection are re-estimated and the algorithm iterates until convergence.

Flowerdew and Green (123) show how the EM algorithm can be used to transfer values from a set of source zones to target zones. This approach is based on the idea that the zones of intersection have, in effect, missing data. The values for these zones can be estimated using the EM algorithm through adding information from ancillary data derived from the target zones (to define control zones) to the basic knowledge of data in the source zones (e.g., variations in population density derived from the source data) and the area of the zones of intersection; the EM algorithm is used to estimate differences in population density between different types of control zones using the source data. That is, the study area is divided into control zones based on ancillary information (for example, whether a zone is urban or rural). Modified areal weighting is used with information about population densities of the control zone and the zone of intersection added. The method of Flowerdew and Green (123) is limited by the pycnophylactic criterion.

Gregory (163) applies the EM algorithm along with other methods. An initial estimate of the population density of a particular control zone, $\lambda_{j(c)}$, can be made through assuming that the population density of urban and rural areas are the same as the average population density of the whole study area.

Gregory's implementation of the EM algorithm can be defined as follows:

E-step: make an estimate of the value of the variable of interest for the zone of intersection between the source zone (parish in this case) and the target zone; this is a modification of the areal weighting formula that includes the densities of each type of control zone:

$$\hat{z}_{pt} = \frac{\lambda_{j(c)} A_{pt}}{\sum_{k=1}^{n} \lambda_{j(k)} A_{pk}} \hat{z}_p \qquad (6.41)$$

where $\hat{z}_{pt}$ is the estimated population of the zone of intersection, and $\lambda_{j(c)}$ is the estimated population density for the type of control zone that the zone of intersection is located within. The population density and area of each zone of intersection k within source zone p are given by $\lambda_{j(k)}$ and A_{pk} respectively.

M-step: new values are estimated based on the estimates of z_{pt} in each control zone. New values of λ_j for each control zone type are calculated with:

$$\hat{\lambda}_j = \sum_{pt \in c(j)} \hat{z}_{pt} / \sum_{pt \in c(j)} A_{pt} \qquad (6.42)$$

The algorithm iterates until it converges. Finally, $\hat{z}_t$ is calculated for each target zone by summing the estimated values of z for each zone of intersection within each target zone. Gregory (163) allocates district-level source data to parish level with Equation 6.37, then the EM algorithm is used as given in Equations 6.41 and 6.42. The procedure is illustrated using an example of allocating population from a single source zone to urban and rural target zones. In the following example, the preliminary stage of estimating the parish population, $\hat{z}_p$, is not required, and so instead z_s is used directly and the index s, rather than p, denotes the source zone.

Initially, it is assumed that the density of the population of the urban and rural areas are the same; thus $\lambda_{jurban} = 1.0$ and $\lambda_{jrural} = 1.0$. The area of the source zone, A_s, is 100 km^2, and the source zone population, z_s, is 60000. The area of the urban zone of intersection is 25 km^2 and the area of the rural zone of intersection is 75 km^2. For both urban and rural areas, the denominator of Equation 6.41 (the E-step) is 100 since $1.0 \times 25 + 1.0 \times 75 = 100$. The numerator of Equation 6.41 for the urban zone of intersection is $1.0 \times 25 = 25$ while the equivalent numerator for the urban zone of intersection is $1.0 \times 75 = 75$. So, the estimated population of the urban zone of intersection is $25/100 \times 60000 = 15000$ people, and the estimated population of the rural zone of intersection is $25/100 \times 60000 = 45000$ people. This is the standard areal weighting solution given Equation 6.27. Given information on population densities in urban and rural areas more accurate reallocations of the population can be performed.

Equation 6.42 (the M-step) is used to estimate population density values for each control zone type. In this theoretical example, the estimated population density of urban areas, λ_{jurban}, given data for the whole study area is 1000 people per km^2, while the equivalent estimated population density for rural areas, λ_{jrural}, is 250 people per km^2.

Given this information, the E-step is repeated. The denominator for the E-step is now: $1000 \times 25 + 250 \times 75 = 43750$. For the urban zone of intersection, the numerator is now $1000 \times 25 = 25000$, and for the rural zone of intersection the numerator is $250 \times 75 = 18750$. Thus, the estimated population for the urban zone of intersection is $25000/43750 \times 60000 = 34286$, and the estimated population for the rural zone of intersection is $18750/43750 \times 60000 = 25714$ (with both figures rounded to produce whole numbers). The effect of altering the urban/rural population balance is clear. Given the new estimates, the M-step is repeated and the procedure continues until convergence. Gregory and Ell (164) give a further worked example of the E-step of the algorithm.

6.6.5 Uncertainty in areal interpolation

As Gregory (163) states, there has been relatively little work in assessing errors in results obtained through areal interpolation. Fisher and Langford (121) use Monte Carlo simulation to compare different regression models in applying the method of Langford et al. (230). Their study indicates that accuracy increases as the number of target zones decreases. Cockings et al. (88) assess mean error for a fixed set of target zones calculated based on interpolation from multiple sets of randomly created source zones, and their study indicates the geometric properties of source zones are important. Sadahiro (327), in a study assessing the accuracy (using a stochastic modelling approach) of areal weighting interpolation outputs, suggests that the size and shape of source zones is important. That is, smaller source zones tend to give the most accurate estimates.

Gregory (163) aggregated observed data to form 'synthetic' versions of parishes and districts and compared six approaches to areal interpolation, as outlined at the end of Section 6.6.4. The differences between the estimated values of the variables and observed values were assessed in various ways. Gregory's study indicated that the combined dasymetric and EM algorithm approaches usually provided the most accurate estimates. Differences were partly a function of the variable being interpolated and the choice of target geography.

6.7 Limitations: Point and areal interpolation

Clearly all of the methods for point and areal interpolation discussed in this chapter have some limitations and different approaches are suitable for different situations. But, which approach is 'best' in a given situation? This is difficult to assess using a global approach but more difficult still for a local approach where factors such as window size are an issue. As with other methods discussed in this book, experimentation with parameters is sensible. Approaches such as cross-validation may be used to guide selection of parameters in the absence of independent validation data. Many studies have shown very similar prediction accuracies where the sample spacing is small relative to the scale of spatial variation. Such case studies help to guide the prediction process, but in most cases comparison of a variety of approaches will be necessary. In some cases, approaches other than interpolation are appropriate. For example, diffusion models may be a more appropriate means of mapping airbourne pollutants than a generic interpolation procedure such as those discussed in this chapter. There are many ways of transferring data values between different zonal units but the approaches to assessing uncertainty that are being developed will hopefully allow more objective comparison of approaches in different situations.

6.8   Overview

In this chapter, a variety of approaches to point interpolation and areal interpolation have been discussed. As stated above, no one method can be considered as inherently 'better' than the others, although some have properties that make them more suitable in certain contexts.

In the next chapter another set of approaches to spatial prediction are discussed, along with associated tools for characterising and modelling spatial variation.

7

Spatial Prediction 2: Geostatistics

Geostatistics is discussed separately to the methods outlined in Chapter 6 on the grounds that geostatistics entails different operational principles, as well as a different underlying philosophy (use of a random function model, as defined below). In practice, as detailed in this chapter, there are similarities between methods outlined in this chapter and some of those discussed in Chapter 6.

Geostatistics is based upon the recognition that in the Earth sciences there is usually a lack of sufficient knowledge concerning how properties vary in space. Therefore, a deterministic model may not be appropriate. If we wish to make predictions at locations for which we have no observations, we must allow for uncertainty in our description as a result of our lack in knowledge. So, the uncertainty inherent in predictions of any property we cannot describe deterministically is accounted for through the use of probabilistic models. With this approach, the data are considered as the outcome of a random process. Isaaks and Srivastava (200) caution that we should remember that use of a probabilistic model is an admission of our ignorance, but it does not mean that any spatially-referenced property varies randomly in reality. For a criticism of geostatistics, see Philip and Watson (317), with responses to this paper given by Journel (209) and Srivastava (340).

In geostatistics, spatial variation is modelled as comprising two distinct parts, a deterministic (trend) component ($m(\mathbf{s})$) and a stochastic ('random') component ($R(\mathbf{s})$):

$$Z(\mathbf{s}) = m(\mathbf{s}) + R(\mathbf{s}) \tag{7.1}$$

This is termed a random function (RF) model. In geostatistics, a spatially-referenced variable, $z(\mathbf{s})$, is treated as the outcome of a RF, $Z(\mathbf{s})$, defined as a set of random variables (RV) that vary as a function of spatial location $\mathbf{s}$. A realisation of a RF is called a regionalised variable (ReV). The Theory of Regionalised Variables (265), (266) is the fundamental framework on which geostatistics is based. Note the common division into two components with the thin plate spline model of Equation 6.6.

Our observations (that is, measurements of the property of interest) are modelled as ReVs. So, if the property of interest was precipitation, the ReV may be viewed as a realisation of the set of RVs 'precipitation' (see Journel and Huijbregts (212) for a clear and detailed overview).

In this chapter, the initial focus is on the theoretical basis of linear geostatistics (Sections 7.1 and 7.2). This is followed by an account of linear geostatistics based on a stationary model (Sections 7.3 through 7.12), and a short review of other approaches is given in Section 7.13. This is necessary to provide a basis upon which the following account of nonstationary (that is, locally-adaptive) geostatistics builds (Sections 7.14 through 7.18). Finally, the chapter summarises some key issues. Illustrated examples of the applications of some key methods are presented in the body of the text.

Many introductions to geostatistics are available (for example, (18), (85), (153), (384)). An introduction to geostatistics is provided by Burrough and McDonnell (70) and others by Atkinson and Lloyd (28) and Lloyd (245). Introductions have been written for users of GISystems (297), physical geographers (298), (299), the remote sensing community (98), and archaeologists (249).

7.1   Random function models

When the variables we are concerned with are continuous (as is usually the case in geostatistics), then we model the variables as continuous RFs. The RF, $Z(\mathbf{s})$, is characterised completely by the cumulative distribution function (cdf), F:

$$F(\mathbf{s}; z) = P\{Z(\mathbf{s}) \leq z\} \text{ for all } z \tag{7.2}$$

The cdf gives the probability, P, that $Z(\mathbf{s})$ is no greater than the specific threshold z. The RF is characterised by the n-variate cdf:

$$F(\mathbf{s}_1, ..., \mathbf{s}_n; z_1, ..., z_n) = P\{Z(\mathbf{s}_1) \leq z_1, ..., Z(\mathbf{s}_n) \leq z_n\} \tag{7.3}$$

This is termed the spatial law of the RF, $Z(\mathbf{s})$. In general, we have only one realisation, $z(\mathbf{s})$, at each location, $\mathbf{s}$. Unfortunately, we cannot know the spatial law of the RF from only one realisation. To overcome this problem we make some assumptions about the way in which a property varies in space. These assumptions, which are encompassed in the hypothesis of stationarity, are discussed in the following section. If we can assume that the property of interest varies in a similar way across the region of interest, and a stationary model is considered appropriate, we can treat each of our observations (ReVs) as realisations of the same RF (212). Following this, we are able to make some statistical inferences about z.

7.2 Stationarity

Stationarity may be divided (for geostatistical purposes) into three classes, for which different parameters of the RF may exist. In turn these are: (i) strict stationarity, (ii) second-order stationarity, and (iii) intrinsic stationarity (212), (283). Myers (283) provides a concise summary of stationarity from a geostatistical viewpoint.

It should be emphasized that stationarity is not a property of the data. It is, rather, a property of the RF model (209), (283). For this reason, it is misleading to equate stationarity with homogeneity of a population (209). As such, stationarity cannot be tested with the data. However, where the objective is prediction, the decision of stationarity or nonstationarity (as any model decision) is testable through some measure of the precision of spatial predictions (210).

7.2.1 Strict stationarity

A RF is strictly stationary if its spatial law is shown to be invariant under translation. That is, the joint distribution of the n RVs $Z(s_i), ..., Z(s_n)$ is identical to that of the joint distribution of $Z(s_i+h), ..., Z(s_n+h)$ (212), (283), where h indicates the lag (distance and direction) with respect to location s_i. In simpler terms, if we move from one area to another all the characteristics of the RF are the same. In practice, a weaker form of stationarity, termed second-order stationarity, is employed in linear geostatistics.

7.2.2 Second-order stationarity

For second-order stationarity only the first and second moments, the mean and the covariance, are required to be constant with location. Therefore, the expected value of the mean should be the same at all locations, s:

$$E\{Z(s)\} = m(s) \text{ for all } s \qquad (7.4)$$

In addition, the covariance, $C(h)$, between the locations s and $s + h$ should depend only on the lag, h, and not on the location, s:

$$\begin{aligned} C(h) &= E[\{Z(s) - m\}\{Z(s + h) - m\}] \\ &= E[\{Z(s)\}\{Z(s + h)\} - m^2] \qquad \text{for all } h \end{aligned} \qquad (7.5)$$

In many cases, second-order stationarity is found to be too strong. In some cases, the variance (or dispersion) may be unlimited. For this reason, Matheron (266), (267) defined the intrinsic hypothesis.

7.2.3   Intrinsic stationarity

For a RF to fulfil the intrinsic hypothesis it is required only that the expected value of the variable should not depend on $\mathbf{s}$:

$$E\{Z(\mathbf{s})\} = m(\mathbf{s}) \text{ for all } \mathbf{s} \qquad (7.6)$$

and the variance of the increments should be finite (212). Thus, the variogram, $\gamma(\mathbf{h})$, defined as half the expected squared difference between paired RVs, exists and depends only on $\mathbf{h}$:

$$\gamma(\mathbf{h}) = \tfrac{1}{2}E[\{Z(\mathbf{s}) - Z(\mathbf{s} + \mathbf{h})\}^2] \text{ for all } \mathbf{h} \qquad (7.7)$$

Second-order stationarity implies the intrinsic hypothesis, but the intrinsic hypothesis does not imply second-order stationarity. Thus, the covariance function and the correlogram (or autocorrelation function) exist only if the RF is second-order stationary, and the variogram is used when intrinsic stationarity only can be assumed (212).

7.2.4   Quasi-intrinsic stationarity

The variogram may be used within limited (moving) neighbourhoods if stationarity is accepted within some (moving) neighbourhood. In this case, the RF is said to be quasi-intrinsic (212).

7.2.5   Nonstationarity

In cases where a stationary model is not realistic then a nonstationary model of some kind may be used. For example, if the mean of a variable varies from place to place then a nonstationary mean model may be necessary. If the variogram differs markedly in form in different areas then similarly a nonstationary variogram model may be utilised. While this chapter necessarily includes a introduction to stationary models (beginning in the following section), the main focus is on cases where a stationary model is inadequate. Sections 7.14 onwards deal with locally-adaptive (i.e., nonstationary) geostatistical models.

7.3   Global models

Sections 7.4 to 7.13 are concerned primarily with the case where spatial variation is assumed to be constant over the region of interest (that is, a stationary model is appropriate). However, ordinary kriging, a widely used geostatistical interpolation algorithm (see Section 7.8), allows the mean of the variable of interest to vary from place to place, although it is assumed to be constant within a moving window. As such, the distinction between global and local models is not always clear. Sections 7.14 through 7.18 focus on models that explicitly allow for local variation in the spatial structure of a variable. Coverage of standard stationary models provides necessary background to the nonstationary models.

7.4 Exploring spatial variation

The determination of regions for which a stationary variogram is appropriate is a central concern of geostatistics. The variogram cloud is useful in assessing whether the division of a region into smaller regions is advisable (209) and for identifying outliers. The variogram cloud plots the semivariance (half the squared difference) between paired data values against their separation distance on a pair-by-pair basis. In contrast, the experimental variogram (discussed below) deals with average semivariances for all pairs separated by a given separation distance, which itself is an average if the data are irregularly distributed. Examination of the variogram cloud provides a means of identifying heterogeneities in spatial variation of a variable (384). This variation is obscured through the summation over lags that occurs when the experimental variogram is estimated. The **h**-scatterplot is another useful summary; it shows all pairs of data values separated by a particular lag and direction (200).

The variogram surface (two-dimensional [2D] variogram) is a graphical summary of the variogram estimated in all directions. The principal use of the variogram surface is the identification of directional trends (anisotropy). The variogram surface is implemented, for example, by Pannatier (307) in his Variowin variography package.

7.4.1 The covariance and correlogram

The variogram is more frequently used than the covariance function and correlogram, although the latter two functions are still frequently encountered in the geostatistical literature. It is, however, considered wise to estimate at least two measures of spatial structure in a geostatistical analysis (107). Furthermore, although the variogram is used primarily because it requires a weaker form of RF model, it has been demonstrated that this is often of no relevance in practical situations (107).

The covariance function ($C(\mathbf{h})$) was defined in Equation 7.5:

$$\begin{aligned} C(\mathbf{h}) &= \mathrm{E}[\{Z(\mathbf{s}) - m\}\{Z(\mathbf{s} + \mathbf{h}) - m\}] \\ &= \mathrm{E}[\{Z(\mathbf{s})\}\{Z(\mathbf{s} + \mathbf{h})\} - m^2] \qquad \text{for all } \mathbf{h} \end{aligned}$$

Where $\mathbf{h}$ is zero the covariance is equal to the variance:

$$C(0) = \mathrm{E}[\{Z(\mathbf{s}) - m\}^2] = \sigma^2 \tag{7.8}$$

The correlogram (or autocorrelation function, signified by $\rho(\mathbf{h})$), is defined by the ratio $C(\mathbf{h})/C(0)$ at lag $\mathbf{h}$. The correlogram has values ranging from 1 to −1. The semivariance, covariance, and autocorrelation function can be related by:

$$\begin{aligned} \gamma(\mathbf{h}) &= C(0) - C(\mathbf{h}) \\ &= \sigma^2\{1 - \rho(\mathbf{h})\} \end{aligned} \tag{7.9}$$

The experimental covariance function is estimated for the $p(\mathbf{h})$ paired observations, $z(\mathbf{s}_i), z(\mathbf{s}_i + \mathbf{h}), i = 1, 2, ..., p(\mathbf{h})$:

$$\hat{C}(\mathbf{h}) = \frac{1}{p(\mathbf{h})} \sum_{i=1}^{p(\mathbf{h})} \{z(\mathbf{s}_i) - m\}\{z(\mathbf{s}_i + \mathbf{h}) - m\} \tag{7.10}$$

For the covariance to exist the variance of the increments must be finite (382). This means that where there is no limit in semivariance, second-order stationarity cannot be assumed and the covariance does not exist.

The correlogram, or autocorrelation function, is the standardised covariance function:

$$\hat{\rho}(\mathbf{h}) = \frac{\hat{C}(\mathbf{h})}{\sigma^2} \tag{7.11}$$

where σ^2 is the variance, which corresponds to the covariance at lag 0, $\hat{C}(0)$.

An alternative form of the covariance function is available to allow for a trend in the data — the mean for different lags varies. The covariance function is then given by:

$$\hat{C}(\mathbf{h}) = \frac{1}{p(\mathbf{h})} \sum_{i=1}^{p(\mathbf{h})} \{z(\mathbf{s}_i) - m_1\}\{z(\mathbf{s}_i + \mathbf{h}) - m_2\} \tag{7.12}$$

where m_1 is the mean of $z(\mathbf{s}_i)$, and m_2 is the mean of $z(\mathbf{s}_i + \mathbf{h})$ (384). The means (where m_1 is termed the mean of the 'heads', and m_2 the mean of the 'tails') do not tend to be the same in practice (200). The correlogram can then be given by:

$$\hat{\rho}(\mathbf{h}) = \frac{\hat{C}(\mathbf{h})}{\sigma_1 \sigma_2} \tag{7.13}$$

where σ_1 is the standard deviation of $z(\mathbf{s}_i)$, and σ_2 is the standard deviation of $z(\mathbf{s}_i + \mathbf{h})$.

7.4.2 The variogram

The variogram, as defined above, is half the expected squared difference between paired RFs, separated by the lag **h**.

$$\gamma(\mathbf{h}) = \frac{1}{2}\mathrm{E}[\{Z(\mathbf{s}) - Z(\mathbf{s} + \mathbf{h})\}^2]$$

This section details estimation of the variogram, variogram models, and model fitting.

7.4.2.1 Experimental variogram

The variogram (or semivariogram) $\gamma(\mathbf{h})$ relates semivariance against lag **h**, the distance and direction between paired RFs. The per-point or block (i.e., positive area) variance is half the variance, thus the term semivariance. Of the three parameters of the RF model mentioned above (variogram, covariance function, and correlogram), the variogram is by far the most frequently used. This is largely due to the weaker form of stationarity required for its existence as discussed in Section 7.2.

The experimental variogram can be estimated for the $p(\mathbf{h})$ paired observations, $z(\mathbf{s}_i), z(\mathbf{s}_i + \mathbf{h}), i = 1, 2, ..., p(\mathbf{h})$ with:

$$\hat{\gamma}(\mathbf{h}) = \frac{1}{2p(\mathbf{h})} \sum_{i=1}^{p(\mathbf{h})} \{z(\mathbf{s}_i) - z(\mathbf{s}_i + \mathbf{h})\}^2 \tag{7.14}$$

In simple terms, the variogram is estimated by calculating the squared differences between all the available paired observations and obtaining half the average for all observations separated by that lag (or within a lag tolerance, where the observations are not on a regular grid). Figure 7.1 gives a simple example of a transect along which observations have been made at regular intervals. Lags (**h**) of 1 and 2 are indicated. So in this case, half the average squared difference between observations separated by a lag of 1 is calculated, and the process is repeated for a lag of 2 and so on. The variogram can be estimated for different directions to enable the identification of directional variation (anisotropy).

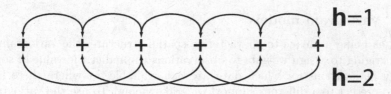

FIGURE 7.1: Transect with observations separated by lag **h**.

Webster and Oliver (383) state that, to estimate a variogram, at least 100 observations are needed. According to Webster and Oliver a variogram based on 150 data might be satisfactory, while using 225 data is usually reliable. Attempts to use smaller datasets, in some cases for as few as 20 observations, may produce reasonable results (for example, (154)) but in such cases there is much potential for instability in terms of results. However, factors such as the sampling strategy are important, and the variogram may be estimated robustly from few data if the sampling strategy is designed well.

If the data on which a variogram is based have a skewed distribution (that is, they are not normally distributed), the variogram will be biased. When the samples are preferentially located, this may also introduce bias. Variograms are also affected adversely by outliers. Standard geostatistical approaches are quite robust, but there are several widely used approaches that may reduce the extent of the problem where it exists. Extreme values may be down-weighted, or the values can be transformed, for instance by using logs. Alternatively, a robust variogram estimator may be used (137), (341). Armstrong (16) recommends the use of the variogram cloud to identify outliers.

The variogram can be used to inform prediction using kriging (as described below), or it may simply be used to summarise and interpret some property. For example, Mulla (282) uses the variogram to explore spatial variation in landform. The variogram has also been used to examine spatial variation in permeability in sedimentary structures formed in different depositional environments (269).

7.4.3   The cross-variogram

At locations where two properties u and v have been measured it is possible to estimate the cross-variogram. This is given as:

$$\hat{\gamma}_{uv}(\mathbf{h}) = \frac{1}{2p(\mathbf{h})} \sum_{i=1}^{p(\mathbf{h})} \{z_u(\mathbf{s}_i) - z_u(\mathbf{s}_i + \mathbf{h})\}\{z_v(\mathbf{s}_i) - z_v(\mathbf{s}_i + \mathbf{h})\} \qquad (7.15)$$

The cross-variogram is used to characterise spatial dependency between two variables. The use of the cross-variogram in spatial prediction using cokriging is outlined in Section 7.9.

7.4.4   Variogram models

As well as being a means to characterise spatial structure, the variogram is used in kriging to assign weights to observations to predict the value of some property at locations for which data are not available, or where there is a desire to predict to a different support or grid spacing. To use the variogram in kriging, a mathematical model must be fitted to it, such that the coefficients may then be used in the system of kriging equations (detailed below).

There are two principal classes of variogram model. Transitive models have a sill (finite variance) and are indicative of second-order stationarity. Unbounded models do not reach a finite variance. Unbounded models fulfil the requirements of intrinsic stationarity only: they are not second-order stationary (268).

With transitive models, the lag at which the sill is reached, the limit of spatial dependence in the variogram, is called the range a. Only pairs of values closer together than the range are spatially dependent. The form of the variogram beyond the range may be considered irrelevant if all the observations used in prediction are always at a distance from the prediction location which is smaller than the range (293).

Theoretically, the semivariance at zero lag is zero. In practice, the value at which the model fitted to the variogram crosses the intercept is usually some positive value, known as the nugget variance (c_0). The nugget effect in most contexts arises from variation that has not been resolved — spatially dependent variation at lags that are shorter than the smallest sampling interval used, any measurement errors, and random variation (212), (298). Variation at small distances is often the main contributor to the nugget effect (294). Atkinson (22) discusses the use of the nugget variance in estimating measurement error in remotely-sensed images. The structured component of the variogram (excluding the nugget variance) is signified by c. The total sill, or combined sill, is given by $c_0 + c$. In Figure 7.2, the coefficients of a bounded variogram model are illustrated.

Where there is no apparent structure in the variogram — that is, there is no increase in semivariance with lag $\mathbf{h}$, and semivariance is similar at all lags, the variogram may be modelled as a pure nugget. A pure nugget effect is indicative of a complete absence of spatial dependence at the scale at which the variable of concern has been measured (212). In such a case a simple average may be an appropriate prediction of values at all locations. In most instances, there is at least some indication of spatial structure, as completely random variation is unlikely to exist in spatial data. A pure nugget effect may indicate that there is spatial dependence at some scale smaller than the smallest sampling interval (294). Therefore, the sampling density may need to be increased if this is feasible.

McBratney and Webster (268) have provided a review of the models most frequently fitted to variograms in the geostatistical literature. A model must be conditional negative semidefinite (CNSD), so that it may be used to model spatial variance in a geostatistical analysis. Use of models that are not CNSD may lead to a prediction variance of less than zero. As a variance of less than zero is impossible, it is necessary to ensure that variances can be only zero or positive (19).

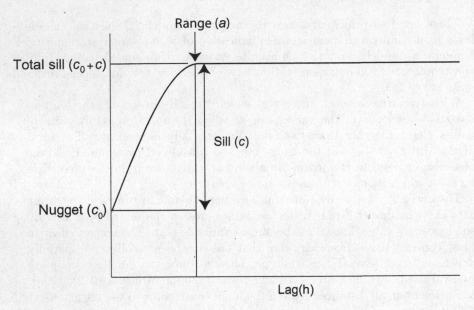

FIGURE 7.2: Bounded variogram.

The stationary RF $Z(\mathbf{s})$ has covariance $C(\mathbf{h})$. The variance of a linear combination Y of outcomes of the RF, $z(\mathbf{s}_i)$, is expressed as a linear combination of covariance values (153):

$$\mathrm{Var}(Y) = \mathrm{Var}\left\{\sum_{i=1}^{n} \lambda_i z(\mathbf{s}_i)\right\} = \sum_{i=1}^{n}\sum_{j=1}^{n} \lambda_i \lambda_j C(\mathbf{s}_i - \mathbf{s}_j) \qquad (7.16)$$

The variance must be non-negative and, to ensure that this is the case, the covariance model must be positive semidefinite. Where the covariance does not exist Equation 7.16 can be rewritten as:

$$\mathrm{Var}(Y) = C(0)\sum_{i=1}^{n}\lambda_i\sum_{j=1}^{n}\lambda_j - \sum_{i=1}^{n}\sum_{j=1}^{n}\lambda_i\lambda_j\gamma(\mathbf{s}_i - \mathbf{s}_j) \qquad (7.17)$$

By making the weights sum to zero, the variance of a linear combination of (intrinsically stationary) random variables can be defined as (212):

$$\mathrm{Var}(Y) = -\sum_{i=1}^{n}\sum_{j=1}^{n}\lambda_i\lambda_j\gamma(\mathbf{s}_i - \mathbf{s}_j) \qquad (7.18)$$

The non-negativity of the variance is guaranteed by ensuring that the weights, λ_i, sum to zero. The process of assessing whether or not a model is CNSD is not straightforward, and in the majority of cases the most practical solution is to use one of the 'authorised' models (268). CNSD models may be combined to model the variogram more accurately (19), (268). In this section, some of the most commonly used authorised models are detailed.

The nugget effect model represents unresolved variation — a mixture of spatial variation at a finer scale than the sample spacing and measurement error. It is given by:

$$\gamma(h) = \begin{cases} 0 \text{ for } h = 0 \\ c_0 \text{ for } |h| > 0 \end{cases} \tag{7.19}$$

The nugget effect model is often used in conjunction with other models, such as those defined below. The two classes of variogram models, transitive (or bounded) models and unbounded models, are detailed below.

7.4.4.1 Bounded models

Three of the most frequently used models are the spherical model, the exponential model and the Gaussian model, each of which is defined in turn.

The exponential model is given by:

$$\gamma(h) = c \cdot \left[1 - \exp\left(-\frac{h}{d} \right) \right] \tag{7.20}$$

where d is the distance parameter. The exponential model reaches the sill asymptotically. The practical range is the distance at which the semivariance equals 95% of the sill variance which is approximately $3d$ (384).

The spherical model (shown in Figure 7.2) is given by:

$$\gamma(h) = \begin{cases} c \cdot \left[1.5\frac{h}{a} - 0.5 \left(\frac{h}{a} \right)^3 \right] \text{ if } h \leq a \\ c \qquad\qquad\qquad\qquad \text{ if } h > a \end{cases} \tag{7.21}$$

The spherical model is very widely used — its structure of almost linear growth near to the range, followed by stabilisation, corresponds with spatial variation that is often observed (18).

The Gaussian model is given by:

$$\gamma(h) = c \cdot \left[1 - \exp\left(-\frac{h^2}{d^2} \right) \right] \tag{7.22}$$

The Gaussian model does not reach a sill at a finite distance, and the practical range is $d\sqrt{3}$ (212). Variograms with parabolic behaviour at the origin are indicative of very regular spatial variation (212). Another model, which like the Gaussian model, is concave upwards at the origin, is the cubic model (18):

$$\gamma(h) = \begin{cases} c \cdot \left[7 \left(\frac{h}{a} \right)^2 - 8.75 \left(\frac{h}{a} \right)^3 + 3.5 \left(\frac{h}{a} \right)^5 - 0.75 \left(\frac{h}{a} \right)^7 \right] \text{ if } h \leq a \\ c \qquad\qquad\qquad\qquad\qquad\qquad\qquad\qquad\qquad \text{ if } h > a \end{cases} \tag{7.23}$$

Further models can be found in, for example, Chilès and Delfiner (80) and Waller and Gotway (375).

A feature often seen in variograms is the nonmonotonic growth of semivariance — there is a peak and valley. A valley following a peak in semivariance is referred to as a 'hole effect'. If semivariance follows a series of peaks and valleys it is known as 'periodicity'. A hole effect is usually the result of pseudo-periodicity in the spatial variability (211).

7.4.4.2 Unbounded models

Where no sill is reached and the variogram model is unbounded, the power model may be used:

$$\gamma(h) = m \cdot h^\omega \qquad\qquad (7.24)$$

where ω is a power $0 < \omega < 2$ with a positive slope, m (107). The linear model is a special case of the power model. Variograms to which a power model fits best may indicate a large scale trend in the data, as discussed in Section 7.15.2.

7.4.4.3 Anisotropic models

One of the advantages of kriging is that it is often fairly straightforward to model anisotropic structures using the variogram. Two primary forms of anisotropy have been outlined in the geostatistical literature.

If the sills for all directions are not significantly different, and the same structural components (for example, spherical or Gaussian) are used, then anisotropy can be accounted for by a linear transformation of the coordinates: this is called geometric or affine anisotropy (382). That is, if the direction which has the smallest range (and greatest proportional increase in semivariance with lag), or gradient for unbounded models, can be related to the direction with the variogram having the maximum range (and least proportional increase in semivariance with lag) by an anisotropy ratio then the anisotropy is geometric. Thus, the anisotropy ratio is computed by dividing the range or gradient in the minor direction (the smallest range or gradient) by the range or gradient in the major direction (the largest range or gradient) for each component (107). Clearly, observations near to the unknown location have more weight in the principal direction (200). In summary, with geometric anisotropy, the sill remains constant, but the range or gradient changes with direction (200). This form of anisotropy can be taken into account comparatively easily for the purposes of modelling variograms, assigning weights for kriging and for optimal sampling strategies. Where the sill changes with direction but the range is similar for all directions, the anisotropy is called zonal (200). Zonal anisotropy may be taken into account by defining a large range in the minor direction so that structure is added only to orthogonal directions (107). However, the modelling of zonal anisotropy is much more problematic than the modelling of geometric anisotropy. In practice, a mixture of geometric and zonal anisotropy has been found to be common (200). As noted above, the variogram surface is a useful tool in the identification and characterisation of anisotropy. Waller and Gotway (375) give a more detailed account of anisotropic models.

7.4.4.4 Fitting variogram models

Examination of the geostatistical literature reveals a variety of different approaches to the fitting of models to variograms. Some practitioners prefer to fit variogram models 'by eye,' on the grounds that it enables the use of personal experience and the possibility of accounting for features or variation that may be difficult to quantify (82), (212). However, such a view would be viewed negatively by many geostatisticians.

Weighted least squares (WLS) has been proposed as a suitable means of fitting models to variograms (94), (310) and the approach has been used by many geostatisticians. The technique is preferred to unweighted ordinary least squares (OLS), as in WLS the weights are proportional to the number of pairs at each lag (94). Thus, lags with many pairs have greater influence in the fitting of a model. WLS is implemented, for example, in the geostatistical package Gstat (313). The use of generalised least squares (GLS; Section 5.3) has also been demonstrated in a geostatistical context (94), (268), (138) (see also Section 7.15.2). Use of maximum likelihood (ML) estimation (268) has become widespread amongst geostatisticians and has been used for WLS. Zimmerman and Zimmerman (405) state the view that by using the simpler least squares algorithms (OLS and WLS) little is lost in terms of results in comparison to the more complex ML estimators. In any case, if the distribution is Gaussian, then OLS is a ML estimator. The use of restricted maximum likelihood (REML) for estimating variogram model parameters is discussed by Dietrich and Osborne (108) and Pardo-Igúzquiza (309).

The goodness of the fit of models to the variogram, and of the relative improvement or otherwise in using different numbers of parameters, may be compared through the examination of the sum of squares of the residuals, or through the use of the Akaike Information Criterion (268), (380) (see Section 5.8.2). Pannatier (307) employs a goodness-of-fit criterion he terms an 'Indicative Goodness of Fit.'

7.4.4.5 Variogram estimation and modelling: An example

Variograms were estimated from the precipitation data described in Section 1.8.1. These data were also used in the previous chapter. The variogram cloud and omnidirectional and directional variograms were estimated from the data. Figure 7.3 shows the variogram cloud. The omnidirectional variogram is shown in Figure 7.4, with a fitted model comprising a nugget effect and two spherical components. The directional variogram is given in Figure 7.5. The variograms were estimated, and the model fitted to the omnidirectional

variogram, using the Gstat software (313). The variogram cloud shows
how the semivariances vary — the semivariance is not simply a function
of the distance (and direction) between observations. The variogram cloud
provides one means of assessing local departures from the average and may,
as suggested above, help in deciding if a nonstationary model is necessary.
The omnidirectional variogram suggests that there are two dominant scales of
spatial variation corresponding to the fitted range components (14 km and 155
km). The directional variogram indicates that spatial structure varies with
direction. That is, the spatial variation is anisotropic although the structure
is very similar in many directions (and particularly for smaller lags).

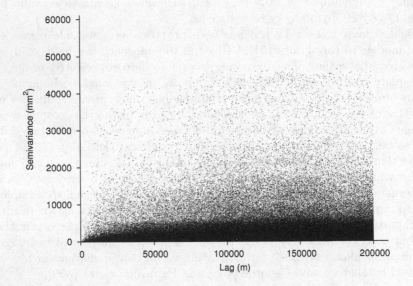

FIGURE 7.3: Precipitation: variogram cloud.

7.5   Kriging

Kriging is the term used by geostatisticians for a family of generalised least-
squares regression algorithms and is named after the South African mining
engineer, Danie Krige (153). Kriging makes fuller use of the data than
methods such as IDW, as with kriging the weights are selected using the
data and the general procedure is outlined here. Kriging is widely known

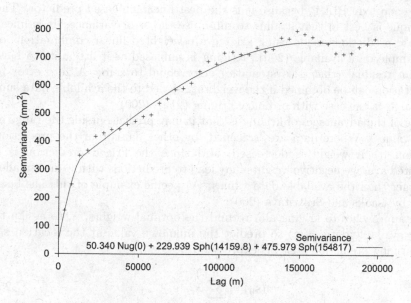

FIGURE 7.4: Precipitation: omnidirectional variogram. Units are decimal degrees clockwise from north.

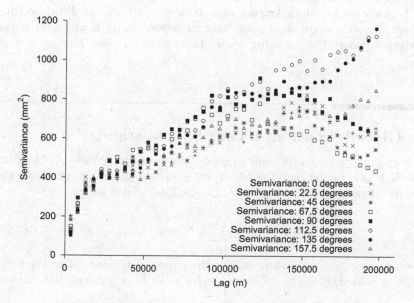

FIGURE 7.5: Precipitation: directional variogram.

by the acronym BLUP, because it is the best linear unbiased predictor. The technique is best in that it aims to minimise the error variance. It is linear because predictions made using kriging are weighted linear combinations of the sample data available (200). Kriging is unbiased as it attempts to have a mean residual (that is, systematic) error equal to zero. A zero error in expectation can be obtained if the weights assigned to the available data sum to one, as is the case with ordinary kriging (212), (200).

One of the advantages of kriging is that it may predict outside the range of data values. Where data are 'screened' by other data from the prediction location, their weight is decreased, and since the kriged prediction is a weighted average negative weights may lead to predictions with values smaller or larger than the available data values. A specific example of this has been given by Isaaks and Srivastava (200).

All approaches to kriging aim to find the optimal weights, λ, to assign to the n available data, $z(\mathbf{s}_i)$ to predict the unknown value at the location $\mathbf{s}_0$. The kriging prediction $\hat{z}(\mathbf{s}_0)$, is expressed as:

$$\hat{z}(\mathbf{s}_0) = \sum_{i=1}^{n} \lambda_i z(\mathbf{s}_i) \tag{7.25}$$

It is usually the case that only the n observations nearest to $\mathbf{s}_0$ are used for kriging.

Punctual (or point) kriging makes predictions on the same support as the sample observations. Where we wish to obtain predictions (averages) over some larger support, block kriging may be used (69), (200). Block kriging produces smoother maps than point kriging, as variability is averaged across the larger support. The following descriptions focus on point kriging.

7.6   Globally constant mean: Simple kriging

Simple kriging (SK) is, as its name suggests, the most simple variant of kriging. With SK it is assumed that the mean, m, is known and constant across the region of interest. The SK prediction, $\hat{z}_{SK}(\mathbf{s}_0)$, is defined as:

$$\hat{z}_{SK}(\mathbf{s}_0) - m = \sum_{i=1}^{n} \lambda_i^{SK}[z(\mathbf{s}_i) - m] \tag{7.26}$$

where the mean, m, is constant across the region of interest. The predictor must be unbiased, thus the prediction error must have an expected value of 0:

$$\mathrm{E}\left\{\hat{z}_{SK}(\mathbf{s}_0) - z(\mathbf{s}_0)\right\} = 0 \tag{7.27}$$

The weights are obtained that minimize the error variance, with the constraint that the predictor is unbiased. The kriging variance, σ_{SK}^2, is expressed as:

$$\sigma_{SK}^2 = \mathrm{E}[\{\hat{z}_{SK}(\mathbf{s}_0) - z(\mathbf{s}_0)\}^2]$$
$$= C(0) - 2\sum_{i=1}^n \lambda_i^{SK} C(\mathbf{s}_i - \mathbf{s}_0) + \sum_{i=1}^n \sum_{j=1}^n \lambda_i^{SK} \lambda_j^{SK} C(\mathbf{s}_i - \mathbf{s}_j) \tag{7.28}$$

The optimal weights are found where the first derivative of the prediction variance is zero:

$$\frac{1}{2}\frac{\partial \sigma_{SK}^2}{\partial \lambda_i^{SK}} = \sum_{j=1}^n \lambda_j^{SK} C(\mathbf{s}_i - \mathbf{s}_j) - C(\mathbf{s}_i - \mathbf{s}_0) = 0 \qquad i = 1, ..., n \tag{7.29}$$

The SK system is given as:

$$\sum_{j=1}^n \lambda_j^{SK} C(\mathbf{s}_i - \mathbf{s}_j) = C(\mathbf{s}_i - \mathbf{s}_0) \qquad i = 1, ..., n \tag{7.30}$$

The SK variance (or minimum error variance) is given by:

$$\sigma_{SK}^2 = C(0) - \sum_{i=1}^n \lambda_i^{SK} C(\mathbf{s}_i - \mathbf{s}_0) \tag{7.31}$$

Using matrix notation the SK system is written as:

$$\mathbf{K}_{SK} \lambda_{SK} = \mathbf{k}_{SK} \tag{7.32}$$

where $\mathbf{K}_{SK}$ is the n by n matrix of covariances:

$$\mathbf{K}_{SK} = \begin{bmatrix} C(\mathbf{s}_1 - \mathbf{s}_1) & \cdots & C(\mathbf{s}_1 - \mathbf{s}_n) \\ \vdots & \vdots & \vdots \\ C(\mathbf{s}_n - \mathbf{s}_1) & \cdots & C(\mathbf{s}_n - \mathbf{s}_n) \end{bmatrix}$$

λ_{SK} are the SK weights, and $\mathbf{k}_{SK}$ are covariances for the observations to prediction location:

$$\lambda_{SK} = \begin{bmatrix} \lambda_1^{SK} \\ \vdots \\ \lambda_n^{SK} \end{bmatrix} \quad \mathbf{k}_{SK} = \begin{bmatrix} C(\mathbf{s}_1 - \mathbf{s}_0) \\ \vdots \\ C(\mathbf{s}_n - \mathbf{s}_0) \end{bmatrix}$$

To obtain the SK weights, the inverse of the data covariance matrix is multiplied by the vector of data to prediction location covariances:

$$\lambda_{SK} = \mathbf{K}_{SK}^{-1} \mathbf{k}_{SK} \tag{7.33}$$

The SK variance is given by:

$$\sigma_{SK}^2 = C(0) - \mathbf{k}_{SK}^T \lambda_{SK} \tag{7.34}$$

SK is usually utilised in its modified form, ordinary kriging (OK). SK has been employed with varying local means derived using secondary information (154). This is discussed in Section 7.16.4.1.

7.7 Locally constant mean models

In many practical contexts it is not reasonable to assume that the mean is constant across the region of interest. As such, the mean is usually assumed to vary from place to place. OK, which is the topic of the following section, entails estimation of the mean in a moving window. As such, it implicitly considers a nonstationary RF model, with stationary assumed within the local neighbourhood of a prediction location (153). The cokriging system outlined in Section 7.9 is an extension of the OK system.

Most of the remainder of this chapter deals with nonstationary models in that the mean is allowed to vary *between* or *within* neighbourhoods or the variogram is assumed to vary from place to place.

7.8 Ordinary kriging

OK has been called the "anchor algorithm of geostatistics" (107, p. 66). OK is used if the mean varies across the region of interest. With OK, the mean is considered constant within a moving window. Since the mean is allowed to vary across the region of interest OK can, in this respect, be considered a nonstationary method. In Section 7.16 the case where the mean is not constant within a moving window will be discussed.

The OK prediction, $\hat{z}_{OK}(\mathbf{s}_0)$, is a linear weighted moving average of the available n observations defined as:

$$\hat{z}_{OK}(\mathbf{s}_0) = \sum_{i=1}^{n} \lambda_i z(\mathbf{s}_i) \tag{7.35}$$

with the constraint that the weights sum to 1:

$$\sum_{i=1}^{n} \lambda_i^{OK} = 1 \tag{7.36}$$

The prediction error must have an expected value of 0:

$$\mathrm{E}\{\hat{z}_{OK}(\mathbf{s}_0) - z(\mathbf{s}_0)\} = 0 \tag{7.37}$$

The weights are obtained that minimize the error variance with the constraint that the predictor is unbiased.

The kriging (or prediction) variance, σ_{OK}^2, is expressed as:

$$\sigma^2_{OK}(\mathbf{h}) = \mathrm{E}[\{\hat{z}_{OK}(\mathbf{s}_0) - z(\mathbf{s}_0)\}^2]$$
$$= 2\sum_{i=1}^{n} \lambda_i^{OK} \gamma(\mathbf{s}_i - \mathbf{s}_0) - \sum_{i=1}^{n}\sum_{j=1}^{n} \lambda_i^{OK}\lambda_j^{OK}\gamma(\mathbf{s}_i - \mathbf{s}_j) \qquad (7.38)$$

That is, we seek the values of $\lambda_1, ..., \lambda_n$ that minimise this expression, with the constraint that the weights sum to one (Equation 7.36). This minimisation may be achieved through the use of Lagrange multipliers, a good introduction to which in geostatistical contexts is provided by Kitanidis (219). The conditions for the minimisation are given by the OK system, comprising $n+1$ equations and $n+1$ unknowns:

$$\begin{cases} \sum_{j=1}^{n} \lambda_j^{OK}\gamma(\mathbf{s}_i - \mathbf{s}_j) + \psi_{OK} = \gamma(\mathbf{s}_i - \mathbf{s}_0) & i = 1, ..., n \\ \sum_{j=1}^{h} \lambda_j^{OK} = 1 \end{cases} \qquad (7.39)$$

where ψ_{OK} is a Lagrange multiplier. Knowing ψ_{OK}, the error variance of OK can be given as:

$$\sigma^2_{OK} = \sum_{i=1}^{n} \lambda_i^{OK}\gamma(\mathbf{s}_i - \mathbf{s}_0) + \psi_{OK} \qquad (7.40)$$

In matrix notation the OK system is written as:

$$\mathbf{K}_{OK}\lambda_{OK} = \mathbf{k}_{OK} \qquad (7.41)$$

where $\mathbf{K}_{OK}$ is the $n+1$ by $n+1$ matrix of semivariances:

$$\mathbf{K}_{OK} = \begin{bmatrix} \gamma(\mathbf{s}_1 - \mathbf{s}_1) & \cdots & \gamma(\mathbf{s}_1 - \mathbf{s}_n) & 1 \\ \vdots & \vdots & \vdots & \vdots \\ \gamma(\mathbf{s}_n - \mathbf{s}_1) & \cdots & \gamma(\mathbf{s}_n - \mathbf{s}_n) & 1 \\ 1 & \cdots & 1 & 0 \end{bmatrix}$$

λ_{OK} are the OK weights, and $\mathbf{k}_{OK}$ are semivariances for the observations to the prediction location:

$$\lambda_{OK} = \begin{bmatrix} \lambda_1^{OK} \\ \vdots \\ \lambda_n^{OK} \\ \psi \end{bmatrix} \quad \mathbf{k}_{OK} = \begin{bmatrix} \gamma(\mathbf{s}_1 - \mathbf{s}_0) \\ \vdots \\ \gamma(\mathbf{s}_n - \mathbf{s}_0) \\ 1 \end{bmatrix}$$

To obtain the OK weights, the inverse of the data semivariance matrix is multiplied by the vector of data to prediction semivariances:

$$\lambda_{OK} = \mathbf{K}_{OK}^{-1}\mathbf{k}_{OK} \qquad (7.42)$$

Finally, the OK variance is the transpose of the vector of data to prediction semivariances multiplied by the OK weights:

$$\sigma^2_{OK} = \mathbf{k}^T_{OK} \lambda_{OK} \tag{7.43}$$

As kriged predictions are weighted averages, predictions are more reliable with normally distributed data. In some cases, if the data are lognormally distributed, logarithmic transformation of the data may be advisable (221). An alternative transformation of positively skewed data is the normal scores transform (153) (107). However, OK is fairly robust.

OK is illustrated below using the same data as used to implement inverse distance weighting (IDW; Section 6.3.6.1) and thin plate splines (TPS; Section 6.3.7.1) in the previous chapter and, as noted in those contexts, the small sample (four observations) is used for ease of illustration. Note that the figures below were rounded for presentation purposes. A further example of OK in practice is given by Waller and Gotway (375) while Chang (78) illustrates OK and kriging with a trend model (KT; detailed below).

First, the distances between the observations and between the observations and the prediction location are fed into the equation for the spherical model (Equation 7.21) and the nugget effect is added. The variogram model coefficients are shown in Figure 7.4. Then the OK system (see equation 7.41) is given with:

$$\begin{bmatrix} 0.00 & 145.524 & 174.745 & 200.908 & 1 \\ 145.524 & 0.00 & 198.419 & 161.713 & 1 \\ 174.745 & 198.419 & 0.00 & 148.571 & 1 \\ 200.908 & 161.713 & 148.571 & 0.00 & 1 \\ 1 & 1 & 1 & 1 & 0 \end{bmatrix} \begin{bmatrix} \lambda_1^{OK} \\ \lambda_2^{OK} \\ \lambda_3^{OK} \\ \lambda_4^{OK} \\ \psi \end{bmatrix} = \begin{bmatrix} 107.383 \\ 127.396 \\ 134.131 \\ 149.088 \\ 1 \end{bmatrix}$$

The OK weights are as follows: $\lambda_1^{OK} = 0.372$, $\lambda_2^{OK} = 0.233$, $\lambda_3^{OK} = 0.244$, and $\lambda_4^{OK} = 0.151$ with $\psi = 0.453$. The OK prediction is obtained with: $(0.372 \times 52.6) + (0.233 \times 40.8) + (0.244 \times 48.3) + (0.151 \times 46.7) = 47.908$.

The OK variance is obtained with: $(0.372 \times 107.383) + (0.233 \times 127.396) + (0.244 \times 134.131) + (0.151 \times 149.088) + (0.453 \times 1) = 125.332$.

A further example of OK, with full working, is provided by Lloyd (245).

7.8.0.6   OK case study

The coefficients of the model fitted to the variogram shown in Figure 7.4 were input into the OK algorithm provided as a part of the Geostatistical Software Library, GSLIB (107). Cross-validation prediction errors are given in Table 7.1. The RMSE values decrease as the number of nearest neighbours increase. A map of precipitation for July 1999 derived using OK (using 64 nearest neighbours) is given in Figure 7.6. The map is quite smooth in appearance, with the most obvious feature being the contrast between large precipitation amounts in western Scotland and small amounts in southern Britain as a whole.

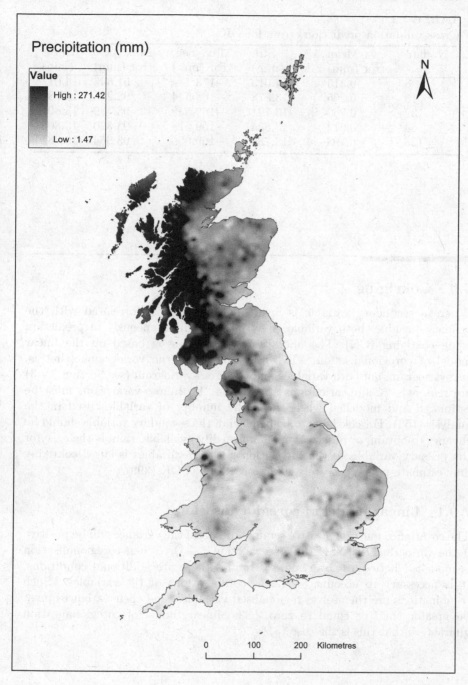

FIGURE 7.6: Precipitation: map derived using OK.

TABLE 7.1
Cross-validation prediction errors for OK.

N. data	Mean error (mm)	Std. dev. (mm)	Max. neg. error (mm)	Max. pos. error (mm)	RMSE (mm)
8	0.315	14.013	−173.714	91.643	14.017
16	0.295	13.878	−168.544	92.235	13.881
32	0.260	13.798	−169.548	93.489	13.801
64	0.212	13.804	−169.518	93.320	13.805
128	0.193	13.780	−169.059	93.964	13.781

7.9   Cokriging

Where a secondary variable is available that is cross-correlated with the primary variable, both variables may be used simultaneously in prediction using cokriging (CK). The operation of cokriging is based on the linear model of coregionalisation. To apply cokriging, the variograms (that is, autovariograms) of both variables and the cross-variogram (see Section 7.4.3) are required. K autovariograms and $K(K-1)/2$ cross-variograms must be estimated and modelled where K is the number of variables used in the analysis (153). For cokriging to be beneficial, the secondary variable should be cheaper to obtain, or observations more readily available, than is the case for the primary variable. If the correlation between variables is large, cokriging may estimate more accurately than, for example, OK (399).

7.9.1   Linear model of coregionalisation

The covariance must be positive semidefinite, and the variance will be positive if the variogram is CNSD (31) (see Section 7.4.4). When coregionalisation is modelled between two or more variables there are additional conditions. It is necessary to account for linear combinations of the variables. Such combinations are themselves regionalised variables, and their variances must be greater than or equal to zero. The linear model of coregionalisation guarantees that this is the case (31).

With the linear model of coregionalisation, each variable is assumed to be a linear sum of L orthogonal (that is, independent) RFs $Y(\mathbf{s})$. Their variograms are then given by $\gamma_l(\mathbf{h}), l = 1, 2, ..., L$. The variograms combine to produce the variogram for any pair of variables u and v with:

$$\gamma_{uv}(\mathbf{h}) = \sum_{l=1}^{L} b_{uv}^l \gamma_l(\mathbf{h}) \tag{7.44}$$

where b_{uv}^l (l being an index and not a power) are the variances and covariances (that is, nugget and sill variances) for the independent components if bounded (nugget variances and gradients if unbounded); $b_{uv}^l = b_{vu}^l$ for all l, and for each value of l the matrix of coefficients:

$$\begin{bmatrix} b_{uu}^l & b_{uv}^l \\ b_{vu}^l & b_{vv}^l \end{bmatrix} \tag{7.45}$$

must be positive definite. As the matrix is symmetric it is only necessary that $b_{uu} \geq 0$ and $b_{vv} \geq 0$ and the determinant of the matrix is positive or zero:

$$|b_{uv}^l| = |b_{vu}^l| \leq \sqrt{b_{uu}^l b_{vv}^l} \tag{7.46}$$

which is termed Schwarz's inequality (31). The implications of this are that, for two variables, all basic structures represented in a cross-variogram must be represented in both autovariograms, whereas structures may be present in the autovariograms without appearing in the cross-variogram (30), (31), (384).

7.9.2 Applying cokriging

The cokriging equations are an extension of the OK system (31). The ordinary cokriging predictor is given by:

$$\hat{z}_{CK}(\mathbf{s}_0) = \sum_{k=1}^{K} \sum_{i=1}^{n_k} \lambda_{ik} z_k(\mathbf{s}_i) \tag{7.47}$$

with $k = 1, 2, ..., K$ variables, and $i = 1, 2, ..., n_k$ observation locations. The variable to be predicted is indicated by u. The λ_{ik} are weights where:

$$\sum_{i=1}^{n_k} \lambda_{ik} = \begin{cases} 1, k = u \\ 0, k \neq u \end{cases}$$

The ordinary CK system is then given by:

$$\begin{cases} \sum_{k=1}^{K} \sum_{i=1}^{n_k} \lambda_{ik} \gamma_{kv}(\mathbf{s}_i - \mathbf{s}_j) + \psi_v = \gamma_{uv}(\mathbf{s}_i - \mathbf{s}_0) \\ \sum_{i=1}^{n_k} \lambda_{ik} = \begin{cases} 1, k = u \\ 0, k \neq u \end{cases} \end{cases} \tag{7.48}$$

for all $v = 1, 2, ..., K$ and all $j = 1, 2, ..., n_v$. Where $k = v$ or $u = v$, $\gamma_{kv}(\mathbf{s}_i - \mathbf{s}_j)$ are the autosemivariances (384).

The prediction variance, $\mathrm{E}[\{\hat{z}_{CK}(\mathbf{s}_0) - z(\mathbf{s}_0)\}^2]$, is obtained with:

$$\sigma_u^2 = \sum_{k=1}^{K} \sum_{j=1}^{n_k} \lambda_{jk} \gamma_{uk}(\mathbf{s}_j - \mathbf{s}_0) + \psi_u \tag{7.49}$$

where ψ_u is the Lagrange multiplier for the uth variable. As before, the matrix form of the equations is given. The case of two variables, u and v will be presented. There is a matrix of semivariances that includes cross-semivariances where $u \neq v$ (384):

$$\mathbf{K}_{uv} = \begin{bmatrix} \gamma_{uv}(\mathbf{s}_1 - \mathbf{s}_1) & \cdots & \gamma_{uv}(\mathbf{s}_1 - \mathbf{s}_{n_v}) \\ \vdots & \cdots & \vdots \\ \gamma_{uv}(\mathbf{s}_{n_u} - \mathbf{s}_1) & \cdots & \gamma_{uv}(\mathbf{s}_{n_u} - \mathbf{s}_{n_v}) \end{bmatrix}$$

the semivariances for variable u and the cross-semivariances are given as:

$$\mathbf{k}_{uu} = \begin{bmatrix} \gamma_{uu}(\mathbf{s}_1 - \mathbf{s}_0) \\ \vdots \\ \gamma_{uu}(\mathbf{s}_{n_u} - \mathbf{s}_0) \end{bmatrix}, \mathbf{k}_{uv} = \begin{bmatrix} \gamma_{uv}(\mathbf{s}_1 - \mathbf{s}_0) \\ \vdots \\ \gamma_{uv}(\mathbf{s}_{n_v} - \mathbf{s}_0) \end{bmatrix}$$

The matrix equation is then:

$$\begin{bmatrix} & & & 10 \\ & & & 10 \\ \mathbf{K}_{uu} & \mathbf{K}_{uv} & \vdots \\ & & & 10 \\ & & & 01 \\ & & & 01 \\ \mathbf{K}_{vu} & \mathbf{K}_{vv} & \vdots \\ & & & 01 \\ 11 \ \cdots \ 1 \ 00 \ \cdots \ 0 \ 00 \\ 00 \ \cdots \ 0 \ 11 \ \cdots \ 1 \ 00 \end{bmatrix} \times \begin{bmatrix} \lambda_{1u} \\ \lambda_{2u} \\ \vdots \\ \lambda_{n_u u} \\ \lambda_{1v} \\ \lambda_{2v} \\ \vdots \\ \lambda_{n_v v} \\ \psi_u \\ \psi_v \end{bmatrix} = \begin{bmatrix} \mathbf{k}_{uu} \\ \\ \mathbf{k}_{uv} \\ 1 \\ 0 \end{bmatrix}$$

The CK system is given by:

$$\mathbf{K}_{CK} \lambda_{CK} = \mathbf{k}_{CK} \tag{7.50}$$

It is solved with:

$$\lambda_{CK} = \mathbf{K}_{CK}^{-1} \mathbf{k}_{CK} \tag{7.51}$$

The prediction variance for a point support is then given by:

$$\sigma_u^2 = \mathbf{k}_{CK}^T \lambda_{CK} \tag{7.52}$$

Various studies have shown the potential benefits of cokriging over methods that make use of only one variable. For example, Goovaerts (155) compared several interpolation techniques including IDW, OK, kriging with an external drift (KED; see Section 7.16.4.3), simple kriging with a locally-varying mean (SKlm; Section 7.16.4.1), ordinary colocated cokriging (OCK), and regression. The focus in that study was on mapping annual and monthly precipitation in a region in the south of Portugal, and elevation was used as a secondary variable. The techniques that used elevation data generally out-performed OK, where the (elevation versus precipitation) correlation coefficient was larger than 0.75. Yates and Warrick (399) found that cokriging was justified if the sample correlation coefficient was greater than 0.5. Dungan (114) identified a correlation coefficient of 0.89 as the limit above which regression provided more accurate estimates than cokriging. Stein et al. (345) made use of secondary variables in the presence of a trend (discussed below) through universal cokriging. The linear model of coregionalisation is summarised and cokriging applied by Atkinson et al. (31). Lloyd (243), in a study concerned with mapping precipitation amount, found that OK outperformed CK (with elevation as the secondary variable) and this was due, at least in part, to the weak global relationship between the two variables.

7.10 Equivalence of splines and kriging

The equivalence of splines and kriging has been demonstrated by several authors (371). The TPS basis function, $d_i^2 \ln d_i$ (see Section 6.3.7), is called the spline covariance model in the geostatistical terminology (371). However, while TPS and kriging are linked in theory the ways in which they are applied are very different in practice (80), (95).

7.11 Conditional simulation

Kriging predictions are weighted moving averages of the available sample data (107). Kriging is thus a smoothing interpolator. Stochastic simulation is not subject to the smoothing evidenced using kriging, as predictions are drawn from "alternative, equally probable, joint realisations of the component RVs from a RF model" (107, p. 18). That is, simulated values are not the expected values but are values taken randomly from a distribution that is a function of the available observations and the modelled spatial variation (115). The simulation is considered 'conditional' if the simulated values honour the observations at their locations (107). Simulation allows the generation of many different possible realisations, which may be used as a guide to potential errors in the construction of a map (210). Perhaps the most widely used form of conditional simulation is sequential Gaussian simulation (SGS) (107).

7.12   The change of support problem

As outlined above, block kriging provides a means of changing support. For example, if the input has a point support, predictions can be made over areas rather than at points. Change of support is a major concern in geostatistics, and these principles can be expanded to addressing the kinds of questions outlined in Section 6.4. Gotway Crawford and Young (158), (159) provide an overview of the change of support problem and some methods that may be used to address it.

Kyriakidis (224) offers a geostatistical framework for such applications. Kyriakidis and Yoo (225) present a method for predicting and simulating from areal data to points. In that paper, the concern is with making predictions or simulations to points using areal data with equal supports. Such a case is a remotely-sensed image, in which all pixels have the same support. In practice, only the area support values are observable, but Kyriakidis and Yoo (225) assume that the point-support covariance model is known. With this assumption, they were able to make predictions that (i) account for the difference between areal data and the point predictions, (ii) are mass preserving (pycnophylactic) predictions, and (iii) include a prediction standard error. Goovaerts (156) presents a methodology for estimating the point variogram from areal data and for area-to-point kriging. This kind of research demonstrates the potential shortcomings of the kinds of methods discussed in Section 6.4 that are used to generate surfaces from areal data.

7.13   Other approaches

The utility of conventional variograms and algorithms such as OK for geostatistical analysis is limited by particular assumptions. Principal problems include concerns with (i) normality (although the techniques are fairly robust), and (ii) the independence of the OK variance on data values. The indicator approach is one means of overcoming both of these limitations. The indicator approach (208) involves the designation of several thresholds, for example the deciles of the sample cumulative distribution (153) which are used to transform the data to a set of binary variables. Disjunctive kriging (381), (323) allows the calculation of the conditional probability of the predicted value of some variable being above or below some critical threshold (321), (400), (401).

There is a growing literature on the use of model-based geostatistics. Diggle et al. (111) are concerned with problems where Gaussian distributional assumptions are inappropriate. The authors detail the use of a Bayesian inference framework which is implemented using Markov chain Monte Carlo (MCMC) methods. It is argued that this approach enables uncertainty in the prediction of model parameters to be accounted for properly.

7.14 Local approaches: Nonstationary models

In the following sections, the main focus is on models that allow for local variation in the mean or the covariance structure of the variable of interest. That is, the concern is with nonstationary models. Note that OK can be considered a nonstationary model in that the local mean is assumed to vary. In the following sections, variation within local neighbourhoods is also considered.

A variety of approaches exist that allow for a nonstationary mean. Section 7.15 introduces some key ideas relating to a nonstationary mean, while Section 7.16 discusses methods for predicting in the presence of a trend (that is, a nonstationary mean). Where a trend is present (there is a systematic change in the mean across the region of interest), a variety of approaches can be used to estimate the trend-free variogram, and this subject is the focus of Section 7.15.2. Approaches like kriging with a trend model, discussed below, only take into account nonstationarity of the mean. That is, the variogram itself is not adapted locally as part of the process (170). In Section 7.17 methods that account for local variation in spatial structure, and thus the variogram, are summarised briefly. Section 7.18 discusses the use of the variogram for texture analysis — such approaches employ locally-estimated variograms.

To set the scene for the following sections, some example one-dimensional data profiles are presented that were generated using unconditional simulation. That is, given a set of variogram model coefficients, profiles were generated without reference to any data. Figures 7.7 and 7.8 represent cases for a single variogram model and no evidence of trend — a stationary model is appropriate. In Figure 7.9, a trend has been added and, for the analysis of such data, an approach like those detailed in Section 7.15 is appropriate. In the case of Figure 7.10, the spatial structure for the first half of the structure is quite different to that in the second half. In this case, a nonstationary variogram model would be appropriate, as is discussed in Section 7.17.

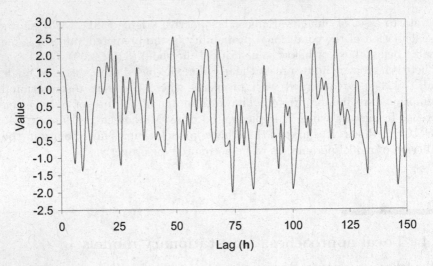

FIGURE 7.7: Simulated data values on a transect: generated using a spherical variogram with $c_0 = 0.1$, $c_1 = 0.9$, and $a = 2.5$.

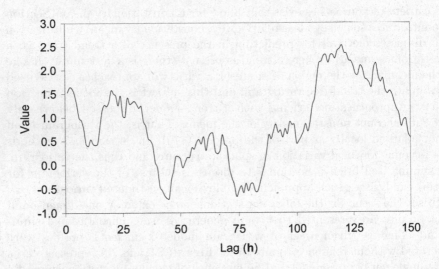

FIGURE 7.8: Simulated data values on a transect: generated using a spherical variogram with $c_0 = 0$, $c_1 = 1$, and $a = 100$.

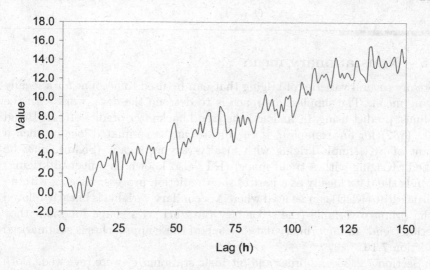

FIGURE 7.9: Simulated data values on a transect: generated using a spherical variogram with $c_0 = 0.1$, $c_1 = 0.9$, and $a = 2.5$ with an to east to west trend added.

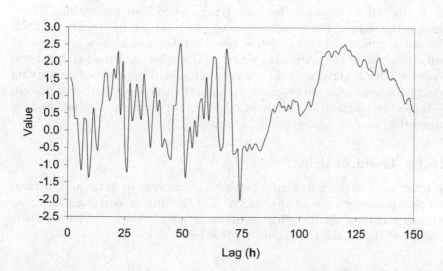

FIGURE 7.10: Simulated data values on a transect: lag $1 - 75$ = model from Figure 7.7, lags $76 - 150$ = model from Figure 7.8.

7.15   Nonstationary mean

There are several varieties of kriging that can be used to account for a locally
varying mean. The simplest approach is to detrend the data, work with the
residuals, predict using SK and add the trend back after prediction (see Hengl
et al. (182) for an example). If the mean can be estimated locally, then a
variant of SK, simple kriging with locally-varying means (SKlm), may be
applied. Kriging with a trend model (KT, also known as universal kriging)
detrends the data locally as a part of the prediction process. Kriging with an
external drift (KED) can be used where a secondary variable is linearly related
to the primary variable (the secondary data act as a shape function; they
describe trends in the primary data). Each of these approaches is summarised
in Section 7.16.

In Section 7.2, second-order and intrinsic stationarity were reviewed. Both
assume that the mean (or expectation) is constant. In this section, some
tools that may be used where this assumption is considered inappropriate are
discussed. There is a variety of published approaches to the separation of
the stochastic and deterministic (that is, the trend) elements of the spatial
variation. One approach, as outlined above, is to compute the variogram
using the residuals from the regional trend (297). Several methods have
been developed to estimate the underlying (trend-free) variogram. These
include estimation of the variogram using the residuals from a trend (287).
Universal kriging (or kriging with a trend model; which does not directly
entail estimation of the underlying variogram) and intrinsic random functions
of order k (IRF-k) kriging (105) are semiautomated approaches to deriving
the trend for prediction purposes. Median polish kriging is a further approach.
The respective methods are summarised here. An overview of nonstationary
geostatistical models is provided by Atkinson (23).

7.15.1   Trend or drift?

The term 'drift' is often used in geostatistical contexts to distinguish it from
the least squares estimate of $m(\mathbf{s})$ (283). That is, drift is used since the mean
is a characteristic of the RF, whereas the term 'trend' refers to a characteristic
of the data. Below, the term trend is preferred.

7.15.2   Modelling and removing large scale trends

Variograms are estimated from the values of the variable itself or from
residuals from a fitted trend model. If the variogram increases more rapidly
than a quadratic for large lags, then the RF is nonstationary (18). In
such cases, it may be necessary to model this nonstationarity. The drift,
a nonstationary mean $m(\mathbf{s})$, may be modelled, for example, as a low-order

polynomial trend. A map of monthly precipitation values obtained when a global first-order polynomial was fitted to monthly precipitation data (see Section 1.8.1) is given in Figure 6.2. Kriging with a trend model, and a commentary on trends, is the subject of Section 7.16.2. The presence of a trend may have a severe and marked effect on the form of the variogram (344). To deal with this bias, where it exists to a significant degree, the trend may be removed and the residuals used for geostatistical analysis.

One approach to detrending data is to use ordinary least squares (OLS) residuals (279), (312) but this may result in biased variograms. In particular, OLS may be markedly affected by outliers. It has, however, been demonstrated that this bias does not usually affect results detrimentally (84), (96), (218), (312). Beckers and Bogaert (35) attempt to quantify bias when OLS is used to detrend. They concluded that variograms computed from OLS residuals tend to underestimate total variance, and may also overestimate the nugget variance.

A set of elevation data are used to illustrate the effect on the variogram of detrending using first-order and second-order polynomials. More information on the application is provided by Lloyd and Atkinson (247). In Figure 7.11 (page 221), a variogram of (i) raw elevation data and residuals from a, (ii) first, and (iii) second-order polynomial trend are shown. The magnitudes of the semivariances decrease with increase in the order of the polynomial used, as more variation in the terrain is explained.

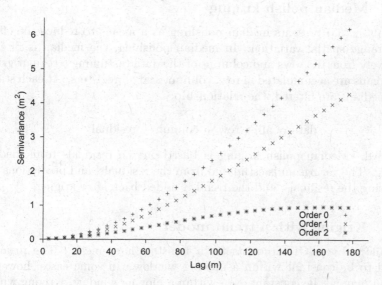

FIGURE 7.11: Variogram of raw data (order 0) and residuals from a first- and second-order polynomial trend.

Another approach involves using residuals from OLS to provide an initial estimate of the trend. Recognising the potential bias in the OLS residuals, this is used as a basis for fitting using generalised least squares (GLS) (32), (182), (244), (287).

Olea (291), (292) presented an approach to the estimation of the variogram where there is a local trend. This was based on the assumption of a linear variogram with no nugget effect, and works only with data on a regular grid. This approach is applied by Lloyd and Atkinson (248).

An alternative approach is to identify the variogram for a direction in which the trend does not dominate modelled spatial variation. The variogram for the selected direction may then be used for all directions under the assumption of isotropic variation (153), (371). An example is given in Section 7.16.4.4.

7.16 Nonstationary models for prediction

This section details several methods for kriging in the presence of a locally-varying mean. These include univariate methods and methods that make use of additional variables (see Section 7.16.4).

7.16.1 Median polish kriging

Cressie (93), (95) presents median polishing as a means to reduce the effect of long-range spatial variation. In median polishing, the median is filtered successively from the rows and columns of the data (assuming a regular grid). The medians are accumulated in row, column, and all registers. At each stage, as the medians are filtered, the relationship:

$$\text{data} = \text{all} + \text{row} + \text{column} + \text{residual}$$

is retained. Median polish kriging is based on the residuals from median polishing. The variogram is estimated from the residuals, and predictions are made using the residuals, and the trend is added back after kriging.

7.16.2 Kriging with a trend model

As detailed above, OK accounts for a nonstationary mean. The mean is assumed to be constant within a moving window. In some cases, however, the mean may not be constant even within a moving window. Kriging with a trend model (KT; or universal kriging) (213) takes account of local trends in data while it minimises the error associated with prediction (107), (153).

The variogram of the trend-free process must be estimated first, but the trend is estimated simultaneously as part of the prediction. For this reason, many researchers do not favour KT, but prefer an approach whereby the trend must be estimated as part of a structural analysis, so that the trend is consciously taken into account (32), (287). Approaches to estimating the trend-free variogram are described above.

With KT, the trend (or more correctly drift), the expected value of $Z(\mathbf{s})$, is modelled as a deterministic function of the coordinates. The unknown parameters are fitted using the data:

$$m(\mathbf{s}) = \sum_{k=0}^{K} a_k f_k(\mathbf{s}) \tag{7.53}$$

where $f_k(\mathbf{s})$ are functions of the coordinates, and a_k are unknown parameters (107), (153). When $k = 0$ the model is equivalent to OK, with the locally constant but unknown mean a_0, since conventionally $f_0(\mathbf{s}) = 1$ (153). The KT prediction is defined as:

$$\hat{z}_{KT}(\mathbf{s}_0) = \sum_{i=1}^{n} \lambda_i z(\mathbf{s}_i) \tag{7.54}$$

with

$$\sum_{i=1}^{n} \lambda_i^{KT} f_k(\mathbf{s}_i) = f_k(\mathbf{s}_0) \ k = 0, ..., K \tag{7.55}$$

The expected error is equal to zero:

$$\mathrm{E}\{\hat{z}_{KT}(\mathbf{s}_0) - z(\mathbf{s}_0)\} = 0 \tag{7.56}$$

The KT variance, σ_{KT}^2 , is given as:

$$\sigma_{KT}^2(\mathbf{s}) = \mathrm{E}[\{\hat{z}_{KT}(\mathbf{s}_0) - z(\mathbf{s}_0)\}^2]$$
$$= 2\sum_{i=1}^{n} \lambda_i^{KT}\gamma(\mathbf{s}_i - \mathbf{s}_0) - \sum_{i=1}^{n}\sum_{j=1}^{n} \lambda_i^{KT}\lambda_j^{KT}\gamma(\mathbf{s}_i - \mathbf{s}_j) \tag{7.57}$$

The weights are determined using $K + 1$ Lagrange multipliers through the KT system:

$$\begin{cases} \sum_{j=1}^{n} \lambda_j^{KT}\gamma(\mathbf{s}_i - \mathbf{s}_j) + \sum_{k=0}^{K} \psi_k^{KT} f_k(\mathbf{s}_i) = \gamma(\mathbf{s}_i - \mathbf{s}_0) & i = 1, ..., n \\ \sum_{j=1}^{n} = 1 \\ \sum_{j=1}^{n} \lambda_j^{KT} f_k(\mathbf{s}_j) = f_k(\mathbf{s}_0) & k = 0, ..., K \end{cases} \tag{7.58}$$

Once we know ψ_{KT}, the error variance of KT can be given as:

$$\sigma_{KT}^2 = \sum_{k=0}^{K} f_k(\mathbf{s}_0)\psi_k^{KT} + \sum_{i=1}^{n} \lambda_i^{KT}\gamma(\mathbf{s}_i - \mathbf{s}_0) \tag{7.59}$$

In matrix notation the KT system is given as:

$$\mathbf{K}_{KT}\lambda_{KT} = \mathbf{k}_{KT} \qquad (7.60)$$

For the case of a linear model in two dimensions, $\mathbf{K}_{KT}$ is the $(n + K + 1)$ by $(n + K + 1)$ matrix of semivariances and functions, f, of location:

$$\mathbf{K}_{KT} = \begin{bmatrix} \gamma(\mathbf{s}_1 - \mathbf{s}_1) & \cdots & \gamma(\mathbf{s}_1 - \mathbf{s}_n) & 1 & f_1(\mathbf{s}_1) & \cdots & f_K(\mathbf{s}_1) \\ \vdots & \vdots & \vdots & \vdots & \vdots & \vdots & \vdots \\ \gamma(\mathbf{s}_n - \mathbf{s}_1) & \cdots & \gamma(\mathbf{s}_n - \mathbf{s}_n) & 1 & f_1(\mathbf{s}_n) & \cdots & f_K(\mathbf{s}_n) \\ 1 & \cdots & 1 & 0 & 0 & \cdots & 0 \\ f_1(\mathbf{s}_1) & \cdots & f_1(\mathbf{s}_n) & 0 & 0 & \cdots & 0 \\ \vdots & \cdots & \vdots & \vdots & \vdots & \cdots & \vdots \\ f_K(\mathbf{s}_1) & \cdots & f_K(\mathbf{s}_n) & 0 & 0 & \cdots & 0 \end{bmatrix}$$

λ_{KT} are the KT weights, and $\mathbf{k}_{KT}$ are semivariances for the observations to prediction location:

$$\lambda_{KT} = \begin{bmatrix} \lambda_1^{KT} \\ \vdots \\ \lambda_n^{KT} \\ \psi_0^{KT} \\ \psi_1^{KT} \\ \vdots \\ \psi_K^{KT} \end{bmatrix} \quad \mathbf{k}_{KT} = \begin{bmatrix} \gamma(\mathbf{s}_1 - \mathbf{s}_0) \\ \vdots \\ \gamma(\mathbf{s}_n - \mathbf{s}_0) \\ 1 \\ f_1(\mathbf{s}_0) \\ \vdots \\ f_K(\mathbf{s}_0) \end{bmatrix}$$

To obtain the KT weights the inverse of the data semivariance matrix is multiplied by the vector of data to prediction semivariances:

$$\lambda_{KT} = \mathbf{K}_{KT}^{-1}\mathbf{k}_{KT} \qquad (7.61)$$

Finally, the KT variance is given as:

$$\sigma_{KT}^2 = \mathbf{k}_{KT}^T \lambda_{KT} \qquad (7.62)$$

It has been suggested that rather than using KT it might often be easier to apply OK within a smaller neighbourhood if there is no marked local trend. Then, as noted above, the (global) trend may be ignored (32), (312), (326). Journel and Rossi (213) are of the same opinion, that OK is preferred in most situations. Furthermore, Journel (209) has written that in his experience KT is necessary only "in cases of extrapolation when the point being predicted is beyond the correlation range of any datum" (p. 127). However, some analyses have indicated that the use of KT may be advantageous in the presence of obvious trends. The choice of neighbourhood, using OK, cannot take into account local trend.

The KT system (see Equation 7.60) is illustrated using the same data used for illustrating OK in Section 7.8. With the variogram model coefficients shown in Figure 7.14 (page 231) and for a first-order polynomial trend we have:

$$
\begin{bmatrix}
0.00 & 167.718 & 194.207 & 218.098 & 1 & 370400 & 415700 \\
167.718 & 0.00 & 215.817 & 182.372 & 1 & 370900 & 412400 \\
194.207 & 215.817 & 0.00 & 170.473 & 1 & 366100 & 414700 \\
218.098 & 182.372 & 170.473 & 0.00 & 1 & 367100 & 411400 \\
1 & 1 & 1 & 1 & 0 & 0 & 0 \\
370400 & 370900 & 366100 & 367100 & 0 & 0 & 0 \\
415700 & 412400 & 414700 & 411400 & 0 & 0 & 0
\end{bmatrix}
\begin{bmatrix}
\lambda_1^{KT} \\
\lambda_2^{KT} \\
\lambda_3^{KT} \\
\lambda_4^{KT} \\
\psi_0^{KT} \\
\psi_1^{KT} \\
\psi_2^{KT}
\end{bmatrix}
=
\begin{bmatrix}
133.348 \\
151.360 \\
157.432 \\
170.941 \\
1 \\
369000 \\
414300
\end{bmatrix}
$$

The KT weights are as follows: $\lambda_1^{KT} = 0.417$, $\lambda_2^{KT} = 0.210$, $\lambda_3^{KT} = 0.272$, and $\lambda_4^{KT} = 0.101$ with $\psi_0^{KT} = -2705.852$, $\psi_1^{KT} = 0.000785$, and $\psi_2^{KT} = 0.00587$. The KT prediction is obtained with: $(0.417 \times 52.6) + (0.210 \times 40.8) + (0.272 \times 48.3) + (0.101 \times 46.7) = 48.357$.

The KT variance is obtained with: $(0.417 \times 133.348) + (0.210 \times 151.360) + (0.272 \times 157.432) + (0.101 \times 170.941) + (-2705.852 \times 1) + (0.000785 \times 369000) + (0.00587 \times 414300) = 161.386$.

Note that four is a small sample size and is used only to illustrate the approach. In this case, the difference between the OK prediction (47.908 mm; illustrated in Section 7.8) and the KT prediction (48.357 mm) is fairly small, but in cases where a marked trend is present we would expect the differences to be larger. Note that the nugget of the variogram used for KT (Figure 7.14) is larger than that of the variogram used for OK (Figure 7.4). This provides an analogy with the case where the nugget of variograms estimated of residuals from a polynomial trend may be overestimated (see Section 7.15.2). One implication of this is that the approach used (selecting the trend-free direction) does not account fully for the trend, while the use of local regression with elevation data appears to remove more of the large-scale spatial variation (see Figure 7.12, page 228).

Zimmerman et al. (404) present a comparison of OK and KT with IDW. They conclude that OK and KT outperform IDW for all cases they explore in terms of surface type, sampling configuration and other factors.

7.16.3 Intrinsic model of order k

KT has been criticised by some researchers because of the difficulty in estimating the unbiased trend-free variogram (17). Dissatisfaction with KT led to the development of an alternative methodology, intrinsic random functions of order k (IRF-k) kriging (105), (267). However, some researchers argue that KT is more flexible and more intuitive than IRF-k kriging (370).

With KT, the coefficients of the trend model are estimated from the data. In IRF-k kriging, the trend is filtered out using linear combinations of the data, termed generalised increments of order k (GI-k). The filtering performed approach is analogous to the ARIMA (autoregressive integrated moving average) model (315). The difference between KT and IRF-k kriging is in how the form of the trend is ascertained and the characterisation of spatial variation rather than in the kriging process itself. The concept of increments of order k is outlined by Chilès and Delfiner (80).

The variance of increments of order k is expressed with the generalised covariance (GCV) of order k, $K(\mathbf{h})$. The variogram is the variance of increments; that is, a difference of order 1 (106). With IRF-k we attempt to filter out higher order differences. The negative of the variogram is equal to the generalised covariance for an IRF of order 0. Several different methods have been developed for inferring the highest order of drift to be filtered and for selecting the optimal GCV. A range of (i) regression-based methods, and (ii) parametric methods have been used. Kitanidis (217) presented a range of parametric methods and algorithms to implement each method have been given by Pardo-Igúzquiza (308). The ISATIS software (50) employs an approach following Delfiner (105).

It has been shown that IRF-k kriging provides identical predictions to KT where the covariance model is identical (83). The IRF-k kriging system and the KT system are essentially the same (209), (283), except that in the IRF-k kriging system there is the generalised covariance, $K(\mathbf{h})$, rather than the variogram.

7.16.4   Use of secondary data

There is a variety of methods whereby secondary data can be used following the assumption that these data, rather than some function of the coordinates (as in KT), explain large scale variation in the primary data. This section details two such approaches.

7.16.4.1   Simple kriging with a locally varying mean

As noted in Section 7.6, SK is the most basic form of kriging. With SK, the mean is assumed to be constant (there is no systematic change in the mean of the property across the region of study) and known. If the mean is not constant and it is possible to estimate the mean at locations in the domain of interest, then this locally varying mean can be used to inform prediction using SK. That is, the local mean is estimated prior to kriging, whereas in OK the mean is estimated as part of the kriging procedure.

The simple kriging with locally-varying means (SKlm) prediction, $\hat{z}_{SKlm}(\mathbf{s}_0)$, is defined as:

$$\hat{z}_{SKlm}(\mathbf{s}_0) - \hat{m}_{SK}(\mathbf{s}_0) = \sum_{i=1}^{n} \lambda_i^{SK}[z(\mathbf{s}_i) - \hat{m}_{SK}(\mathbf{s}_i)] \qquad (7.63)$$

Where $\hat{m}_{SK}$ is a known locally-varying mean. The weights are found through the SKlm system:

$$\sum_{j=1}^{n} \lambda_j^{SKlm} C_R(\mathbf{s}_i - \mathbf{s}_j) = C_R(\mathbf{s}_i - \mathbf{s}_0) \qquad i = 1, ..., n \qquad (7.64)$$

where $C_R(\mathbf{h})$ is the covariance function of the residuals from the local mean. Where the mean is not constant, or the variance is unbounded, then the variogram is used (so, for example, the variogram is used in the OK system). The locally-varying mean can be estimated in various different ways. One approach is to use regression to predict the value of the primary variable at (i) all observation locations, and (ii) all locations where SKlm predictions will be made. The variogram is then estimated using the residuals from the regression predictions at the data locations. Goovaerts (155) applies SKlm for mapping precipitation in Portugal. Lloyd (240) uses local regressions of precipitation and elevation to estimate the local mean of precipitation amount, followed by the application of SKlm.

7.16.4.2 SKlm case study

The variogram estimated from residuals of a local regression of elevation against monthly precipitation (using the data described in Section 1.8.1) was estimated for prediction using SKlm. The variogram of locally detrended precipitation values is given in Figure 7.12. SKlm cross-validation prediction errors are given in Table 7.2. The SKlm derived map, generated using 128 nearest neighbours, is given in Figure 7.13. Using smaller data neighbourhoods with SKlm was invalid as moving window regression (MWR, used to derive the local mean) residuals for smaller neighbourhoods were unstructured — all spatial structure had been removed. Precipitation amounts are large in the north and west. The map derived using SKlm is quite similar in appearance to that derived using OK (Figure 7.6). SKlm was conducted using the GSLIB software (107).

TABLE 7.2
Cross-validation prediction errors for SKlm.

N. data	Mean error (mm)	Std. dev. (mm)	Max. neg. error (mm)	Max. pos. error (mm)	RMSE (mm)
128	0.144	13.785	−176.042	79.444	13.786

7.16.4.3 Kriging with an external drift model

Kriging with an external drift (KED) (107), (153), (154) exploits relationships between primary and secondary variables. The method is used under the

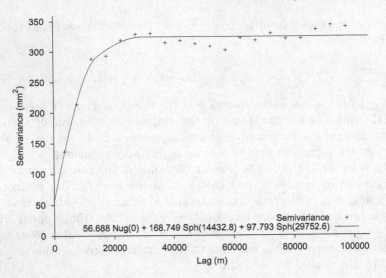

FIGURE 7.12: Precipitation: variogram estimated from residuals of a local regression.

assumption that the drift is due to a secondary variable. The primary advantage over cokriging is that it is not necessary to model the covariation. KED has also been extended to the use of a categorical external drift (280). This enables integration of thematic information in the prediction process.

To apply KED, it is necessary to have secondary data that are (i) available at the primary data locations as well as at all locations for which predictions are desired, and (ii) linearly related to the primary variable. KED predictions are a function of (i) the form of the variogram model, (ii) the neighbouring primary data, and (iii) the (modelled) relationship between the primary variable and the secondary variable locally. More specifically, the local mean of the primary variable is derived using the secondary information and simple kriging (SK) is carried out on the residuals from the local mean. The KED prediction is defined as:

$$\hat{z}_{KED}(\mathbf{s}_0) = \sum_{i=1}^{n} \lambda_i^{KED} z(\mathbf{s}_i) \tag{7.65}$$

The weights are determined through the KED system of $n + 2$ linear equations (107), (153), where the secondary variable acts as a constraint:

$$\begin{cases} \sum_{j=1}^{n} \lambda_j^{KED} \gamma(\mathbf{s}_i - \mathbf{s}_j) + \psi_0^{KED} + \psi_1^{KED} y(\mathbf{s}_i) = \gamma(\mathbf{s}_i - \mathbf{s}_0) \ i = 1, ..., n \\ \sum_{j=1}^{h} \lambda_j^{KED} = 1 \\ \sum_{j=1}^{h} \lambda_j^{KED} y(\mathbf{s}_j) = y(\mathbf{s}_0) \end{cases}$$

$$\tag{7.66}$$

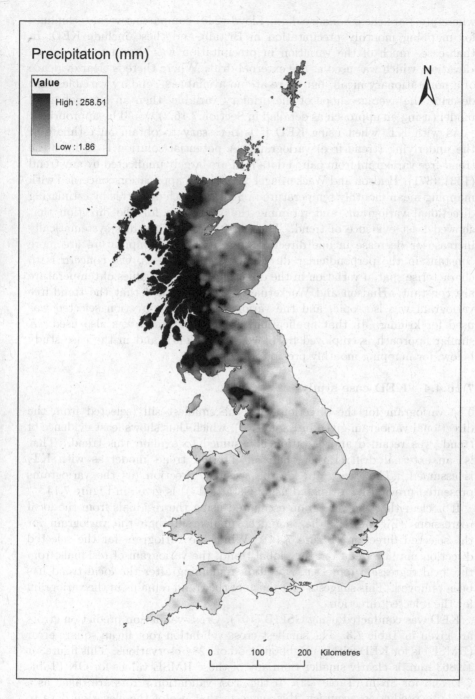

FIGURE 7.13: Precipitation: map derived using SKlm.

where $y(\mathbf{s}_i)$ are the secondary data. Lloyd (240) compares several techniques for mapping monthly precipitation in Britain and these include KED. In that case, much of the variation in precipitation was shown to be due to elevation, which was used as the external drift. Where there is clear evidence of a nonstationary mean, but there are no available secondary variables that describe the average shape of the primary variable, then an internal trend model (using an approach as detailed in Section 7.16.2) would be appropriate.

As with KT, when using KED it is necessary to obtain an estimate of the underlying (trend-free) variogram. A potential solution is to infer the trend-free variogram from paired data that are largely unaffected by any trend (153), (371). Hudson and Wackernagel (191), in an application concerned with mapping mean monthly temperature in Scotland, achieved this by estimating directional variograms and retaining the variogram for the direction that showed least evidence of trend. That is, temperature values systematically increase or decrease in one direction, but values of temperature are more constant in the perpendicular direction. In such cases, the concern is to characterise spatial variation in the direction for which values of temperature are constant. Hudson and Wackernagel (191) assumed that the trend-free variogram was isotropic, and the variogram for the direction selected was used for kriging. In that application, an internal trend was also used. A similar approach is employed by Lloyd (239), (240), and in the case study below, for mapping monthly precipitation.

7.16.4.4   KED case study

The variogram for the direction with the smallest sill (selected from the directional variogram given in Figure 7.5), which thus shows least evidence of trend, was retained and elevation is assumed to explain this trend. That is, an external drift, rather than an internal trend model as with KT, is assumed. The variogram for the selected direction (as the variograms presented previously, estimated using Gstat (313)) is given in Figure 7.14.

It is clear that the variogram estimated using the residuals from the local regressions (Figure 7.12) has a much smaller sill than the variogram for the selected direction (Figure 7.14). While the variogram for the selected direction may be free from the global trend, the variogram of residuals from the local regression represents variation remaining after the local trend has been removed. This suggests that much of the trend remains in the variogram for the selected direction.

KED was conducted using GSLIB (107). Cross-validation prediction errors are given in Table 7.3. The smallest cross-validation root mean square error (RMSE) is for KED with a neighbourhood of 128 observations. This figure, at 12.863 mm, is clearly smaller than any of those RMSE values for OK (Table 7.1) and for SKlm (Table 7.2). If the cross-validation errors are taken as a guide to prediction accuracy, this suggests that use of the elevation data to inform prediction of monthly precipitation is justified.

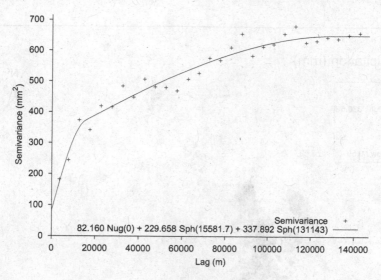

FIGURE 7.14: Precipitation: variogram for selected direction.

TABLE 7.3
Cross-validation prediction errors for KED.

N. data	Mean error (mm)	Std. dev. (mm)	Max. neg. error (mm)	Max. pos. error (mm)	RMSE (mm)
8	0.116	14.014	−162.579	88.635	14.015
16	0.119	13.527	−138.897	86.234	13.528
32	0.143	13.014	−123.399	85.681	13.014
64	0.110	12.973	−133.962	92.204	12.973
128	0.110	12.862	−133.786	91.139	12.863

The KED map, generated using 64 nearest neighbours, is given in Figure 7.15. The KED derived map shows clearly the form of the terrain (the external drift). The KED maxima is larger than that of the other methods. KED can extrapolate outside the range of the data where values of the exhaustively sampled secondary variable (here, elevation) are outside the range of the values of the secondary variable at locations where the primarily variable has also been sampled (371). That is, in this case the maximum values of elevation represented in the digital elevation model (DEM) are larger than the maximum values of elevation at the locations where precipitation amount is recorded.

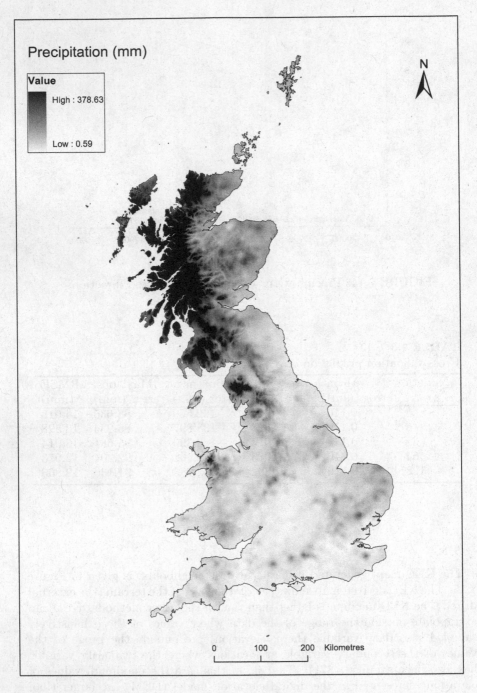

FIGURE 7.15: Precipitation: map derived using KED.

7.16.5 Locally varying anisotropy

An interesting recent development is the derivation of locally varying anisotropy fields as presented by Boisvert et al. (51). With this approach, there is a consistent variogram model with respect to the nested structures and variance components, but the range and direction parameters are allowed to vary. Such an approach acknowledges the sometimes locally-marked directional variations in environmental phenomena. The following section focuses on the case where the variogram model as a whole is allowed to vary in some way between different locations.

7.17 Nonstationary variogram

Various approaches have been developed for estimating the variogram where the spatial structure of a given property varies across the region of interest. The simplest approach is to split the data into subsets within which spatial variation is considered similar at all locations. Alternatively, the variogram can be estimated within a moving window, and a model can be fitted automatically for the purposes of kriging (170), (171), (172), (246). Haas (170) has presented an approach where the variogram is estimated for a moving window, and the model fitted to the local variogram used to inform kriging. In addition, the data are detrended locally to take into account what Haas terms mean nonstationarity (the trend) in addition to allowing for covariance nonstationarity (change in the form of the variogram, or other function, across the region of interest) through use of a moving window. Harris et al. (178) compute local variograms using geographical weights.

The size of window is crucial in local variogram estimation — a sufficient number of data are required to estimate the variogram, but the number should be small enough to capture local variation. Haas (171), in a study concerned with acid deposition, presents what he terms a simple heuristic determination of window size. The scheme of Haas (171) was specific to his study, but he generalises the scheme. Haas states that a lag interval and number of observations in the window should be specified so that at least two paired observations ('couples') are usually found at the shortest lag and at least two at the longest lag.

Haas (171) used the Levenberg-Marquardt algorithm (see Press et al. (319)) to fit the variogram models (specifically, a spherical model was selected). Where a negative nugget was fitted given an initial unconstrained search, the nugget was fixed at zero and the sill and range subspace was searched again. Haas details a set of procedures for determining final model parameters, with the initial value for the nugget set at 85% of the median of the semivariances at the first three lags. The starting sill value was set as the median of the semivariances at the four largest lags. Given these starting values for the nugget and sill, ten starting values of the range were specified starting at one-tenth of the maximum lag and up to the maximum lag in equal increments. The parameters corresponding to the model with the smallest residual sum of squares (RSS) for the ten fits were selected for kriging. Where any of the ten fits did not converge, the global search was repeated but instead using the median of the semivariances at the first three lags as the starting nugget, rather than using the 85% criterion. After the global search, if the nugget was larger than the sill then both were set equal to the sample variance in the window. In terms of assessing convergence, Haas accepted the model fit when, for 20 consecutive steps, the difference between the RSS for two consecutive steps did not change or decreased by no more than 10^{-5} times the RSS for the earlier stage.

A limitation of moving window variogram approaches include difficulties with estimating uncertainty, but such approaches provide the main focus here because of their potential for widespread application. Lloyd and Atkinson (246) fitted models, using GLS, to variograms estimated within a moving window for the purposes of predicting elevation values at unsampled locations. In that study, the global model was used where the standardised weighted sum of squares for the local model fit exceeded some fixed amount. In another application, Lloyd (244) estimated variograms for a moving window and determined models automatically using maximum likelihood (ML). The Vesper software environment can be used to estimate local variograms and fit models to them automatically (392).

To illustrate local variogram modelling the precipitation data detailed in Section 1.8.1 were analysed using the same procedure detailed by Lloyd (244). Using this procedure, variogram models were derived using ML with the local elevation-precipitation relationship modelled using GLS.

Figures 7.16, 7.17, and 7.18 show, respectively, the nugget effect (c_0), structured component (c_1), and range (a) of spherical models fitted automatically given the 128 nearest neighbours to each location. Local variograms were also estimated without accounting for the local elevation-precipitation relation. The models fitted to these latter variograms were used for prediction with OK, while the models summarised in Figures 7.16, 7.17 and 7.18 were used for prediction with KED. Judging by the cross-validation RMSE values given in Table 7.4, local variogram OK is less accurate than global variogram OK (see Table 7.1; page 212), whereas local variogram KED results in a slightly smaller RMSE than global variogram KED (see Table 7.3; page 231).

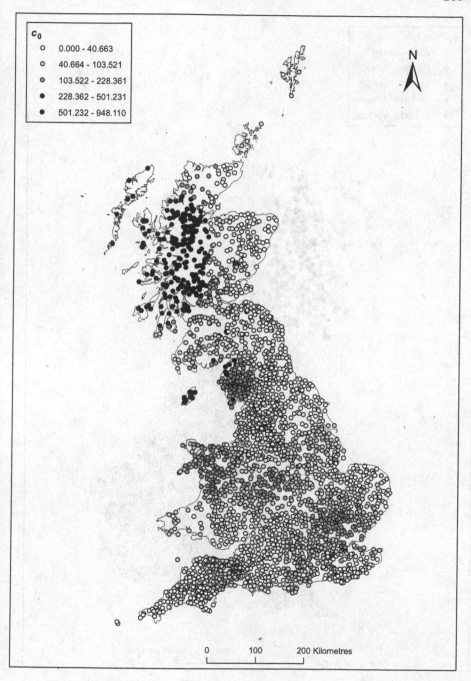

FIGURE 7.16: Local nugget effect for 128 nearest neighbours.

FIGURE 7.17: Local structured component for 128 nearest neighbours.

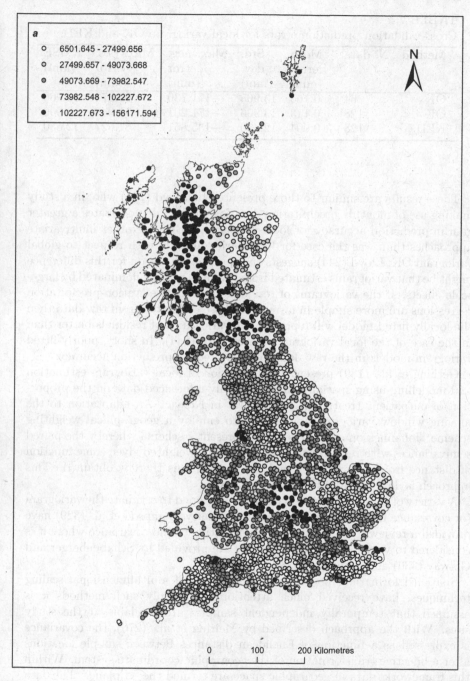

FIGURE 7.18: Local range for 128 nearest neighbours.

TABLE 7.4
Cross-validation prediction errors for local variogram OK and KED.

Method	N. data	Mean error (mm)	Std. dev. (mm)	Max. neg. error (mm)	Max. pos. error (mm)	RMSE (mm)
OK	64	0.165	13.965	−171.730	69.973	13.966
OK	128	0.113	13.906	−171.201	71.046	13.906
KED	128	−0.004	12.850	−135.851	87.027	12.850

These results are similar to those presented by Lloyd (244) who, in a study making use of monthly precipitation data for the UK, demonstrates a greater gain in prediction accuracy for local variogram KED over other multivariate approaches than was the case for local variogram OK with respect to global variogram OK. Lloyd (244) suggests that part of the reason for this difference might be that variograms estimated from raw data may be dominated by large-scale effects. If the variograms of residuals from local elevation-precipitation regressions are more simple in form than those estimated from raw data then the locally fitted model will represent the variograms of residuals better than in the case of the local variograms of raw data (244). In short, poorly-fitted variogram models in the raw data case may reduce prediction accuracy.

Gething et al. (139) present an application of local variogram estimation and modelling using spatially and temporally referenced data on the proportion of outpatient treatments for malaria in Kenya. An adaptation to the moving window variogram approach is to employ a geographical weighting scheme. Johannesson and Cressie (203) present a schema whereby the paired semivariances with respect to location s_i can be weighted given some function of distance from that location and local variograms thereby obtained. This approach is developed by Harris et al. (178).

A variety of other techniques have been developed to estimate the variogram (or covariance function) where it is nonstationary. Sampson et al. (329) have provided a review of methods that aim to estimate the covariance where it is considered to vary locally. Other reviews are provided by Schabenberger and Gotway (330) and Sampson (328).

Spatial deformation models, which make use of multidimensional scaling techniques, have received much attention. To apply such methods it is assumed that temporally independent samples are available at the study sites. With the approach described by Meiring et al. (270), the covariance is expressed as a function of Euclidean distances between sample locations after a bijective transformation of the geographic coordinate system. Within this framework, data in geographic space are termed the 'G plane.' The data locations are transformed, and the result is termed the 'D plane' (note that the approach itself is global). The variogram of data in the 'D plane' is stationary (329).

Løland and Høst (254), in a study concerned with modelling in a coastal domain, consider the use of a deformation approach for calculating a Euclidean approximation to water distances. With this approach, approximate water distances are derived and multidimensional scaling is applied to the grid points using these approximate water distances. The data locations in original geographic space are then mapped to the transformed Euclidean space (Euclidean distances which approximate water distance in the original geographic space) using linear interpolation. The general topic of non-Euclidean distance measures in a geostatistical context is discussed by Curriero (99).

Another important class of approaches is based on process convolution. Such approaches are based on the principle that a Gaussian process, z, can be represented as the moving average (convolution) of a white noise process (i.e., zero-mean and spatially independent), x:

$$z(\mathbf{s}) = \int_A k(\mathbf{s} - \mathbf{u}) x(\mathbf{u}) \ d\mathbf{u} \qquad (7.67)$$

where the kernel is centred on $\mathbf{s}$, and A indicates the study region.

Higdon (184) and co-workers (185), (351) have presented approaches whereby the convolution kernels are allowed to differ between spatial locations. In this framework, the covariance structure is defined indirectly. Instead of estimating the covariance parameters, we estimate the kernel parameters and the kernel functions lead to different autocorrelation functions. For example, the Gaussian kernel leads to the Gaussian correlation function model (330). As noted by Schabenberger and Gotway (330), as the covariance function depends on the choice of the kernel function, a nonstationary covariance function can be built through varying the kernel spatially. In short, using the approach of Higdon (184), kernels with different bandwidths can be identified at individual locations. Also, the variances of the kernels may be unequal in different directions and, therefore, geometric anisotropy can be modelled.

Higdon et al. (185) illustrate the application of this approach to the exploration, and spatial prediction, of dioxin concentrations. The method is implemented in a Bayesian framework. A prior distribution on the parameters of the kernel ellipse is specified as a Gaussian random field with a Gaussian spatial covariance function (see Sampson et al. (329), for another summary), and thus the kernels are constrained to vary smoothly across space. Higdon et al. (185) present local kernel standard deviation ellipses and posterior mean estimates for the dioxin concentrations. Paciorek and Schervish (305) implement a similar approach to Higdon and colleagues, but, in addition to a fully Bayesian Gaussian process model, they also assess the application of an *ad hoc* nonstationary kriging approach based on nonstationary correlation functions.

The above approaches assume that data are available at the appropriate support. Atkinson and Jeganathan (26) employ a deconvolution–convolution procedure to estimate the local punctual variogram given data with a coarse spatial resolution relative to the desired finer spatial resolution.

7.18   Variograms in texture analysis

The variogram has been used in characterising image texture on a moving window basis (72), (272), (273). A summary of the subject is provided by Tso and Mather (364). Lark (231), uses square-root differences rather than the semivariance. Various researchers have estimated the range of the variogram locally (320), (201). Lloyd and Atkinson (248) use the double log variogram to estimate the fractal dimension of topography in a moving window. Additionally, locally-computed variograms have been used for automated seafloor classification (183). Franklin et al. (133) use the variogram to determine appropriate window sizes for texture analysis. Such an approach could be used to identify areas for which the variogram may be stationary — regions for which the semivariance is comparatively homogeneous for each lag. As noted above, Haas (170) uses a moving window approach to estimate the variogram locally for the purposes of kriging. Similarly, Lloyd and Atkinson (246) fit models automatically to variograms estimated in a moving window. In that case, the spherical model was fitted, but such an approach could be adapted so that a range of different models could be fitted and the best fitting model retained.

An application of region-growing segmentation (see Section 3.7) of the mapped range of the variogram has been presented by St-Onge and Cavayas (342) and was concerned with identifying forest stands. Where kriging is the aim, segmentation could be the first step for what is termed stratified kriging (153). Another approach is to utilise prior data and estimate the variogram within subsets. Berberoglu et al. (39) estimated the variogram of digital number (DN) values contained by areas represented by boundaries digitised from paper maps. These per-field variograms were used as inputs to maximum likelihood (ML) and artificial neural network (ANN) classifiers. Lloyd et al. (251) followed up this work by using other geostatistical structure functions and statistics derived from the grey-level cooccurrence matrix. Atkinson and Lewis (27) provide an overview of geostatistical methods for classification of remotely-sensed imagery.

7.19 Summary

Kriging is theoretically superior to the other prediction approaches outlined since it makes use of spatial variation in the data. However, if this spatial variation is complex and varies substantially locally, it may be very difficult to use this information effectively. A lot of interaction is required for the application of geostatistical prediction in the usual manner, and this can be viewed as both a strength and a disadvantage. For example, the user must make several decisions such as the stationarity decision.

There is now a wide range of geostatistical software available. As well as the various packages mentioned in the text, a variety of routines have been written in the R language (see, for example, Bivand et al. (48)). Techniques like OK and KT can be implemented using popular GISystems. However, at present there is little software available to allow estimation of the nonstationary variogram. Important developments have taken place in recent years, although some of these seem unlikely to benefit the majority of users of GISystems in the near future.

The following chapter moves to a different focus from all of the previous chapters and it discusses the analysis of point patterns.

8

Point Patterns and Cluster Detection

In this chapter, a variety of methods for exploring the configuration of point events are outlined. The chapter first discusses some key principles and commonly used approaches are then detailed. Global summaries (e.g., the K function) are outlined first to provide a context for local approaches. In Section 8.3, intensity estimation and distance methods are defined. Section 8.4 is concerned with statistical tests of point patterns. Sections 8.5 through 8.8 focus on global summaries of point pattern structure. Local measures of point pattern intensity are discussed in Sections 8.9 through 8.11, while a local measure of point pattern structure is discussed in Section 8.12. Section 8.13 deals with the identification of clusters or clustering. The application of selected methods is illustrated in the text using the point data described in Section 1.8.3.

8.1 Point patterns

A point pattern comprises a set of point event locations, s_i, $i = 1, 2, ..., n$ (with spatial coordinates x, y) within a specific study area. In common with previous discussions about point pattern analysis (e.g., (32), (109)), here an 'event' refers to an observation, whereas a 'point' refers to an arbitrary location in the study region. That is, some methods are used to analyse the spatial distribution of events, while the intensity of a point pattern may be estimated at any location in the region of interest. There is often a desire to assess how clustered or dispersed events are. Alternatively, we may be concerned with assessing how similar two point patterns are. This chapter presents some key methods for analysing the spatial configuration of samples (that is, point patterns). More specifically, there are two areas of focus: (i) the analysis of point patterns, and (ii) the identification of clusters (or clustering) of events.

With a global approach to point pattern analysis, statistics summarise the degree of clustering or dispersion across the entire region of interest. Clearly, if the concern is to assess the possibility of variation in the spatial configuration of the data across the region of interest, then it is necessary to apply some form of locally-adaptive measure. The basic principles and tools of point pattern analysis are outlined below, but the focus is on local measures.

Detailed accounts of methods for the analysis of spatial point patterns are provided by a range of authors, including Ripley (322), Bailey and Gatrell (32), Gatrell et al. (134), Boots and Getis (55), Cressie (95), Diggle (109), Fotheringham et al. (128), and O'Sullivan and Unwin (304). Much of the development of methods for point pattern analysis took place in ecology, and there are many applications in physical geography. Bailey and Gatrell (32) list several examples of spatial point patterns, including locations of craters in a volcanic field and locations of granite tors. Diggle (109) has three case studies concerned with analysis of Japanese black pine saplings, redwood seedlings, and biological cells.

Conventionally, introductions to point pattern analysis outline ways in which point patterns can be summarised with, first, measures of intensity and, second, event separation (distance measures). In this case, the concern is with division into global and local approaches, and the chapter is arranged to reflect this. First, visual examination of point patterns is discussed. Next, intensity and distance measures are outlined and measures such as the mean centre and standard distance, which provide global summaries, are detailed. Some other background is then provided. The remainder of the chapter focuses on (i) global measures of event intensity (Section 8.6), (ii) distance-based measures that provide global summaries (Section 8.7), (iii) local measures of event intensity (Sections 8.10 and 8.11), (iv) local distance-based methods (Section 8.12), and (v) methods for detection of clusters or clustering in point patterns (Section 8.13). The other sections deal with a variety of specific issues not addressed in the main body of the text.

8.2   Visual examination of point patterns

The examination of point patterns in the form of scatter plots provides a sensible first step in the analysis of a point pattern. This enables a first impression of the tendency towards clustering or regularity, and also local variation in the point pattern. However, to gain a better impression of how clustered a point pattern is, it is necessary to derive measures that summarise in various ways properties of the point pattern. Where several events occur at one location, perhaps because the locational coordinates are imprecise and events separated in reality appear to overlap, different symbols may be used to represent these multiple event locations (128).

An additional concern may be to account for the underlying population at risk. For example, symbols of different sizes could be used that are inversely proportional to the population intensity at a particular location (32). An alternative approach is to generate a density-equalised point pattern (a cartogram), such that the point pattern in geographic space is transformed into a point pattern with a homogeneous population at risk.

8.3   Measuring event intensity and distance methods

There are two main approaches to the description of point patterns. These are concerned with (i) point event intensity, and (ii) point separation. Where first-order effects are most obvious, location is important, and measures of point pattern intensity are most appropriate. In other words, they are concerned with expected values of points over a study area. In cases where second-order effects are most marked, measures of point separation are most appropriate. In these cases, interaction between events and the relative location of observations are important (304). However, in practice it is very difficult to distinguish between first- and second-order effects.

The terms 'density' and 'intensity' are often used interchangeably, but Waller and Gotway (375) define the density function as the probability of observing an event at location **s**, whereas the intensity function is defined as the number of events expected per unit area at location **s**. A density integrates to one over the study area while the intensity estimate integrates to the overall mean number of events per unit area (375). The mean intensity of a point pattern is given by:

$$\lambda = \frac{\#(A)}{|A|} \tag{8.1}$$

where $\#(A)$ is the number of events in region A (which is a subregion within $\Re$), and $|A|$ is the area of that region.

A point process may be represented as a set of random variables (RVs) $\{\#(A), A \subseteq \Re\}$. A spatial point pattern describes a stochastic process where RVs represent the spatial locations of events, and a realisation of this process is a set of locations generated under the point process model, whether observed or simulated (375). For an area centred on **s**, the intensity is given as the mathematical limit:

$$\lambda(\mathbf{s}) = \lim_{d \longrightarrow 0} \frac{\mathrm{E}(\#(C(\mathbf{s}, d)))}{\pi d^2} \tag{8.2}$$

where $\mathrm{E}(\#(C(\mathbf{s}, d)))$ is the expected number of events in the circle $C(\mathbf{s}, d)$, with radius d around the location **s**. If $\lambda(\mathbf{s})$ is constant across the entire study area $\Re$, then $E(\#(A)) = \lambda |A|$, where $|A|$ is the area of A, and the point process is termed stationary.

The second-order properties of a point pattern can be examined using the covariances between events in two areas. The second-order intensity provides a way of describing the relationship between paired points and it is given as:

$$\gamma(\mathbf{s}_i, \mathbf{s}_j) = \lim_{d_i \longrightarrow 0, d_j \longrightarrow 0} \frac{E(\#(C(\mathbf{s}_i, d_i)) \#(C(\mathbf{s}_j, d_j)))}{\pi^2 (d_i d_j)^2} \qquad (8.3)$$

If the point process is stationary and $\gamma(\mathbf{s}_i, \mathbf{s}_j)$ is a function of only the distance and direction $\mathbf{h}$ separating events and not their location, then the second-order intensity can be written as $\gamma(\mathbf{s}_i - \mathbf{s}_j)$ or $\gamma(\mathbf{h})$. An isotropic process is a function only of distance and not direction.

8.4 Statistical tests of point patterns

The measures discussed in this chapter provide various ways of assessing the degree of clustering or dispersion of a point pattern. To consider how clustered or how dispersed a point pattern is, there is a need to use a statistical test of some kind. This is often achieved by considering if an observed process is likely to be an outcome of some hypothesized process. Models of complete spatial randomness (CSR), whereby events are the outcome of a homogeneous Poisson process, are used widely. For a CSR process, the intensity is constant (that is, each location has an equal chance of containing an event) and events are independently distributed. So, we may want to assess if an observed point pattern shows any departure from CSR, in that events are either clustered or regularly distributed (134). An alternative problem might be to ascertain if two separate point patterns are realisations of the same process. Other models are discussed by Bailey and Gatrell (32). Testing for CSR is a sensible starting point in any point pattern analysis (55). Some tests for CSR using different point pattern measures are outlined below.

8.5 Global methods

The focus of the next section is on global summaries of the structure of point patterns. The measures do not allow for spatial variation in point patterns. In other words, when these approaches are applied, the assumption is made that the point process is stationary. A summary of the first-order properties of a point pattern would be provided by simply measuring the average intensity over the whole study area. In practice, most measures of first-order properties are locally-derived (see Section 8.10).

8.6 Measuring event intensity

Local variation in intensity can be explored by partitioning a set of points into subsets with equal area (quadrats). Quadrat analysis is a core means of assessing the intensity of point patterns. Although quadrats are local spaces, most applications of quadrats have global summaries as their outputs, and this approach is the focus here.

8.6.1 Quadrat count methods

Quadrat counts are conducted either as an exhaustive census of quadrats, or by placing quadrats randomly across the area of interest (this latter approach is often used in fieldwork). In either case, the output comprises the counts in each cell. Once this information is available it is possible to compare the frequency distribution with the expected distribution. The origin, orientation and size of cells is crucial and experimentation is a practical way of selecting an appropriate scheme. When cells are too large a danger is that much variation will be lost (averaged out). Conversely, when cells are too small, fine scale spatial variation may be captured, but there may be many empty cells. The use of (overlapping) moving windows provides one way to overcome this problem, but a problem then is that no information on the relative location of points is available (32). Kernel estimation (KE) is one way forward, and this approach is discussed in Section 8.10.2.

Boots and Getis (55) summarise an application in which quadrats were located randomly (as opposed to forming a complete grid covering the study region). In the study outlined, the quadrats were located according to CSR (see Section 8.4) assumptions — each location in the study area has an equal chance of receiving a quadrat centre, while the location of an existing quadrat centre does not influence the location of further quadrat centres. A simpler approach is to superimpose a grid on the study area and select randomly x,y coordinates to represent the centres of quadrats. In some instances, where quadrats are large, they will overlap. In this situation, the quadrats are not independent of each other and this violates the condition of independence for generating a CSR point process (55).

Quadrat analysis is usually conducted using quadrats placed on a regular grid. A key issue in this context is to select appropriate shapes and sizes for the quadrats (and possibly their orientation). These are usually selected through experimentation. One general rule of thumb that has been suggested is that an appropriate quadrat size is approximated as twice the size of the area per point. So, for square quadrats, the length of the side of a quadrat would be $\sqrt{(2A/N)}$ where A is the area and there are N points (55). In practice, a variety of different quadrats are often used and results compared.

8.6.2   Quadrat counts and testing for complete spatial randomness (CSR)

The expected probability distribution for a quadrat count of a point pattern can be given by the Poisson distribution:

$$P(k) = \frac{e^{-\lambda}\lambda^k}{k!} \text{ for } k = 0, 1, 2... \tag{8.4}$$

where e is the mathematical constant 2.178282, λ is the expected number of points per sample area (which can be estimated by the mean number of points per quadrat), and there are k points in a quadrat. To ascertain how well a null hypothesis of CSR explains the observed point pattern, a quadrat count distribution is compiled, and it may then be compared to the Poisson distribution with λ estimated from the point pattern. As an example, following Boots and Getis (55), if 78 events are found in 30 quadrats the intensity, λ, is 78/30=2.6. In the case study summarised in Table 2.1 of Boots and Getis (55), the probability of a quadrat with zero points is given as 0.0743 and for one point it is 0.1931: the probability of an empty quadrat, that is when $k = 0$, is obtained (where 0! is defined as 1) with

$$P(k) = \frac{e^{-2.6}2.6^0}{0!} = e^{-2.6} = 0.0743 \tag{8.5}$$

and $P(k)$ when $k = 1$ is obtained with:

$$P(k) = \frac{e^{-2.6}2.6^1}{1!} = e^{-2.6}2.6 = 0.1931 \tag{8.6}$$

Probabilities for larger values of k are obtained in the same way. The probability of finding more than the observed maximum number of points must also be calculated. In the case study of Boots and Getis (55) the maximum observed count was 8. $P(k > 8)$ is obtained with:

$$P(k > 8) = 1 - \sum_{k=0}^{8} P(k) \tag{8.7}$$

In this case $P(k > 8) = 1 - 0.9985 = 0.0015$. Once all the probabilities have been calculated, the next step is to calculate the expected frequencies of quadrats that contain k points. This is achieved by multiplying the relevant probability by the total number of quadrats. In the example of Boots and Getis (55) for empty quadrats this gives $30 \times 0.0743 = 2.23$, while for quadrats with one point it is $30 \times 0.1931 = 5.79$, and the remaining expected frequencies are calculated in the same way.

A simple statistical test to assess how well the observed distribution fits the Poisson distribution is obtained using the variance/mean ratio — the mean is the mean average number of events in a quadrat, and the variance is the mean squared difference between each quadrat count and the mean quadrat count. The variance/mean ratio (VMR) is expected to be 1 if the distribution is Poisson. If the VMR is greater than one, then the point pattern is clustered, whereas if the VMR is less than one, the pattern tends towards being evenly spaced. Lloyd (245) provides a worked example of the VMR.

The index of dispersion test provides one approach for testing for CSR. A test can be conducted following the idea (as considered above) that if counts in quadrats follow a Poisson distribution the mean and variance of the counts would be expected to be equal. The test statistic is given as:

$$X^2 = \frac{(m-1)s^2}{\mu_k} = \frac{\sum_{i=1}^{m}(k_i - \mu_k)^2}{\mu_k} \tag{8.8}$$

where $k_1, ..., k_m$ are the number of events in m quadrats, μ_k is the mean of the counts, and s^2 is their variance. Under CSR, the theoretical probability distribution of X^2 approximates χ^2_{m-1} well, assuming $m > 6$ and $\mu_k > 1$. The observed values can be compared with the percentage points of the χ^2_{m-1} distribution. Significantly large values correspond to point patterns that are clustered departures from CSR, while small values indicate regular departures from CSR. s^2/μ_k is referred to as the index of dispersion, while $(s^2/\mu_k) - 1$ is termed the index of cluster size (ICS) (32).

Another approach is to apply a test of spatial autocorrelation for quadrat counts. That is, a CSR point pattern would be expected to exhibit no spatial autocorrelation, so it provides a useful means of comparing the observed point pattern to a CSR pattern (245). Tests of spatial autocorrelation are discussed in Section 4.3.1.

The following section is concerned with global measures of second-order properties, and it details nearest neighbour methods and the K function.

8.7 Distance methods

The second-order properties of a point pattern are described using measures based on distances between events. The nearest neighbour is an important concept in this context, and a variety of measures are based on the nearest neighbouring event to some location. The measures discussed in this section are global summaries, but they are locally derived. Local variants of some of these measures have been developed and, in the case of the K function, such an approach is summarised in Section 8.12.

8.7.1   Nearest neighbour methods

The distance between event $\mathbf{s}_i$ and event $\mathbf{s}_j$ is obtained using Pythagoras' theorem:

$$d(\mathbf{s}_i, \mathbf{s}_j) = \sqrt{(x_i - x_j)^2 + (y_i - y_j)^2} \qquad (8.9)$$

The mean nearest neighbour distance is given by:

$$\overline{d}_{min} = \frac{\sum_{i=1}^{n} d_{min}(\mathbf{s}_i)}{n} \qquad (8.10)$$

where d_{min} is the distance from the nearest event, and there are n events.

There are several approaches that enable exploration of events separated by different distances. Event-event nearest neighbour distances can be summarised using the G function. G, the cumulative frequency distribution of the nearest neighbour distances, is obtained with:

$$G(d) = \frac{\#(d_{min}(\mathbf{s}_i) < d)}{n} \qquad (8.11)$$

where, as before, $\#$ means 'the number of.' G for distance d indicates the proportion of nearest neighbour distances in the point pattern that are less than d. Where events are clustered, G rises rapidly at short distances. Where the events are distributed evenly, there is a gradual increase in G, which then increases rapidly at the distance by which most events are separated.

Point-event distances can be summarised using the F function, whereby distances are computed between randomly selected locations and the nearest-neighbour events. The F function is the cumulative frequency distribution of nearest neighbours for the m randomly selected locations:

$$F(d) = \frac{\#(d_{min}(\mathbf{s}_i, X) < d)}{m} \qquad (8.12)$$

where $d_{min}(\mathbf{s}_i)$ is the minimum distance from location $\mathbf{s}_i$ in the randomly selected set of locations to the nearest event in the point pattern, X (304), and there are m random points sampled. F and G can be modified such that values close to the boundary of the region are effectively ignored and edge effects are, therefore, minimised.

One advantage of F over G is the capacity to increase the sample size, which has the effect of smoothing the cumulative frequency distribution curve. G and F can provide different information — for example, a small cluster leads to a rapid rise in G, but most of the study area may be empty, so F may rise slowly, while the reverse is true for regular point patterns.

The mean nearest neighbour distribution can be tested for conformance with CSR using the R statistic of Clark and Evans (86). The expected value for the mean nearest neighbourhood distance is:

$$\mathrm{E}(d) = \frac{1}{2\sqrt{\lambda}} \tag{8.13}$$

The ratio R of the observed mean nearest neighbour distance to this value may be used to assess the point pattern relative to CSR.

$$R(d) = \frac{\overline{d}_{min}}{1/2\sqrt{\lambda}} = 2\overline{d}_{min}\sqrt{\lambda} \tag{8.14}$$

R of less than 1 indicates a tendency towards clustering while R of more than 1 indicates a tendency towards evenly-spaced events.

Expected values of G and F under CSR are given with:

$$\mathrm{E}[G(d)] = 1 - e^{-\lambda \pi d^2} \tag{8.15}$$

$$\mathrm{E}[F(d)] = 1 - e^{-\lambda \pi d^2} \tag{8.16}$$

The expected value is the same for both functions; for a clustered pattern G and F may lie on opposite sides of the curve, representing expected values (304).

8.7.2 The K function

Like the methods discussed above, the K function is normally used as a global summary, but a local variant is outlined in Section 8.12. G and F are based on the nearest neighbour for each event. In contrast, the K function (the reduced second moment measure) uses distances between all events in the study area. The K function for distance d is given as:

$$K(d) = \frac{\#(C(\mathbf{s}, d))}{\lambda} \tag{8.17}$$

where $\#(C(\mathbf{s}, d))$ is, as before, the number of events in the circle $C(\mathbf{s}, d)$, with radius d centred on location $\mathbf{s}$, and λ is the average intensity of the process. In essence, the K function measures the degree of spatial dependence in the arrangement of events (134). Generally, spatial dependence is examined over small spatial scales, as homogeneity and isotropy are assumed over the scale of analysis. The K function can be obtained by (i) finding all points within radius d of an event, (ii) counting these and calculating the mean count for all events (that is, count all events within d at all event locations) after which the mean count is divided by the overall study area event intensity which gives $K(d)$, (iii) increasing the radius by some fixed step, and (iv) repeating steps (i), (ii), and (iii) to the maximum desired value of d:

$$\hat{K}(d) = \frac{|A|}{n^2} \sum_{i=1}^{n} \#(C(\mathbf{s}_i, d)) \tag{8.18}$$

where $|A|$ is the area of zone A. In other words, the estimation of the K function could be summarised as follows (where the order of procedures differs to that in the description above):

1. An event is visited,

2. around the event a set of concentric circles is constructed with a small spacing,

3. the cumulative number of events in each distance band is counted,

4. the other events are visited in the same way,

5. the cumulative number of events in distance bands around all events (to the radius d) are counted and scaled by $|A|/n^2$. This is the estimate of $K(d)$.

To account for edge effects, counts of events centred around events that are less than distance d from the boundary may be adjusted. The count is divided by w_i, the proportion of $C(\mathbf{s}_i, d)$ that is located in the study area. This gives (128):

$$\hat{K}(d) = \frac{|A|}{n^2} \sum_{i=1}^{n} \frac{\#(C(\mathbf{s}_i, d))}{w_i} \tag{8.19}$$

The K function can be used to characterise scales of point patterns. That is, it may be possible to distinguish separation of individual events and separation of clusters. For a clustered point process there will be more events than expected at short distances. The use of the K function is illustrated below.

As stated above, the definition of the study area may affect results markedly. One way of minimising these effects is to use a guard zone. Points within the inner boundary of the guard zone are considered a part of the point pattern for all purposes. Points within the guard zone are used in the determination of interevent distances for the G and K functions, or point to event distances for F, but they are not considered as a part of the point pattern.

The K function describes the average number of events in a circle with radius d which centres on a particular event. For CSR, the probability of occurrence of an event at any location in the study region is independent of other events, and it is equally likely across the whole study area (134). The area of each circle is given by πd^2, and λ is the mean intensity of events per unit area. So, the expected value of $K(d)$ is given by:

$$\mathrm{E}[K(d)] = \frac{\lambda \pi d^2}{\lambda} = \pi d^2 \tag{8.20}$$

If a point pattern is clustered $K(d) > \pi d^2$, while if it is regular $K(d) < \pi d^2$ (134). Since $K(d)$ is based on squared distances, it can become very large for large values of d. To see small differences, the expected value of $K(d)$ can be converted to zero. With one approach, $K(d)$ is divided by π, the square root is taken and finally d is subtracted. The transformed values of $K(d)$ should be close to zero if the pattern conforms to CSR. This function is termed the L function:

$$\hat{L}(d) = \sqrt{\frac{\hat{K}(d)}{\pi}} - d \qquad (8.21)$$

$\hat{L}(d)$ is plotted against d. For a given value of d, the expected value is zero. Where $\hat{L}(d)$ is above zero this indicates clustering, if it is less than zero this indicates regularity. Figure 8.1 shows two synthetic point patterns and their corresponding K function and L function.

The K function is illustrated through the analysis of the granite data described in Chapter 1. As detailed in Chapter 1, the boundary of the region is an arbitrarily-selected rectangle. The analysis was conducted using the S+ SpatialStats™ software (214). The K function for the distribution of mafic minerals is given in Figure 8.2. It is expected that $K(d) = \pi d^2$ where a distribution is homogeneous with no spatial dependence (32). Figure 8.2 shows as a solid line the expected value of K under a CSR process. The expected values are very close to the observed values (indicated by circles), so the K function does not indicate marked spatial dependence. In other words, the K function is close to what would be expected for a CSR process. This does not mean that the point process was generated randomly, but rather that its pattern may be regarded as similar to that of a CSR process (55). While the K function suggests a lack of spatial structure in the point pattern, this may obscure local variation (see Section 8.12).

$\hat{L}(d)$ provides a way of relating the K function to CSR. However, conducting a test of significance necessitates knowing about the sampling distribution of $\hat{L}(d)$ and $\hat{K}(d)$ under CSR. But, as Bailey and Gatrell (32) state, due to the edge correction built into $\hat{K}(d)$ this is unknown and complex. One way of overcoming such difficulties is to use a simulation approach. A Monte Carlo simulation approach can be used to generate multiple realisations that are subject to the same edge effects as the observed data. Therefore, the sampling distribution obtained accounts for edge effects. Bailey and Gatrell (32) outline an approach by which simulations are generated under the hypothesis of CSR. The upper and lower simulation envelopes are obtained and placed on the plot of $\hat{L}(d)$ against d, to allow assessment of departures from CSR. Waller and Gotway (375) detail an application of this approach for the analysis of medieval grave site locations. Using the example of disease events, it is possible to use the at risk population (see Section 8.11) to simulate point patterns and then this can be compared to the observed disease data.

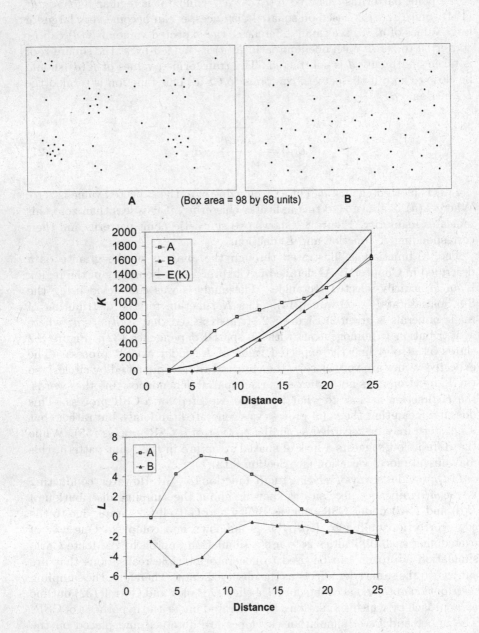

FIGURE 8.1: Point patterns with corresponding K function and L function.

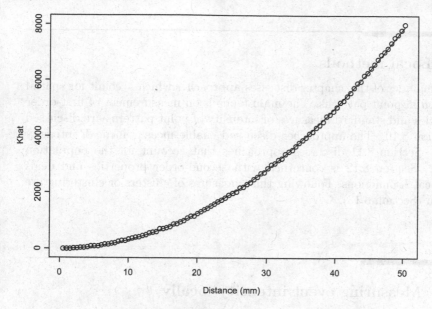

FIGURE 8.2: K function: mafic mineral locations. The expected value of K under a CSR process: solid line; observed values: circles.

8.8 Other issues

There are various other issues which are outside the scope of this introduction to point pattern analysis. In a case where there are two separate sets of events, cross functions can be computed that relate distances from each event in the first pattern to events in the second pattern. The cross K function is defined by Bailey and Gatrell (32). The cross K function is obtained using counts of the number of events in one pattern within different distances of each event in another pattern. If the concern is to assess when in time events were clustered, one approach that can be used is to compute the D function (110). This is a modified form of the K function, and it comprises two K functions that are computed based on the number of events within distance d and time t of an arbitrary event.

The main focus in this chapter is on point events which have no attached attributes. A point process with attached attributes is termed a marked point process and can be distinguished from geostatistical data in that in the case of the latter the domain is continuous and the data are a sample. In contrast, the marked point pattern represents all events (330). Schabenberger and Gotway (330) provide an introduction to marked point patterns.

8.9   Local methods

The remainder of this chapter discusses approaches which account for spatial variation in point patterns. The main focus is on measurement of first-order properties and standard measures of intensity of point patterns are discussed in Section 8.10. The approaches discussed enable measurement of intensity locally. Section 8.11 discusses approaches that account for the population at risk. Section 8.12 is concerned with second-order properties and deals with local K functions. Following this, measures of clusters or clustering are detailed (Section 8.13).

8.10   Measuring event intensity locally

This section details methods for exploring local variations in point pattern intensity. The specific concern is with kernel estimation. Intensity measures are extremely sensitive to the definition of the study area, and edge effects are an important consideration. There are several ways of dealing with this problem and some relevant approaches are outlined below.

8.10.1   Quadrat count methods

The concept of quadrat count methods was outlined in Section 8.6.1. In that section, the focus was on global summaries. However, counts in quadrats allow, of course, for consideration of local variations in event intensity. Methods like local Moran's I (see Section 4.4) could be used to assess clustering in quadrat cell counts.

8.10.2   Kernel estimation

Whereas quadrat counts provide local estimates of the intensity of a point pattern, kernel estimation allows estimation of intensity at all locations and not just at those locations where there is an event. For a circular kernel the intensity estimate at point $\mathbf{s}$ is given by:

$$\hat{\lambda}(\mathbf{s}) = \frac{\#(C(\mathbf{s}, d))}{\pi d^2} \qquad (8.22)$$

where, following previous definitions, $\#(C(\mathbf{s}, d))$ is the number of events in the circle $C(\mathbf{s}, d)$ with radius d centred on location $\mathbf{s}$. This is sometimes called the naive estimator. The selection of d is important as it may affect markedly the intensity surface. For example, if d is very large in relation to the study area then estimates may be similar to the average intensity for the point pattern as a whole. The mapped intensity estimates provide a useful means of identifying visually 'hot spots.'

A more sophisticated approach than the naive estimator is to use a spatial kernel that weights events as a function of their distance from the kernel centre (see Section 2.2). Conventionally, the size of the kernel is fixed at all locations, but alternative approaches exist, as discussed below. If the kernel function is selected appropriately then KE can be used to derive a surface that encloses a volume equal to the total number of events in the pattern. This is important as otherwise the intensity surface may indicate that there are more events than exist in reality. The KE of intensity is given as:

$$\hat{\lambda}_k(\mathbf{s}) = \sum_{i=1}^{n} \frac{1}{\tau^2} k\left(\frac{\mathbf{s} - \mathbf{s}_i}{\tau}\right) \tag{8.23}$$

where $k()$ is the kernel function, with $i = 1, ..., n$ events with location $\mathbf{s}_i$ around location $\mathbf{s}$, and $\tau > 0$ is the bandwidth which determines the width of the kernel and, therefore, the amount of smoothing. The individual intensity estimates can be displayed as a raster grid. Adapting the kernel and bandwidth allows exploration of changes in intensity across A.

KEs near to the boundary of the study area tend to be distorted as there can be no neighbours outside of the boundary. One way to deal with this problem is to use a guard zone near the boundary. The points in the guard zone are used in kernel estimation at locations outside the guard zone, but estimates are not conducted within the guard zone. An alternative approach is to add an edge correction factor (32):

$$\delta_\tau(\mathbf{s}) = \int_A \frac{1}{\tau^2} k\left(\frac{\mathbf{s} - \mathbf{u}}{\tau}\right) d\mathbf{u} \tag{8.24}$$

This is the volume under the scaled kernel centred on $\mathbf{s}$ which lies in the region of interest, A. The KE including this edge correction factor is then:

$$\hat{\lambda}_k(\mathbf{s}) = \frac{1}{\delta_\tau(\mathbf{s})} \sum_{i=1}^{n} \frac{1}{\tau^2} k\left(\frac{\mathbf{s} - \mathbf{s}_i}{\tau}\right) \tag{8.25}$$

A range of different kernel forms have been employed. Previous analyses indicate that similar results are likely to be achieved for different kernel forms if the bandwidth is constant. A common form of kernel is the quartic kernel (32):

$$k(\mathbf{u}) = \begin{cases} \frac{3}{\pi}(1 - \mathbf{u}^T\mathbf{u})^2 & \text{for } \mathbf{u}^T\mathbf{u} \leq 1 \\ 0 & \text{otherwise} \end{cases} \tag{8.26}$$

where $\mathbf{u}$ is d_i/τ, and d_i is the distance from the centre of the kernel. The quartic kernel is illustrated in Figure 8.3, and it was also shown in Figure 2.5

where it was superimposed on a hypothetical study area. Using the quartic kernel, the KE (without the edge correction factor) is given as:

$$\hat{\lambda}_k(\mathbf{s}) = \sum_{d_i \leq \tau} \frac{3}{\pi\tau^2} \left(1 - \frac{d_i^2}{\tau^2}\right)^2 \tag{8.27}$$

where d_i is the distance between the point $\mathbf{s}$, and the event location $\mathbf{s}_i$ and summation takes place only for distances d_i that are less than or equal to the bandwidth, τ. At location $\mathbf{s}$ the weight is $3/\pi\tau^2$, while it is zero at a distance of τ. For large values of τ a smooth map will be generated, while for small values of τ then the output will appear as a series of spikes centred on $\mathbf{s}$ (32). Waller and Gotway (375) include a graphic illustration of kernel intensity estimates for the same set of locations, but for different bandwidths.

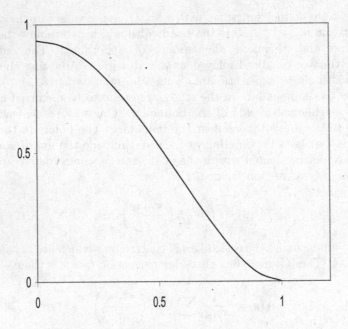

FIGURE 8.3: Section through the quartic kernel.

KE is illustrated using the granite data described in Chapter 1. The intensity of the point pattern was estimated at grid cells of 0.2 mm by 0.2 mm using ArcGIS™ Spatial Analyst. The kernel used by the software follows Silverman (333, p. 76, Equation 4.5 — i.e., the same form as Equation 8.26).

The intensity for mafic minerals (that is, mafic minerals per unit square) is shown in Figure 8.4 (page 260) for bandwidths of 2.5, 5, 7.5, and 10 mm. There are, as expected, clear concentrations of mafic minerals locally.

Distinguishing between first-order and second-order effects is, as noted in this chapter, problematic. The K function for this point pattern was given in Figure 8.2. While derived locally, it is a global summary and as such it is not surprising that it indicates a lack of structure (clustering or regularity) globally. An expanded analysis could be used to explore hypotheses such as the assumption that quartz and feldspar tend to cooccur.

There are a variety of ways of identifying a suitable bandwidth. Examination of $\hat{\lambda}_k(\mathbf{s})$ for different values of τ is a common approach. A value of $\tau = 0.68n^{-0.2}$ has been suggested for the unit square A with n events in A (32). Another approach to informing bandwidth selection is based on the mean integrated squared error (MISE) (128), (375). The standard approach to KE is to use the same bandwidth locally. An alternative approach is to adapt the bandwidth locally. Adaptive kernel estimation provides a means to adapt to spatial variation in point pattern intensity (32), (128). For example, where there is a dense point pattern locally, a small bandwidth is sensible as it prevents excessive smoothing of fine scale spatial variation (32). One adaptive kernel approach is given by Brunsdon (62). Allowing τ to vary, with no edge correction, gives:

$$\hat{\lambda}_k(\mathbf{s}) = \sum_{i=1}^{n} \frac{1}{\tau^2(\mathbf{s}_i)} k\left(\frac{(\mathbf{s} - \mathbf{s}_i)}{\tau(\mathbf{s}_i)}\right) \tag{8.28}$$

The adaptive bandwidth $\tau(\mathbf{s}_i)$ can be specified in various ways. One way of doing this is to make intensity estimates using a non-adaptive bandwidth which gives a pilot estimate, $\widetilde{\lambda}(\mathbf{s})$. Then the geometric mean $\widetilde{\lambda}_g$ of the pilot estimates is obtained for each location $\mathbf{s}_i$. The adaptive bandwidths are then given as (32):

$$\tau(\mathbf{s}_i) = \tau_0 \left(\frac{\widetilde{\lambda}_g}{\widetilde{\lambda}(\mathbf{s}_i)}\right)^r \tag{8.29}$$

The sensitivity parameter $0 \leq r \leq 1$ determines the degree of local variation in the bandwidth, where a value of 0 represents no local adjustment, and 1 represents maximum adjustment (32). Brunsdon (62) illustrates adaptive bandwidth kernel estimation using the widely-applied California Redwood seedling location dataset (see also Diggle (109)). The case study highlights benefits of the adaptive bandwidth approach including the presence of sharper edges between densely clustered and more sparsely clustered areas than is the case for the fixed bandwidth approach.

In some contexts, comparison of KEs is of interest. In an epidemiological application, for example, one set of estimates may refer to cases, and the other set to controls (134). KE is a standard means of exploring spatial variations in event intensity and, in the context of Geographical Information Systems, is more widely used than quadrat counts. The ArcGIS™ software, for example, includes functions for KE.

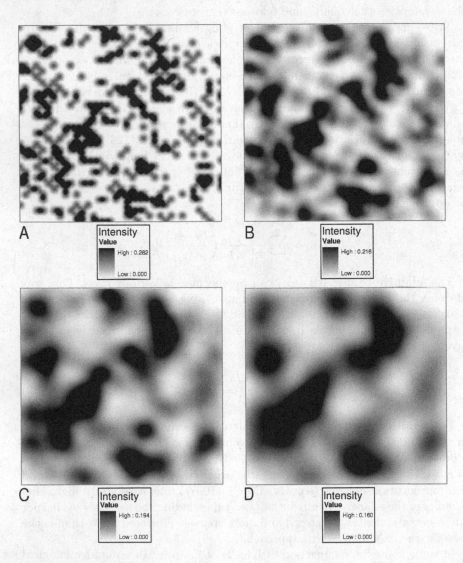

FIGURE 8.4: Mapped intensity of mafic minerals, bandwidth of: A. 2.5 mm, B. 5 mm, C. 7.5 mm, D. 10 mm.

8.11 Accounting for the population at risk

In many cases, the hypothesis of CSR will not be sensible since there is spatial variation in the expected intensity of the process (32). For example, there will tend to be more cases of a disease in a city than in a village since there are likely to be more people in a city than in a village. This has been termed environmental heterogeneity (55). If we have information about spatial variation in the population at risk, for example population counts for specific administrative areas, it is possible to make an estimate of events per unit population, $\hat{\rho}_\tau(\mathbf{s})$. If the population values, y_j associated with point location $\mathbf{s}'_j$, are located at centroids (or population centres) of areas, we can make an estimate of population intensity, $\hat{\lambda}'_\tau(\mathbf{s})$, with (32):

$$\hat{\lambda}'_k(\mathbf{s}) = \sum_{j=1}^{m} \frac{1}{\tau^2} k\left(\frac{(\mathbf{s} - \mathbf{s}'_j)}{\tau}\right) y_j \qquad (8.30)$$

for population values at m locations around $\mathbf{s}$. This allows an estimate based on the ratio of the kernel estimates for event intensity and population intensity:

$$\hat{\rho}_k(\mathbf{s}) = \frac{\sum_{i=1}^{n} k\left(\frac{(\mathbf{s}-\mathbf{s}_i)}{\tau}\right)}{\sum_{j=1}^{m} k\left(\frac{(\mathbf{s}-\mathbf{s}'_j)}{\tau}\right) y_j} \qquad (8.31)$$

However, interpolating population values from zone centroids is problematic. One way around this problem is to use a second spatial process (a control process) which is considered to represent population variations. The use of such a control process is described by Bailey and Gatrell (32). The idea of reallocating population values across areas links to Section 6.4, which discusses a variety of approaches to areal interpolation.

8.12 The local K function

A local variant of the K function is described by Getis (140) and Getis and Franklin (143). The local K function, $K_i(d)$, is similar to the global K function described above, but only pairs of points that have a given point i as one of the members of the pair are included. Using the same notation as the global K function (Equation 8.19), for point i it can be given by:

$$\hat{K}_i(d) = \frac{|A|}{n} \frac{\#(C(\mathbf{s}_i, d))}{w_i} \tag{8.32}$$

where counts are of all points within distance d of point i. As before, $|A|$ is the area of the study area A. Getis and Franklin (143), set the denominator on the left hand side of the equation to $n-1$. The local L function can be given by:

$$\hat{L}_i(d) = \sqrt{\frac{\hat{K}_i(d)}{\pi}} - d \tag{8.33}$$

Figure 8.5 shows unstandardised local K, for point pattern A of Figure 8.1, given a radius of 5 units. In Figure 8.6 (page 263) local L is given for (A) 5 units, and (B) 12.5 units. Note how some of the points at the edges of clusters move from negative values in (A) to positive values in (B).

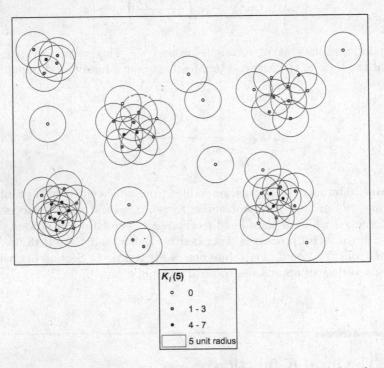

FIGURE 8.5: Point pattern with radii of 5 units superimposed and unstandardised K for each point.

An example follows using the mafic mineral point data described in Chapter 1. The area covered by the point data is 5040 square mm, and there are

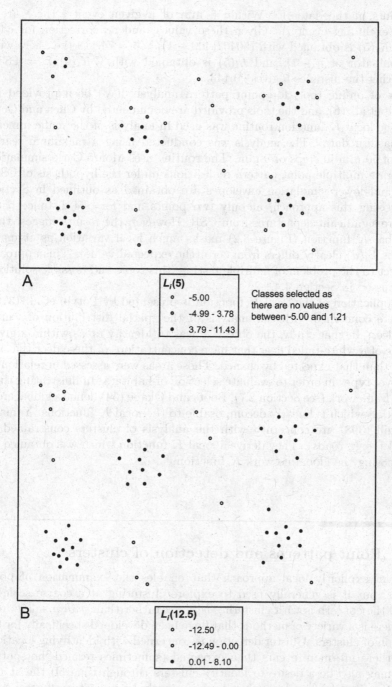

A

$L_i(5)$

○	-5.00
◐	-4.99 - 3.78
●	3.79 - 11.43

Classes selected as there are no values between -5.00 and 1.21

B

$L_i(12.5)$

○	-12.50
◐	-12.49 - 0.00
●	0.01 - 8.10

FIGURE 8.6: L_i for radii of (A) 5 units and (B) 12.5 units.

409 points in the dataset. Within 5 mm of a given event i there are 6 points (excluding event i). Given these values, and no correction for edge effects, $\hat{K}_i(5)$ is obtained with $(5040/(409-1)) \times 6 = 74.118$ (i.e., here with a denominator of $n-1$), and $\hat{L}_i(5)$ is obtained with $\sqrt{K_i(5)/\pi} = 4.857$; subtracting the distance 5 gives –0.143.

A set of online tools for point pattern analysis have been provided by Aldstadt et al. (6), and the tools provided are documented by Chen and Getis (79). The local K function routine was used in analysis of the mafic mineral point pattern data. The analysis was conducted using a maximum search radius of 50 mm in steps of 5 mm. The routine uses Monte Carlo simulation to generate multiple point pattern realisations under the hypothesis of CSR. Upper and lower simulation envelopes are obtained as outlined in Section 8.7.2. Using this approach, at only two points and for one distance band are there significant departures from CSR. However, the results suggest that the global K function (Figure 8.2) masks much local variation as at many locations $\hat{L}_i(d)$ clearly differs from zero (the expected value). This approach is subject to the problem of multiple testing; this topic and possible solutions are discussed in Section 4.4.1.2.

An application of the local K function is presented by Potvin et al. (318), who were concerned with the analysis of the spatial distribution of white-tailed deer. In that study, the objective was to identify areas within a given distance of a white-tailed deer that have concentrations of these deer that are greater than that expected by chance. These areas were assessed in relation to land cover types in order to evaluate selection of habitats. In illustrating their LoSSA framework (see Section 2.7), Boots and Okabe (54) define a global cross K function which is linearly decomposable to the local K functions. Yamada and Thill (398) are concerned with the analysis of clusters constrained to networks (e.g., roads). They derive a local K function which was obtained by decomposing the global network K function.

8.13   Point patterns and detection of clusters

KE is an explicitly local approach that enables the examination of point clusters, but it is generally used to explore clustering at a coarse scale — that is, the concern is with clustering *between* rather than *within* small areas (7). There is a variety of methods that have been developed specifically for the detection of clusters. Cluster detection is concerned with identifying locations where there are more events than expected (sometimes termed 'hotspots'). There may also be a desire to identify clusters through time. If the at risk population is unknown, local clusters may simply be indicative of areas with a high population intensity. If the at risk population is known, then it is possible to distinguish meaningful clusters. In this case, local point pattern analysis is concerned with identifying samples that are clustered relative to the population (131). The theme of spatial clustering analysis is introduced by Aldstadt (5).

In many instances, traditional point pattern analysis, as discussed previously in this section, may not be suitable. Openshaw (302) stated that there was a need for automated approaches that would enable users to extract the maximum possible information from a growing range of large datasets. Such approaches, Openshaw et al. (302) noted, were made feasible by increases in the computational power of computers.

The need for methods for automated local analysis of point patterns has led to the development of a range of methods. Alexander and Boyle (8) includes chapters which outline a variety of ways of investigating clusters in incidences of cancers. Other comparisons of methods are provided by Hill et al. (186) and by Waller and Gotway (375).

Methods may be divided into those designed for testing for clustering and those designed for identifying clusters (7), (40). Another distinction is between methods based on quadrat counts and methods based on distances. Also, tests may be either focused or general in nature (40). That is, focused tests concentrate on regions around one or more locations that are thought *a priori* to be linked to, for example, disease incidence. General tests are applied to all regions in the dataset with no such prior considerations. Distance-based methods include those of Openshaw et al. (302), Besag and Newell (40), and Cuzick and Edwards (100) and such approaches are the main focus here.

The geographical analysis machine (GAM) was developed by Openshaw et al. to identify areas which deviate from a null hypothesis (302). That is, the concern was to identify areas where there were more observed points within a specific radius of a given location than expected. A test statistic was then applied to assess the significance of a cluster.

The core components of the GAM are four-part:

1. A spatial hypothesis generator.

2. A procedure for assessing significance.

3. A GISystem for retrieval of spatial data.

4. A means of displaying the subregions with anomalous point patterns.

In (1) the concern is to ascertain if there is an excess of events within distance d of a particular location. Part (1) of the GAM operates as follows:

1. Set up a grid over the area of interest. Define the minimum, maximum, and the increment value of radii of the circles that will be located at the intersections of the grid. The grid intersection spacing and the radii of the circles should be such that the circles overlap.

2. Move the circle so that it is located at each grid intersection systematically. Compute the test statistic for each circle. If the test statistic passes the significance test, then the circle is tagged for visualisation.

3. Increase the circle radius by the predefined increment.

4. Repeat steps 2 and 3 until the radius reaches the maximum value.

For each (circular) subregion a test statistic is computed to ascertain whether the circle contains a statistically significant excess of events. Monte Carlo simulation was used by Openshaw et al. (302) to assess the significance of clusters. A count of events within a circle is made, and this can be compared to the number expected under a particular null hypothesis (e.g., comparison of the observed point pattern with what would be expected if the pattern was generated by a Poisson process). The Monte Carlo procedure entails ranking the value of the observed test statistic, u_1, amongst a set of $n - 1$ values generated by sampling randomly from the null distribution of u. Under the null hypothesis each of the n possible rankings of u_1 is equally likely. So, the significance level of the test is obtained by determining the rank of u_1 amongst the set of values $u_i, i = 1, ..., n$. Openshaw et al. (302) state that a set of 100 simulated values is sufficient, although more simulations are required if significance levels of lower than 0.01 are needed. The key advantage in using the Monte Carlo procedure is that it allows the use of any test statistic and null hypothesis, and it is not constrained by known distribution theory.

The circles with a greater than expected number of points are mapped. In other words, all locations where the null hypothesis breaks down are visualised. Openshaw et al. (302) considered that a genuine cluster will persist across a range of scales while, in contrast, spurious clusters are likely to exist at a more limited range of scales.

Openshaw et al. (302) illustrated the application of the GAM by analysing the spatial distribution of two forms of cancer in the Newcastle and Manchester cancer registries. The two types of cancer were acute lymphoblastic leukaemia and Wilms' tumour; the former is believed to cluster, while the latter shows little evidence of clustering. The analysis using GAM confirmed, to some degree, these expectations, in that there was evidence of intense clustering of acute lymphoblastic leukaemia but much less obvious clusters of Wilms' tumour. In the case of acute lymphoblastic leukaemia, the presence of an unexpected large cluster, focused on Gateshead in Tyneside, provided grounds for further research.

Besag and Newell (40) highlight a number of criticisms that had been made of the GAM of Openshaw et al. (302). The most serious criticism concerned the lack of control for multiple testing locally — the change in radii and location of circles is not taken into account in calculation of significance levels. Besag and Newell (40) present an approach which, like the GAM, is intended to help detect possible clusters of a rare disease. However, Openshaw (301) notes that the method of Besag and Newell suffers from similar problems to the GAM. Also, Openshaw (301) has argued that the GAM was intended to identify areas for further analysis, rather than as a statistical test to identify clusters in themselves.

The approach of Besag and Newell (40) is presented through a case study which is based on enumeration district (ED) centroids, to which disease cases have been assigned with the population at risk, estimated from census population figures. Note that, therefore, the technique, unlike most of those considered in this chapter, is not based on raw point event data (but, rather, events within zones). Each centroid then visited and a significance test is used to ascertain if the centroid forms the centre of a cluster of a given size $k + 1$. The null hypothesis usually states that the observed total number of cases, n, is distributed at random amongst the population at risk. That is, any given case occurs in zone i with probability t_i/N where t_i is the corresponding population and N is the total population at risk.

A case is selected and the label A_0 is assigned to the zone (zone centroids in the study by Besag and Newell (40)) within which it occurs. The cases around A_0 are labelled $A_i, i = 1, ...$ sequentially according to their distance from A_0. Then y_i is the observed number of cases in zone i, and t_i is the population at risk in zone i. The accumulated number of other cases in $A_0, A_1, A_2... (D_i)$ and the corresponding accumulated number of the population at risk (u_i) are given respectively by:

$$D_i = \left(\sum_{j=0}^{i} y_j \right) - 1 \tag{8.34}$$

$$u_i = \left(\sum_{j=0}^{i} t_j \right) - 1 \tag{8.35}$$

and let

$$M = \min(i : D_i \geq k) \tag{8.36}$$

where k is a preselected number of observed cases, and M is such that zones $A_0, ..., A_m$ contain at least k other cases. The observed number of zones to contain k cases is given by m. When m is small this indicates a cluster around A_0. The probability of there being exactly y cases of the disease among u_m individuals (the population at risk) is approximated by:

$$\frac{e^{-\lambda} \lambda^y}{y!} \quad \text{for} \quad y = 1, 2, 3... \tag{8.37}$$

where $\lambda = \overline{P} \times u_m$, $\overline{P}$ is n/N (n is the total observed cases (sum of ys), and N, as defined previously, is the total population at risk (sum of ts)). If m is the observed value of M, the significance level of the test is $P(M \leq m)$:

$$P(M \le m) = 1 - P(M > m) = 1 - \sum_{y=0}^{k-1} \frac{e^{-\lambda}\lambda^y}{y!} \qquad (8.38)$$

The circles with a particular significance level (e.g., 0.05) may then be mapped and used as a diagnostic tool. The procedure is illustrated with an example using the North Carolina sudden infant death syndrome (SIDS) data used by Symons et al. (355) and described by Bivand (45). The R package DCluster (48), (148) allows the application of the method of Besag and Newell (in addition to other methods for detecting clusters) and it was applied in this example. Data on births in 1974 (BIR74) and SIDS events in 1974 (SID74) were utilised. The sum of BIR74 (N) was 329,962, and the sum of SID74 (n) was 667. Given a value of $k = 20$, five locations were identified as the centres of clusters. As an example, at one of these five locations the accumulated SID74 values for the four nearest zones were required to exceed 20 cases. In this example, $\overline{P} = n/N = 667/329,962 = 0.002021$. For the selected location, u_m (the accumulated population at risk) = 4617. Therefore, $\lambda = 0.002021 \times 4617 = 9.333011$. The significance level is then obtained with Equation 8.38 with summation over $y = 0, ..., k - 1$, thus $y = 0, ..., 19$ in this case. For $y = 0$, this leads to:

$$\frac{e^{-9.333011} - 9.333011^0}{0!} = 0.000088455 \qquad (8.39)$$

The sum of probabilities (for $y = 0, ..., 19$) is 0.001605653. As Fotheringham and Zhan (131) note, k is selected in an ad hoc manner, and generating results for different values of k is desirable to assess persistence of clusters. The method of Besag and Newell is also implemented in the ClusterSeer® software*.

Fotheringham and Zhan (131) present a further approach for detecting point clusters, and they compare results obtained using (i) the original GAM, (ii) the method of Besag and Newell (40), and (iii) their own approach. The Fotheringham and Zhan (131) method differs from the original GAM in two ways: rather than using a regular grid, the location and sizes of circles are determined randomly within pre-defined ranges, and the significance of clusters is assessed directly using the Poisson probability distribution. Fotheringham and Zhan (131) note that the application of Besag and Newell (40) is based on aggregated data (i.e., numbers of events within zones), whereas their approach makes use of disaggregated data (the original point events).

If the total population at risk in the region of interest is N, and the total number of observed cases with a particular attribute is n, then the mean probability of observing a case in the whole area is as defined above:

*www.terraseer.com/products_clusterseer.php

$$\overline{P} = \frac{n}{N} \tag{8.40}$$

For a circle with a randomly-determined location and radius, the number of cases (y) and the population at risk (t) within the circle can be extracted. The expected number of cases in the circle, λ, is then calculated as:

$$\lambda = \overline{P} \times t \tag{8.41}$$

The probability of observing y cases in a circle with λ expected cases can be determined using the Poisson distribution:

$$P(y, \lambda) = \frac{e^{-\lambda} \lambda^y}{y!} \quad \text{for} \quad y = 1, 2, 3... \tag{8.42}$$

The significance can be measured directly using $P(y, \lambda)$. If $P(y, \lambda) < \rho$, where ρ is a given level of significance, the relevant circle is stored. The method of Besag and Newell (40), as summarised above, can also be used to assess significance.

An alternative to basing the radius size on a fixed number of cases is to base it on the number of persons at risk. The cluster evaluation permutation procedure (CEPP) of Turnbull et al. (366) is based on such an approach. Given a fixed population radius containing t persons, Turnbull et al. are concerned with the maximum number of cases observed in any radius containing t persons, and they seek to test whether the largest observed count is larger than would be anticipated under the constant risk hypothesis. A Monte Carlo test is conducted whereby the maximum observed count is compared to the distribution of maxima from the constant risk simulations (375). The maximum from any simulation may occur at any location and there is a single test, thus avoiding the multiple testing problem (375).

The spatial scan method of Kulldorff (222) utilises a variety of radii. The approach is based on assessing a null hypothesis of constant risk as against alternatives where the (disease) rate with the scanning window is greater than the rate outside the window. Kulldorff's method utilises the Monte Carlo approach proposed by Turnbull et al. (366) (as outlined above), and an estimate of the distribution of the test statistic is obtained in this way. The approach provides a single probability value and, in this sense, is a test of clustering, although there is also the capacity to map the most likely cluster (375). The radii range from the smallest distance between any pair of cases and some maximum such as half the study area extent. The spatial scan method can be implemented using the SaTScan™ software*.

---

*www.satscan.org

8.14   Overview

This chapter has summarised some widely used approaches to analysing spatial point patterns and for identifying spatial clusters. Global summaries were described for context and because they provide sensible first steps in point pattern analysis. Methods like KE are long-established local approaches, while the methods outlined in Section 8.13 provide the means to conduct automated analyses of very large sets of data. While such methods may be of minimal value if applied in isolation, they are a powerful means of identifying local anomalies that can be explored further.

The following chapter summarises the contents of the book and pulls together some of the key themes discussed throughout the text.

9

Summary: Local Models for Spatial Analysis

The methods outlined in this book have been developed in many different disciplines and with a diversity of problems in mind. Also, applications illustrated are wide ranging. Nonetheless, there are many common themes that link these approaches. In the following sections, the contents of the book are reviewed, some key issues are outlined and availability of software is discussed, before considering possible future developments in local modelling.

9.1 Review

This book describes methods which can be applied to solve problems including characterising single properties and multivariate relationships locally for various applications, cluster detection, image compression, and spatial prediction. The substantive chapters concern analyses of single variables on grids (Chapter 3), spatial patterning in single variables represented as points or areas (Chapter 4), multiple variables (Chapter 5), deterministic and thin plate spline approaches to spatial prediction (Chapter 6), geostatistical prediction (Chapter 7), and point patterns (Chapter 8). Chapter 3 discusses methods for the analysis of gridded data (e.g., moving window statistics in general and wavelets in particular). Chapter 4 provides outlines of geographically weighted summary statistics and measures of spatial autocorrelation, while Chapter 5 considers recent developments in techniques such as geographically weighted regression for exploration of spatial relations. Spatial prediction is the subject of Chapters 6 and 7. In Chapter 8, widely used standard methods for point pattern analysis, both global and local, are outlined, in addition to some more specialised methods for the detection of spatial clusters or clustering. In the following section, some key issues raised in the book are summarised.

9.2 Key issues

In the first two chapters of the book, the rationale behind the book was established and some general approaches to dealing with spatial variation were outlined. One issue that was outlined was that, as spatially detailed datasets covering large areas become more widely available, it is increasing likely that global models will be inappropriate. In addition, concern with spatial scale is likely to increase as data become available at a very fine spatial resolution (or small sample spacing). The book has detailed methods that offer potential solutions to both of these problems. Many of the techniques are computed automatically using moving windows (often using a spatial kernel; e.g., kernel estimation, geographically weighted regression, spatial prediction routines like thin plate splines). In addition, window bandwidths can be varied. Thus, a different model could be readily obtained for any location. Also, some methods, such as wavelets and characterisation of spatial variation using the variogram, provide ways of identifying different scales of variation in large datasets. The wavelet transform is by definition local, while the variogram can be estimated for a moving window, and there are various other approaches for estimating the variogram in the nonstationary case. These approaches can be used to identify appropriate scales of observation — for example, the spatial resolution of an image that provides the optimal balance between the amount of data and the information it contains can be identified. In short, there are at least partial answers to the questions that have been raised in this book.

Most of the methods detailed in this book are likely to be used by a wide body of GISystems users. With respect to Chapter 7, for example, techniques like kriging are used widely, but there is still experimentation with, for example, different ways of modelling the nonstationary variogram. In some cases, the practical implications of new developments may, at least in the short term, be minimal for most users of GISystems. But, in addition to outlining techniques that can be readily applied, it is hoped that this book will alert readers to some of the possibilities and encourage exploration of the methods and their application.

9.3 Software

There are many software environments available for the analysis of spatial data. For example, there is a range of free software packages for the analysis of point patterns, for spatial regression, wavelet transforms, and geostatistical prediction, as entering the relevant terms in a web search engine will reveal.

At present, to utilise the full range of kinds of methods detailed in this book it is necessary to use several different software packages. Many packages with extensive functionality are free or available at a low cost (including some of the packages used in case studies presented in this book) so, depending on the tools required, cost may not be a major limiting factor. Many routines for spatial analysis are provided in the R programming language[*], and increasingly it is possible to use one software environment to apply a wide range of tools for the analysis of spatial data. The book by Bivand et al. (48) provides a detailed account of the analysis of spatial data using R, and a shorter discussion about exploratory spatial data analysis is presented by Bivand (47). Popular GISystems now offer tools for the analysis of point patterns, for the measurement of spatial autocorrelation, for spatial filtering of images and for spatial prediction. So, while it is still true that the array of available functions may be limited in many software environments, it is at least possible to find software relatively easily to implement most of the kinds of approaches detailed in this book.

A summary list of software environments and some of the methods they implement is provided in Appendix A. However, software environments change rapidly and, rather than providing a very detailed list of relevant available functions in the text itself, the website associated with this book[†] provides fuller information on software that can be used to implement the methods outlined in the book. At the time of writing, almost all of the methods detailed here can be implemented using routines in the R language. In addition, a large proportion of the methods are available in ArcGIS™ and its associated extensions.

9.4 Future developments

A wide range of tools is now available for the analysis of local variation in spatially-referenced data. The need to provide tools that can be applied without extensive expertise has led to the development of fully automated approaches for processing and analysing large datasets. Automation is necessary to apply particular approaches to large datasets — it is not realistically possible to estimate interactively the parameters of a model at thousands of locations. So, there may be no choice but to apply automated procedures. One downside of the automation of some widely used procedures is that the human element is removed to a large degree, and this perhaps contributes to a perceived reduction in the need to attempt to *understand*

---

[*]www.r-project.org
[†]http://www.crcpress.com/product/isbn/9781439829196

spatial processes. That is, in a modern WYSIWYG (What You See Is What You Get) computer environment, a bewildering array of complex, but easy to apply, methods may be presented to the user. But, if such methods are used blindly, potential benefits may be lost, as users become detached from their data and decision-making is transferred to automated routines. On balance, the availability of methods to a wide user base that overcome many of the limitations of traditional methods, must be considered a positive development.

Extensive research is ongoing, and it is anticipated that many of the limitations of existing approaches will be resolved in the near future. In addition, as more applications are published, the costs and benefits of applying local models in different situations will be understood more fully.

9.5   Summary

Whatever developments take place, the selection of any one (local) model is likely to be on a largely subjective basis, since spatial properties tend to vary at a range of scales and at scales smaller than that at which we are likely to measure them (e.g., it may be considered unlikely that there will be much demand to characterise topography at the molecular level for the majority of users). With increased availability of high spatial resolution data over very large areas, problems such as inappropriateness of global statistics and variations at different spatial scales arise. However, as hopefully this book has shown, geographers, and those in cognate disciplines, now have access to a powerful array of methods which can be used to help tackle the problems that these technological advances bring. The development of such approaches and their use must come hand-in-hand with the recognition that we should balance complexity with understanding — there is no point in using a complex method if the results cannot be interpreted and are not useful. This book was intended to, first, contribute to the appreciation of when such approaches may be suitable and, second, provide guidance as to which methods may be appropriate and how they are applied. Spatial data sources are abundant for most parts of the Earth (and, increasingly, other planets...) and the advent of Global Positioning Systems means that it is easier than ever to collect our own spatial data. It is the hope that this book will guide readers in maximising the potential of such data, and in producing new and innovative outputs.

A

Software

This appendix includes a summary list of selected software environments which can be used to implement most of the methods detailed in the book. Table A.1 details functions available in ArcGIS™ (version 9.3) and R and associated extensions or packages. Note that there is a standalone version of gstat (313).

TABLE A.1
Software and selected methods: ArcGIS™ and R.

Software	Package or extension	Methods
ArcGIS™	Geostatistical Analyst	Interpolation (IDW, TPS, kriging, etc.)
ArcGIS™	Spatial Analyst	Spatial filters, elevation derivatives, kernel estimation, interpolation
ArcGIS™	Spatial Statistics toolbox	Global and local autocorrelation, GWR
R	DCluster	Cluster analysis
R	geoR	Geostatistical methods
R	gstat	Geostatistical methods
R	fields	Thin plate splines
R	spatialkernel	Kernel estimation
R	spdep	Global and local autocorrelation measures
R	spgwr	Geographically weighted statistics including GWR
R	spatstat	Point pattern analysis
R	splancs	Point pattern analysis
R	wavelets	Wavelet analysis

Other packages with related functionality include the free GRASS GIS (with, for example, extensive thin plate spline functionality) and GeoDa™, which includes facilities for the global and local analysis of spatial autocorrelation and for spatial regression. A variety of specialist geostatistical software exists including GSLIB (geostatistical software library). The VESPER software enables estimation and modelling of local variograms. A wide range of Bayesian methods can be implemented in WinBUGS, and these include a Bayesian implementation of GWR. Web addresses for sites providing more details about selected packages are given in Table A.2.

TABLE A.2
Software environments: web addresses.

Software	Web address
ArcGIS™	http://www.esri.com/software/arcgis/index.html
GeoDa™	http://geodacenter.asu.edu/
Gstat	http://www.gstat.org/
GRASS	http://grass.itc.it/
GSLIB	http://www.gslib.com/
R	http://www.r-project.org/
VESPER	http://www.usyd.edu.au/agriculture/acpa/software/vesper.shtml
WinBUGS	http://www.mrc-bsu.cam.ac.uk/bugs/

References

[1] F. Abramovich, T. C. Bailey, and T. Sapatinas. Wavelet analysis and its statistical applications. *The Statistician*, **49**:1–29, 2000.

[2] M. Abramowitz and I. A. Stegun. *Handbook of Mathematical Functions*. Dover Publications, New York, 1964.

[3] P. S. Addison. *The Illustrated Wavelet Transform Handbook: Introductory Theory and Applications in Science, Medicine and Finance*. Institute of Physics Publishing, Bristol, 2002.

[4] J. Albert. *Bayesian Computation with R*. Springer, New York, 2009. Second Edition.

[5] J. Aldstadt. Spatial clustering. In M. M. Fischer and A. Getis, editors, *Handbook of Applied Spatial Analysis: Software Tools, Methods and Applications*, pages 279–300. Springer, Heidelberg, 2010.

[6] J. Aldstadt, D. Chen, and A. Getis. *Point Pattern Analysis. Version 1.0a*. San Diego State University, 2006. http://www.nku.edu/~longa/cgi-bin/cgi-tcl-examples/generic/ppa/ppa.cgi; last accessed 2/10/2009.

[7] F. E. Alexander and P. Boyle. Introduction. In F. E. Alexander and P. Boyle, editors, *Methods for Investigating Localized Clustering of Disease*. International Agency for Research on Cancer, Lyon, 1996. IARC Scientific Publications No. 135.

[8] F. E. Alexander and P. Boyle, editors. *Methods for Investigating Localized Clustering of Disease*. International Agency for Research on Cancer, Lyon, 1996. IARC Scientific Publications No. 135.

[9] L. Anselin. *Spatial Econometrics: Methods and Models*. Kluwer, Dordrecht, 1988.

[10] L. Anselin. Local indicators of spatial association — LISA. *Geographical Analysis*, **27**:93–115, 1995.

[11] L. Anselin. The Moran scatterplot as an ESDA tool to assess local instability in spatial association. In M. M. Fischer, H. Scholten, and D. Unwin, editors, *Spatial Analytical Perspectives on GIS*, pages 111–125. Taylor and Francis, London, 1996.

[12] L. Anselin. *GeoDa™ 0.9 User's Guide*. Center for Spatially Integrated Social Science, Santa Barbara, CA, 2003. http://sal.agecon.uiuc.edu/csiss/geoda.html.

[13] L. Anselin, I. Syabri, and Y. Kho. GeoDa: An introduction to spatial data analysis. *Geographical Analysis*, **38**:5–22, 2006.

[14] L. Anselin, I. Syabri, and Y. Kho. GeoDa: an introduction to spatial data analysis. In M. M. Fischer and A. Getis, editors, *Handbook of Applied Spatial Analysis: Software Tools, Methods and Applications*, pages 73–89. Springer, Heidelberg, 2010.

[15] L. Anselin, I. Syabri, and O. Smirnov. *Visualizing Multivariate Spatial Correlation with Dynamically Linked Windows*. University of California, Santa Barbara, Santa Barbara, 2002. http://www.real.uiuc.edu/d-paper/01/01-t-10.pdf.

[16] M. Armstrong. Improving the estimation and modelling of the variogram. In G. Verly, M. David, A. G. Journel, and A. Marechal, editors, *Geostatistics for Natural Resources Characterization*, pages 1–19. Reidel, Dordrecht, 1984.

[17] M. Armstrong. Problems with universal kriging. *Mathematical Geology*, **16**:101–108, 1984.

[18] M. Armstrong. *Basic Linear Geostatistics*. Springer-Verlag, Berlin, 1998.

[19] M. Armstrong and R. Jabin. Variogram models must be positive definite. *Mathematical Geology*, **13**:455–459, 1981.

[20] R. Assunção and E.A. Reis. A new proposal to adjust Moran's I for population density. *Statistics in Medicine*, **18**:2147–2162, 1999.

[21] B. W. Atkinson and P. A. Smithson. Precipitation. In T. J. Chandler and S. Gregory, editors, *The Climate of the British Isles*, pages 129–182. Longman, London, 1976.

[22] P. M. Atkinson. On estimating measurement error in remotely-sensed images with the variogram. *International Journal of Remote Sensing*, **18**:3075–3084, 1997.

[23] P. M. Atkinson. Geographical information science: geocomputation and nonstationarity. *Progress in Physical Geography*, **25**:111–122, 2001.

[24] P. M. Atkinson. Spatially weighted supervised classification for remote sensing. *International Journal of Applied Earth Observation and Geoinformation*, **5**:277–291, 2004.

[25] P. M. Atkinson and P. Aplin. Spatial variation in land cover and choice of spatial resolution for remote sensing. *International Journal of Remote Sensing*, **25**:3687–3702, 2004.

[26] P. M. Atkinson and C. Jeganathan. Estimating the local small support semivariogram for use in super-resolution mapping. In P. M. Atkinson and C. D. Lloyd, editors, *GeoENV VII: Geostatistics for Environmental Applications*, pages 279–294. Springer, Dordrecht, 2010.

[27] P. M. Atkinson and P. Lewis. Geostatistical classification for remote sensing: an introduction. *Computers and Geosciences*, **26**:361–371, 2000.

[28] P. M. Atkinson and C. D. Lloyd. Geostatistics and spatial interpolation. In A. S. Fotheringham and P. A. Rogerson, editors, *The SAGE Handbook of Spatial Analysis*, pages 159–181. SAGE Publications, London, 2009.

[29] P. M. Atkinson and N. J. Tate. Spatial scale problems and geostatistical solutions: a review. *Professional Geographer*, **52**:607–623, 2000.

[30] P. M. Atkinson, R. Webster, and P. J. Curran. Cokriging with ground-based radiometry. *Remote Sensing of Environment*, **41**:45–60, 1992.

[31] P. M. Atkinson, R. Webster, and P. J. Curran. Cokriging with airborne MSS imagery. *Remote Sensing of Environment*, **50**:335–345, 1994.

[32] T. C. Bailey and A. C. Gatrell. *Interactive Spatial Data Analysis*. Longman Scientific and Technical, Harlow, 1995.

[33] S. Banerjee, B. P. Carlin, and A. E. Gelfand. *Hierarchical Modeling and Analysis for Spatial Data*. Chapman and Hall/CRC Press, Boca Raton, FL, 2004. Monographs on Statistics and Applied Probability 101.

[34] D. M. Bates, M. J. Lindstrom, G. Wahba, and B. S. Yandell. GCVPACK — routines for generalized cross validation. *Communications in Statistics — Simulation and Computation*, **16**:263–297, 1987.

[35] F. Beckers and P. Bogaert. Nonstationarity of the mean and unbiased variogram estimation: extension of the weighted least-squares method. *Mathematical Geology*, **30**:223–240, 1998.

[36] D. A. Belsley, E. Kuh, and R. E. Welsch. *Regression Diagnostics: Identifying Influential Data and Sources of Collinearity*. Wiley, New York, 1980.

[37] Y. Benjamini and Y. Hochberg. Controlling the false discovery rate: A practical and powerful approach to multiple testing. *Journal of the Royal Statistical Society, Series B*, **57**:289–300, 1995.

[38] S. Berberoglu and P. J. Curran. Merging spectral and textural information for classifying remotely sensed images. In S. M. de Jong and F. D. van der Meer, editors, *Remote Sensing Image Analysis: Including the Spatial Domain*, pages 113–136. Kluwer Academic Publishers, Dordrecht, 2004.

[39] S. Berberoglu, C. D. Lloyd, P. M. Atkinson, and P. J. Curran. The integration of spectral and textural information using neural networks for land cover mapping in the Mediterranean. *Computers and Geosciences*, **26**:385–396, 2000.

[40] J. Besag and J. Newell. The detection of clusters in rare diseases. *Journal of the Royal Statistical Society*, **A 154**:143–155, 1991.

[41] L. Bian. Multiscale nature of spatial data in scaling up environmental models. In D. A. Quattrochi and M. F. Goodchild, editors, *Scale in Remote Sensing and GIS*, pages 13–26. CRC Press, Boca Raton, FL, 1997.

[42] S. D. Billings, R. K. Beatson, and G. N. Newsam. Interpolation of geophysical data using continuous global surfaces. *Geophysics*, **67**:1810–1822, 2002.

[43] S. D. Billings, G. N. Newsam, and R. K. Beatson. Smooth fitting of geophysical data using continuous global surfaces. *Geophysics*, **67**:1823–1834, 2002.

[44] D. Birkes and Y. Dodge. *Alternative Methods of Regression*. John Wiley & Sons, New York, 1993.

[45] R. Bivand. Introduction to the North Carolina SIDS data set (revised), 2009. http://cran.r-project.org/web/packages/spdep/vignettes/sids.pdf; last accessed 22/6/09.

[46] R. Bivand, W. G. Müller, and M. Reder. Power calculations for global and local Moran's *I*. *Computational Statistics and Data Analysis*, **53**:2859–2872, 2007.

[47] R. S. Bivand. Exploratory spatial data analysis. In M. M. Fischer and A. Getis, editors, *Handbook of Applied Spatial Analysis: Software Tools, Methods and Applications*, pages 219–254. Springer, Heidelberg, 2010.

[48] R. S. Bivand, E. J. Pebesma, and V. Gómez-Rubio. *Applied Spatial Data Analysis with R*. Springer, New York, 2008.

[49] J. T. Bjørke and S. Nilsen. Wavelets applied to simplification of digital terrain models. *International Journal of Geographical Information Science*, **17**:601–621, 2002.

[50] C. Bleines, S. Perseval, F. Rambert, D. Renard, and Y. Touffait. *Isatis Software Manual*. Geovariances and Ecole des Mines de Paris, Avon, 2000.

[51] J. B. Boisvert, J. G. Manchuk, and C. V. Deutsch. Kriging in the presence of locally varying anisotropy using non-Euclidean distances. *Mathematical Geosciences*, **41**:585–601, 2009.

[52] B. Boots. Local measures of spatial association. *Ecoscience*, **9**:168–176, 2002.

[53] B. Boots. Developing local measures of spatial association for categorical data. *Journal of Geographical Systems*, **5**:139–160, 2003.

[54] B. Boots and A. Okabe. Local statistical spatial analysis: inventory and prospect. *International Journal of Geographical Information Science*, **21**:355–375, 2007.

[55] B. N. Boots and A. Getis. *Point Pattern Analysis*. SAGE Publications, London, 1988.

[56] I. Bracken and D. Martin. Linkage of the 1981 and 1991 UK Censuses using surface modelling concepts. *Environment and Planning A*, **27**:379–390, 1995.

[57] G. A. Bradshaw and T. A. Spies. Characterizing canopy gap structure in forests using wavelet analysis. *Journal of Ecology*, **80**:205–215, 1992.

[58] I. C. Briggs. Machine contouring using minimum curvature. *Geophysics*, **39**:39–48, 1974.

[59] E. O. Brigham. *The Fast Fourier Transform and its Applications*. Prentice Hall, Englewood Cliffs, NJ, 1988.

[60] L. A. Brown and J. P. Jones III. Spatial variation in migration processes and development: a Costa Rican example of conventional modelling augmented by the expansion method. *Demography*, **22**:327–352, 1985.

[61] A. Bruce and H.-Y. Gao. *Applied Wavelet Analysis with S-Plus*. Springer, New York, 1996.

[62] C. Brunsdon. Estimating probability surfaces for geographical point data: an adaptive algorithm. *Computers and Geosciences*, **21**:877–894, 1995.

[63] C. Brunsdon. Exploratory spatial data analysis and local indicators of spatial association with XLISP-STAT. *The Statistician*, **47**:471–484, 1998.

[64] C. Brunsdon. Inference and spatial data. In J. P. Wilson and A. S. Fotheringham, editors, *The Handbook of Geographic Information Science*, pages 337–351. Blackwell Publishing, Maldon, MA, 2008.

[65] C. Brunsdon, M. Aitkin, A. S. Fotheringham, and M. E. Charlton. A comparison of random coefficient modelling and geographically weighted regression for spatially non-stationary regression problems. *Geographical and Environmental Modelling*, **3**:47–62, 1999.

[66] C. Brunsdon and M. Charlton. Local trend statistics for directional data — a moving window approach. *Computers, Environment and Urban Systems*, **30**:130–142, 2006.

[67] C. Brunsdon, A. S. Fotheringham, and M. E. Charlton. Geographically weighted summary statistics — a framework for localised exploratory data analysis. *Computers, Environment and Urban Systems*, **26**:501–524, 2002.

[68] C. Brunsdon, J. McClatchey, and D. J. Unwin. Spatial variations in the average rainfall-altitude relationship in Great Britain: an approach using geographically weighted regression. *International Journal of Climatology*, **21**:455–466, 2001.

[69] T. M. Burgess and R. Webster. Optimal interpolation and isarithmic mapping of soil properties. II. Block kriging. *Journal of Soil Science*, **31**:333–341, 1980.

[70] P. A. Burrough and R. A. McDonnell. *Principles of Geographical Information Systems*. Oxford University Press, Oxford, 1998.

[71] C. S. Burrus, R. A. Gopinath, and H. Guo. *Introduction to Wavelets and Wavelet Transforms: A Primer*. Prentice Hall, Upper Saddle River, NJ, 1998.

[72] J. R. Carr. Spectral and textural classification of single and multiple band digital images. *Computers and Geosciences*, **22**:849–865, 1996.

[73] E. Casetti. Generating models by the expansion method: applications to geographic research. *Geographical Analysis*, **4**:81–91, 1972.

[74] E. Casetti. Expansion method, dependency, and multimodeling. In M. M. Fischer and A. Getis, editors, *Handbook of Applied Spatial Analysis: Software Tools, Methods and Applications*, pages 461–486. Springer, Heidelberg, 2010.

[75] K. R. Castleman. *Digital Image Processing*. Prentice Hall, Upper Saddle River, NJ, 1996.

[76] G. Catney. *Internal Migration, Community Background and Residential Segregation in Northern Ireland*. 2008. Unpublished Ph.D. Thesis, School of Geography, Archaeology and Palaeoecology, Queen's University, Belfast.

[77] A. K. Chan and C. Peng. *Wavelets for Sensing Technologies*. Artech House, Boston, 2003.

[78] K. Chang. *Introduction to Geographic Information Systems*. McGraw-Hill, Boston, 2004. Second Edition.

[79] D. Chen and A. Getis. *Point Pattern Analysis*. Department of Geography, San Diego State University, San Diego, 1998. http://www.geog.ucsb.edu/~dongmei/ppa/ppa.html; last accessed 2/10/2009.

[80] J. P. Chilès and P. Delfiner. *Geostatistics: Modeling Spatial Uncertainty*. John Wiley & Sons, New York, 1999.

[81] Y.-H. Chou. *Exploring Spatial Analysis in Geographic Information Systems*. OnWord Press, Albany, NY, 1996.

[82] G. Christakos. On the problem of permissible covariance and variogram models. *Water Resources Research*, **20**:251–265, 1984.

[83] R. Christensen. The equivalence of predictions from universal kriging and intrinsic random-function kriging. *Mathematical Geology*, **22**:655–664, 1990.

[84] R. Christensen. Quadratic covariance estimation and equivalence of predictions. *Mathematical Geology*, **25**:541–558, 1993.

[85] I. Clark and W. V. Harper. *Practical Geostatistics 2000*. Ecosse North America Llc, Columbus, OH, 1979.

[86] P. J. Clark and F. C. Evans. Distance to nearest neighbor as a measure of spatial relationship in populations. *Ecology*, **34**:445–453, 1954.

[87] A. D. Cliff and J. K. Ord. *Spatial Autocorrelation*. Pion, London, 1973.

[88] S. Cockings, P. F. Fisher, and M. Langford. Parameterization and visualisation of the errors in areal interpolation. *Geographical Analysis*, **29**:314–328, 1997.

[89] P. Congdon. *Applied Bayesian Modelling*. John Wiley & Sons, Chichester, 2003.

[90] P. Congdon. Modelling spatially varying impacts of socioeconomic predictors on mortality outcomes. *Journal of Geographical Systems*, **5**:161–184, 2003.

[91] P. Congdon. *Bayesian Statistical Modelling*. John Wiley & Sons, Chichester, 2006. Second Edition.

[92] J. W. Cooley and J. W. Tukey. An algorithm for the machine calculation of complex Fourier series. *Mathematics of Computation*, **19**:297–301, 1965.

[93] N. A. C. Cressie. Kriging nonstationary data. *Journal of the American Statistical Association*, **81**:625–634, 1983.

[94] N. A. C. Cressie. Fitting variogram models by weighted least squares. *Mathematical Geology*, **17**:563–586, 1985.

[95] N. A. C. Cressie. *Statistics for Spatial Data*. John Wiley & Sons, New York, 1993. Revised Edition.

[96] N. A. C. Cressie and D. L. Zimmerman. On the stability of the geostatistical method. *Mathematical Geology*, **24**:45–59, 1992.

[97] G. P. Cressman. An operational objective analysis system. *Monthly Weather Review*, **87**:367–374, 1959.

[98] P. J. Curran and P. M. Atkinson. Geostatistics in remote sensing. *Progress in Physical Geography*, **22**:61–78, 1998.

[99] F. C. Curriero. On the use of non-Euclidean distance measures in geostatistics. *Mathematical Geology*, **38**:907–926, 2006.

[100] J. Cuzick and R. Edwards. Clustering methods based on k nearest neighbour distributions. In F. E. Alexander and P. Boyle, editors,

Methods for Investigating Localized Clustering of Disease, pages 53–67. International Agency for Research on Cancer, Lyon, 1996. IARC Scientific Publications No. 135.

[101] C. Daly, R. P. Neilson, and D. L. Phillips. A statistical-topographical model for mapping climatological precipitation over mountainous terrain. *Journal of Applied Meteorology*, **31**:140–158, 1994.

[102] I. Daubechies. Orthonormal bases of compactly supported wavelets. *Communications on Pure and Applied Mathematics*, **41**:909–996, 1988.

[103] I. Daubechies. *Ten Lectures on Wavelets*. Philadelphia, PA, Society for Industrial and Applied Mathematics, 1992. CBMS-NSF Regional Conference Series in Applied Mathematics No. 61.

[104] M. J. de Smith, M. F. Goodchild, and P. A. Longley. *Geospatial Analysis: A Comprehensive Guide to Principles, Techniques and Software Tools*. Matador, Leicester, 2007. Second Edition.

[105] P. Delfiner. Linear estimation of non stationary spatial phenomena. In M. Guarascio, M. David, and C. Huijbregts, editors, *Advanced Geostatistics in the Mining Industry*, pages 49–68. Reidel Publishing Corp., Dordrecht, 1976.

[106] J. P. Delhomme. Kriging in the hydrosciences. *Advances in Water Resources*, **1**:251–266, 1978.

[107] C. V. Deutsch and A. G. Journel. *GSLIB: Geostatistical Software Library and User's Guide*. Oxford University Press, New York, 1998. Second Edition.

[108] C. R. Dietrich and M. R. Osborne. Estimation of covariance parameters in kriging via restricted maximum likelihood. *Mathematical Geology*, **23**:119–135, 1991.

[109] P. J. Diggle. *Statistical Analysis of Spatial Point Patterns*. Arnold, London, 2003. Second Edition.

[110] P. J. Diggle, A. G. Chetwynd, R. Haggkvist, and S. E. Morris. Second-order analysis of space-time clustering. *Statistical Methods in Medical Research*, **4**:124–136, 1995.

[111] P. J. Diggle, J. A. Tawn, and R. A. Moyeed. Model-based geostatistics. *Applied Statistics*, **47**:299–350, 1998.

[112] R. Dubin. Spatial weights. In A. S. Fotheringham and P. A. Rogerson, editors, *The SAGE Handbook of Spatial Analysis*, pages 125–157. SAGE Publications, London, 2009.

[113] H. Duchon. Splines minimizing rotation invariant seminorms in Sobolev spaces. In W. Schempp and K. Zeller, editors, *Constructive Theory of Functions of Several Variables. Proceedings of a Conference held*

at Oberwolfach, April 25 – May 1, 1976, pages 85–100. Springer-Verlag, Berlin, 1994. Lecture Notes in Mathematics, Vol. 571.

[114] J. Dungan. Spatial prediction of vegetation quantities using ground and image data. *International Journal of Remote Sensing*, **19**:267–285, 1998.

[115] J. L. Dungan, D. L. Peterson, and P. J. Curran. Alternative approaches for mapping vegetation quantities using ground and image data. In W. K. Michener, J. W. Brunt, and S. G. Stafford, editors, *Environmental Information Management and Analysis: Ecosystem to Global Scales*, pages 237–261. Taylor & Francis, London, 1994.

[116] J. Dykes. Cartographic visualization: exploratory spatial data analysis with local indicators of spatial association using Tcl/Tk and cdv. *The Statistician*, **47**:485–497, 1998.

[117] I. S. Evans. General geomorphology, derivatives of altitude and descriptive statistics. In R. J. Chorley, editor, *Spatial Analysis in Geomorphology*, pages 17–90. Harper & Row, New York, 1972.

[118] I. S. Evans. An integrated system of terrain analysis and slope mapping. *Zeitschrift für Geomorphologie*, Supplement **36**:274–295, 1980.

[119] I. S. Evans. What do terrain statistics really mean? In S. N. Lane, K. S. Richards, and J. H. Chandler, editors, *Landform Monitoring, Modelling and Analysis*, pages 119–138. John Wiley & Sons, Chichester, 1998.

[120] F. F. Feitosa, G. Câmara, A. M. V. Monteiro, T. Koschitzki, and M. P. S. Silva. Global and local spatial indices of urban segregation. *International Journal of Geographical Information Science*, **21**:299–323, 2007.

[121] P. F. Fisher and M. Langford. Modelling the errors in areal interpolation between zonal systems by Monte Carlo simulation. *Environment and Planning A*, **27**:211–224, 1995.

[122] I. V. Florinsky. Accuracy of local topographic variables derived from digital elevation models. *International Journal of Geographical Information Science*, **12**:47–61, 1998.

[123] R. Flowerdew and M. Green. Data integration: statistical methods for transferring data between zonal systems. In I. Masser and M. Blakemore, editors, *Handling Geographical Information: Methodology and Potential Applications*, pages 38–54. Longman Scientific and Technical, Harlow, 1991.

[124] S. A. Foster and W. L. Gorr. An adaptive filter for estimating spatially varying parameters: application to modelling police hours spent in response to calls for service. *Management Science*, **32**:878–889, 1986.

[125] A. S. Fotheringham. Trends in quantitative methods I: stressing the local. *Progress in Human Geography*, **21**:88–96, 1997.

[126] A. S. Fotheringham. "The problem of spatial autocorrelation" and local spatial statistics. *Geographical Analysis*, **41**:398–403, 2009.

[127] A. S. Fotheringham and C. Brunsdon. Local forms of spatial analysis. *Geographical Analysis*, **31**:340–358, 1999.

[128] A. S. Fotheringham, C. Brunsdon, and M. Charlton. *Quantitative Geography: Perspectives on Spatial Data Analysis*. SAGE Publications, London, 2000.

[129] A. S. Fotheringham, C. Brunsdon, and M. Charlton. *Geographically Weighted Regression: The Analysis of Spatially Varying Relationships*. John Wiley & Sons, Chichester, 2002.

[130] A. S. Fotheringham, M. Charlton, and C. Brunsdon. The geography of parameter space: an investigation of spatial non-stationarity. *International Journal of Geographical Information Systems*, **10**:605–627, 1996.

[131] A. S. Fotheringham and F. Zhan. A comparison of three exploratory methods for cluster detection in spatial point patterns. *Geographical Analysis*, **28**:200–218, 1996.

[132] R. Franke. Thin plate splines with tension. *Computer Aided Geometric Design*, **2**:87–95, 1982.

[133] S. E. Franklin, M. A. Wulder, and M. B. Lavigne. Automated derivation of geographic window sizes for use in remote sensing digital image texture analysis. *Computers and Geosciences*, **22**:665–673, 1996.

[134] A. C. Gatrell, T. C. Bailey, P. J. Diggle, and B. S. Rowlington. Spatial point pattern analysis and its application in geographical epidemiology. *Transactions of the Institute of British Geographers*, **21**:256–274, 1996.

[135] A. Geddes, A. Gimona, and D. A. Elston. Estimating local variation in land use statistics. *International Journal of Geographical Information Science*, **17**:299–319, 2003.

[136] A. E. Gelfand, H.-J. Kim, C. F Sirmans, and S. Banerjee. Spatial modeling with spatially varying coefficient processes. *Journal of the American Statistical Association*, **98**:387–396, 2003.

[137] M. G. Genton. Highly robust variogram estimation. *Mathematical Geology*, **30**:213–221, 1998.

[138] M. G. Genton. Variogram fitting by generalized least squares using an explicit formula for the covariance function. *Mathematical Geology*, **30**:323–345, 1998.

[139] P. W. Gething, P. M. Atkinson, A. M. Noor, P. W. Gikandi, S. I. Hay, and M. S. Nixon. A local space-time kriging approach applied to a national outpatient malaria data set. *Computers and Geosciences*, **33**:1337–1350, 2007.

[140] A. Getis. Interaction modeling using second-order analysis. *Environment and Planning A*, **16**:173–183, 1984.

[141] A. Getis. Spatial autocorrelation. In M. M. Fischer and A. Getis, editors, *Handbook of Applied Spatial Analysis: Software Tools, Methods and Applications*, pages 255–278. Springer, Heidelberg, 2010.

[142] A. Getis and J. Aldstadt. Constructing the spatial weights matrix using a local statistic. *Geographical Analysis*, **36**:90–104, 2004.

[143] A. Getis and J. Franklin. Second-order neighbourhood analysis of mapped point patterns. *Ecology*, **68**:473–477, 1987.

[144] A. Getis and J. K. Ord. The analysis of spatial association by use of distance statistics. *Geographical Analysis*, **24**:189–206, 1992.

[145] A. Getis and J. K. Ord. Local spatial statistics: an overview. In P. Longley and M. Batty, editors, *Spatial Analysis: Modelling in a GIS Environment*, pages 261–277. GeoInformation International, Cambridge, 1996.

[146] W. R. Gilks. Full conditional distributions. In W. R. Gilks, S. Richardson, and D. J. Spiegelhalter, editors, *Markov Chain Monte Carlo in Practice*, pages 75–88. Chapman & Hall/CRC, Boca Raton, 1998.

[147] H. Goldstein. *Multilevel Statistical Models*. Arnold, London, 2003. Third Edition.

[148] V. Gómez-Rubio, J. Ferrándiz-Ferragud, A. López-Quílez, and R. Bivand. Package 'DCluster', 2009. http://cran.r-project.org/web/packages/DCluster/DCluster.pdf; last accessed 02/10/09.

[149] M. F. Goodchild. Models of scale and scales of modelling. In N. J. Tate and P. M. Atkinson, editors, *Modelling Scale in Geographical Information Science*, pages 3–10. John Wiley & Sons, Chichester, 2001.

[150] M. F. Goodchild, L. Anselin, and U. Deichmann. A framework for the areal interpolation of socioeconomic data. *Environment and Planning A*, **25**:383–397, 1993.

[151] M. F. Goodchild and N. S.-N. Lam. Areal interpolation: a variant of the traditional spatial problem. *Geo-Processing*, **1**:297–312, 1980.

[152] M. F. Goodchild and D. A. Quattrochi. Scale, multiscaling, remote sensing and GIS. In D. A. Quattrochi and M. F. Goodchild, editors, *Scale in Remote Sensing and GIS*, pages 1–11. CRC Press, Boca Raton, FL, 1997.

[153] P. Goovaerts. *Geostatistics for Natural Resources Evaluation*. Oxford University Press, New York, 1997.

[154] P. Goovaerts. Using elevation to aid the geostatistical mapping of rainfall erosivity. *Catena*, **34**:227–242, 1999.

[155] P. Goovaerts. Geostatistical approaches for incorporating elevation into the spatial interpolation of rainfall. *Journal of Hydrology*, **228**:113–129, 2000.

[156] P. Goovaerts. Kriging and semivariogram deconvolution in the presence of irregular geographical units. *Mathematical Geosciences*, **40**:101–128, 2008.

[157] W. L. Gorr and A. M. Olligschlaeger. Weighted spatial adaptive filtering: Monte Carlo studies and application to illicit drug market modelling. *Geographical Analysis*, **26**:67–87, 1994.

[158] C. A. Gotway and L. J. Young. Combining incompatible spatial data. *Journal of the American Statistical Association*, **97**:632–648, 2002.

[159] C. A. Gotway Crawford and L. J. Young. Change of support: an inter-disciplinary challenge. In P. Renard, H. Demougeot-Renard, and R. Froidevaux, editors, *Geostatistics for Environmental Applications: Proceedings of the Fifth European Conference on Geostatistics for Environmental Applications*, pages 1–13. Springer, Berlin, 2005.

[160] A. Graps. An introduction to wavelets. *IEEE Computational Sciences and Engineering*, **2**:50–61, 1995.

[161] P. J. Green and B. W. Silverman. *Nonparametric Regression and Generalized Linear Models: A Roughness Penalty Approach*. Chapman and Hall, London, 1994.

[162] I. N. Gregory. An evaluation of the accuracy of the areal interpolation of data for the analysis of long-term change in England and Wales. In *Proceedings of the 5th International Conference on GeoComputation, University of Greenwich, 23–25 August 2000*. GeoComputation CD-ROM, Greenwich, 2000.

[163] I.N. Gregory. The accuracy of areal interpolation techniques: standardising 19th and 20th century census data to allow long-term comparison. *Computers, Environment and Urban Systems*, **26**:293–314, 2002.

[164] I.N. Gregory and P.S. Ell. Breaking the boundaries: geographical approaches to integrating 200 years of the census. *Journal of the Royal Statistical Society, Series A*, **168**:419–437, 2005.

[165] D. A. Greiling, G. M. Jacquez, A. M. Kaufmann, and R. G. Rommel. Space-time visualization and analysis in the Cancer Atlas Viewer. *Journal of Geographical Systems*, **7**:67–84, 2005.

[166] D. A. Griffith. *Spatial Autocorrelation: A Primer.* Association of American Geographers, Washington, DC, 1987.

[167] D. A. Griffith. *Advanced Spatial Statistics: Special Topics in the Exploration of Quantitative Spatial Data Series.* Kluwer, Dordrecht, 1988.

[168] D. A. Griffith. Spatial-filtering-based contributions to a critique of geographically weighted regression (GWR). *Environment and Planning A*, **40**:2571–2769, 2008.

[169] D. A. Griffith. Spatial filtering. In M. M. Fischer and A. Getis, editors, *Handbook of Applied Spatial Analysis: Software Tools, Methods and Applications*, pages 301–318. Springer, Heidelberg, 2010.

[170] T. C. Haas. Kriging and automated variogram modeling within a moving window. *Atmospheric Environment*, **24A**:1759–1769, 1990.

[171] T. C. Haas. Lognormal and moving window methods of estimating acid deposition. *Journal of the American Statistical Association*, **85**:950–963, 1990.

[172] T. C. Haas. Local prediction of a spatio-temporal process with an application to wet sulfate deposition. *Journal of the American Statistical Association*, **90**:1189–1199, 1995.

[173] R. Haining. *Spatial Data Analysis: Theory and Practice.* Cambridge University Press, Cambridge, 2003.

[174] R. M. Haralick, K. Shanmugam, and I. Dinstein. Textural features for image classification. *IEEE Transactions on Systems, Man, and Cybernetics*, **3**:610–621, 1973.

[175] R. M. Haralick and L. G. Shapiro. Image segmentation techniques. *Computer Vision, Graphics, and Image Processing*, **29**:100–132, 1985.

[176] J. Harris, D. Dorling, D. Owen, M. Coombes, and T. Wilson. Lookup tables and new area statistics for the 1971, 1981, and 1991 Censuses. In P. Rees, D. Martin, and P. Williamson, editors, *The Census Data System*, pages 67–82. John Wiley & Sons, Chichester, 2002.

[177] P. Harris and C. Brunsdon. Exploring spatial variation and spatial relationships in a freshwater acidification critical load data set for Great Britain using geographically weighted summary statistics. *Computers and Geosciences*, **36**:54–70, 2010.

[178] P. Harris, M. Charlton, and A. S. Fotheringham. Moving window kriging with geographically weighted variograms. *Stochastic Environmental Research and Risk Assessment*, 2010. In press.

[179] J. Haslett, R. Bradley, P. Craig, A. Unwin, and G. Wills. Dynamic graphics for exploring spatial data with applications to locating

global and local anomalies. *The American Statistician*, **45**:234–242, 1991.

[180] T. Hastie, R. Tibshirani, and J. Friedman. *The Elements of Statistical Learning: Data Mining, Inference, and Prediction*. Springer, New York, 2001.

[181] T. Hengl. *A Practical Guide to Geostatistical Mapping*. Office for Official Publications of the European Communities, Luxembourg, 2009. Second Edition.

[182] T. Hengl, G. B. M. Heuvelink, and D. G. Rossiter. About regression-kriging: from equations to case studies. *Computers and Geosciences*, **33**:1301–1315, 2007.

[183] U. C. Herzfeld and C. A. Higginson. Automated geostatistical seafloor classification — principles, parameters, feature vectors, and discrimination criteria. *Computers and Geosciences*, **35**:35–52, 1996.

[184] D. Higdon. A process-convolution approach to modelling temperatures in the North Atlantic Ocean. *Journal of Environmental and Ecological Statistics*, **5**:173–190, 1998.

[185] D. M. Higdon, J. Swall, and J. Kern. Non-stationary spatial modelling. In J. M. Bernardo, J. O. Berger, A. P. David, and A. F. M. Smith, editors, *Bayesian Statistics 6*, pages 761–768. Oxford University Press, Oxford, 1999.

[186] E. G. Hill, L. Ding, and L. A. Waller. A comparison of three tests to detect general clustering of a rare disease in Santa Clara County, California. *Statistics in Medicine*, **19**:1363–1378, 2000.

[187] M. E. Hodgson. What cell size does the computed slope/aspect angle represent? *Photogrammetric Engineering and Remote Sensing*, **61**:513–517, 1995.

[188] B. K. P. Horn. Hill shading and the reflectance map. *Proceedings of the IEEE*, **69**:14–47, 1981.

[189] S. Hoshino. Multilevel modeling on farmland distribution in Japan. *Land Use Policy*, **18**:75–90, 2001.

[190] B. B. Hubbard. *The World According to Wavelets*. A. K. Peters, Natick, MA, 1998. Second Edition.

[191] G. Hudson and H. Wackernagel. Mapping temperature using kriging with external drift: theory and an example from Scotland. *International Journal of Climatology*, **14**:77–91, 1994.

[192] M. F. Hutchinson. A new procedure for gridding elevation and stream line data with automatic removal of spurious pits. *Journal of Hydrology*, **106**:211–232, 1989.

[193] M. F. Hutchinson. The application of thin-plate smoothing splines to content-wide data assimilation. In J. D. Jasper, editor, *BMRC Research Report No. 27, Data Assimilation Systems*, pages 104–113. Bureau of Meteorology, Melbourne, 1991.

[194] M. F. Hutchinson. A locally adaptive approach to the interpolation of digital elevation models. In *Proceedings of the Third International Conference / Workshop on Integrating GIS and Environmental Modeling. Santa Fe, New Mexico, USA, January 21–25, 1996.* University of California, National Center for Geographic Information and Analysis: CD-ROM and WWW, Santa Barbara, CA, 1996.

[195] M. F. Hutchinson. ANUDEM software, 2000. The Australian National University Centre for Resource and Environmental Studies, Canberra. http://cres.anu.edu.au/outputs/anudem.php.

[196] M. F. Hutchinson. *ANUSPLIN Version 4.1 User Guide.* The Australian National University Centre for Resource and Environmental Studies, Canberra, 2000.

[197] M. F. Hutchinson and T. I. Dowling. A continental hydrological assessment of a new grid-based digital elevation model of Australia. *Hydrological Processes*, 5:45–58, 1991.

[198] M. F. Hutchinson and P. E. Gessler. Splines — more than just a smooth interpolator. *Geoderma*, 62:45–67, 1994.

[199] M. F. Hutchinson, H. A. Nix, J. P. McMahon, and K. D. Ord. The development of a topographic and climate database for Africa. In J. P. Wilson and J. Gallant, editors, *Proceedings of the Third International Conference / Workshop on Integrating GIS and Environmental Modeling. Santa Fe, New Mexico, USA, January 21–25, 1996.* University of California, National Center for Geographic Information and Analysis: CD-ROM and WWW, Santa Barbara, CA, 1996.

[200] E. H. Isaaks and R. M. Srivastava. *An Introduction to Applied Geostatistics.* Oxford University Press, New York, 1989.

[201] A. Jakomulska and M. N. Stawiecka. Integrating spectral and textural information for land cover mapping. In G. Begni, editor, *Observing Our Environment from Space: New Solutions for a New Millennium — Proceedings of the 21st EARSel Symposium, Paris, France, 14–16 May 2001*, pages 347–354. A. A. Balkema, Lisse, 2002.

[202] K. Jayaraman and J. Lappi. Estimation of height-diameter curves through multilevel models with special reference to even-aged teak stands. *Forest Ecology and Management*, 142:155–162, 2001.

[203] G. Johannesson and N. Cressie. Finding large-scale spatial trends in massive, global, environmental datasets. *Environmetrics*, 15:1–44, 2004.

[204] W. C. Johnson, M. D. Dixon, R. Simons, S. Jenson, and K. Larson. Mapping the response of riparian vegetation to possible flow reductions in the Snake River, Idaho. *Geomorphology*, **13**:159–173, 1995.

[205] K. Johnston, J. M. Ver Hoef, K. Krivoruchko, and N. Lucas. *Using ArcGIS^{TM} Geostatistical Analyst*. ESRI, Redlands, CA, 2001.

[206] K. Jones. *Multilevel models for Geographical Research*. Environmental Publications, Norwich, 1991. Concepts and Techniques in Modern Geography 54.

[207] K. Jones. Specifying and estimating multilevel models for geographical research. *Transactions of the Institute of British Geographers*, **16**:148–159, 1991.

[208] A. G. Journel. Nonparametric estimation of spatial distributions. *Mathematical Geology*, **15**:445–468, 1983.

[209] A. G. Journel. Geostatistics: Models and tools for the Earth sciences. *Mathematical Geology*, **18**:119–140, 1986.

[210] A. G. Journel. Modelling uncertainty and spatial dependence: Stochastic imaging. *International Journal of Geographical Information Systems*, **10**:517–522, 1996.

[211] A. G. Journel and R. Froidevaux. Anisotropic hole-effect modelling. *Mathematical Geology*, **14**:217–239, 1982.

[212] A. G. Journel and C. J. Huijbregts. *Mining Geostatistics*. Academic Press, London, 1978.

[213] A. G. Journel and M. E. Rossi. When do we need a trend model in kriging? *Mathematical Geology*, **21**:715–738, 1989.

[214] S. P. Kaluzny, S. C. Vega, T. P. Cardoso, and A. A. Shelly. *S+ Spatial Stats: User's Manual for Windows® and UNIX®*. MathSoft, Seattle, WA, 1998.

[215] I. Kaplan. Wavelets and signal processing, 2003. http://www.bearcave.com/misl/misl_tech/wavelets/; last accessed 09/12/2009.

[216] R. E. J. Kelly and P. M. Atkinson. Modelling and efficient mapping of snow cover in the UK for remote sensing validation. In P. M. Atkinson and N. J. Tate, editors, *Advances in Remote Sensing and GIS Analysis*, pages 75–95. John Wiley & Sons, Chichester, 1999.

[217] P. K. Kitanidis. Statistical estimation of polynomial generalised covariance functions and hydrologic applications. *Water Resources Research*, **19**:909–921, 1983.

[218] P. K. Kitanidis. Generalized covariance functions in estimation. *Mathematical Geology*, **25**:525–540, 1993.

[219] P. K. Kitanidis. *Introduction to Geostatistics: Applications to Hydrogeology.* Cambridge University Press, Cambridge, 1997.

[220] I. Kreft and J. De Leeuw. *Introducing Multilevel Modeling.* SAGE Publications, London, 1998.

[221] D. G. Krige and E. J. Magri. Studies of the effects of outliers and data transformation on variogram estimates for a base metal and a gold ore body. *Mathematical Geology,* 14:557–564, 1982.

[222] M. Kulldorff. A spatial scan statistic. *Communications in Statistics — Theory and Methods,* 26:1481–1496, 1997.

[223] P. Kumar and E. Foufoula-Georgiou. Wavelet analysis for geophysical applications. *Reviews of Geophysics,* 35:385–412, 1997.

[224] P. C. Kyriakidis. A geostatistical framework for area-to-point spatial interpolation. *Geographical Analysis,* 36:259–289, 2004.

[225] P. C. Kyriakidis and E.-H. Yoo. Geostatistical prediction/simulation of point values from areal data. In *Proceedings of the 7th International Conference on GeoComputation, University of Southampton, UK, 8–10 September 2003.* GeoComputation CD-ROM, Greenwich, 2003.

[226] J. LaGro. Assessing patch shape in landscape mosaics. *Photogrammetric Engineering and Remote Sensing,* 57:285–293, 1991.

[227] N. S. -N. Lam. Spatial interpolation methods: a review. *The American Cartographer,* 10:129–149, 1983.

[228] N. S. -N. Lam. Fractals and scale in environmental assessment and monitoring. In E. Sheppard and R. B. McMaster, editors, *Scale and Geographic Inquiry: Nature, Society, and Method,* pages 23–40. Blackwell Publishing, Malden, MA, 2004.

[229] I. H. Langford, A. H. Leyland, J. Rasbash, and H. Goldstein. Multilevel modelling of the geographical distributions of diseases. *Journal of the Royal Statistical Society, Series C,* 48:253–268, 1999.

[230] M. Langford, D. J. Maguire, and D. J. Unwin. The areal interpolation problem: estimating population using remote sensing in a GIS framework. In I. Masser and M. Blakemore, editors, *Handling Geographical Information: Methodology and Potential Applications,* pages 55–77. Longman Scientific and Technical, Harlow, 1991.

[231] R. M. Lark. Geostatistical description of texture on an aerial photograph for discriminating classes of land cover. *International Journal of Remote Sensing,* 17:2115–2133, 1996.

[232] A. B. Lawson and S. Banerjee. Bayesian spatial analysis. In A. S. Fotheringham and P. A. Rogerson, editors, *The SAGE Handbook of Spatial Analysis,* pages 321–342. SAGE, Los Angeles, 2009.

[233] A. B. Lawson, W. J. Browne, and C. L. Vidal Rodeiro. *Disease Mapping with WinBUGS and MLwiN.* John Wiley & Sons, Chichester, 2003.

[234] J. P. LeSage. A family of geographically weighted regression models. In L. Anselin, R. J. G. M. Florax, and S. J. Rey, editors, *Advances in Spatial Econometrics. Methodology, Tools and Applications*, pages 241–264. Springer, Berlin, 2004.

[235] Y. Leung, C.-L. Mei, and W.-X. Zhang. Statistical tests for spatial nonstationarity based on the geographically weighted regression model. *Environment and Planning A*, **32**:9–32, 2000.

[236] M. D. Levine and S. I. Shaheen. A modular computer vision system for picture segmentation and interpretation. *IEEE Transactions on Pattern Analysis and Machine Intelligence*, **3**:540–556, 1981.

[237] A. H. Leyland, I. H. Langford, J. Rasbash, and H. Goldstein. Multivariate spatial models for event data. *Statistics in Medicine*, **19**:2469–2478, 1999.

[238] Z. Li, Q. Zhu, and C. Gold. *Digital Terrain Modeling: Principles and Methodology*. Boca Raton, CRC Press, 2004.

[239] C. D. Lloyd. Increasing the accuracy of predictions of monthly precipitation in Great Britain using kriging with an external drift. In G. M. Foody and P. M. Atkinson, editors, *Uncertainty in Remote Sensing and GIS*, pages 243–267. John Wiley & Sons, Chichester, 2002.

[240] C. D. Lloyd. Assessing the effect of integrating elevation data into the estimation of monthly precipitation in Great Britain. *Journal of Hydrology*, **308**:128–150, 2005.

[241] C. D. Lloyd. Analysing population characteristics using geographically weighted principal components analysis: a case study of Northern Ireland in 2001. *Computers, Environment and Urban Systems*, **34**:389–399, 2010.

[242] C. D. Lloyd. Exploring population spatial concentrations in Northern Ireland by community background and other characteristics: an application of geographically weighted spatial statistics. *International Journal of Geographical Information Science*, **24**:1193–1221, 2010.

[243] C. D. Lloyd. Multivariate interpolation of monthly precipitation amount in the United Kingdom. In P. M. Atkinson and C. D. Lloyd, editors, *GeoENV VII: Geostatistics for Environmental Applications*, pages 27–39. Springer, Dordrecht, 2010.

[244] C. D. Lloyd. Nonstationary models for exploring and mapping monthly precipitation in the United Kingdom. *International Journal of Climatology*, **30**:390–405, 2010.

[245] C. D. Lloyd. *Spatial Data Analysis: An Introduction for GIS Users*. Oxford University Press, Oxford, 2010.

[246] C. D. Lloyd and P. M. Atkinson. Interpolating elevation with locally adaptive kriging. In P. M. Atkinson and D. J. Martin, editors,

Innovations in GIS 7: GIS and Geocomputation, pages 241–253. Taylor & Francis, London, 1999.

[247] C. D. Lloyd and P. M. Atkinson. Deriving DSMs from LiDAR data with kriging. *International Journal of Remote Sensing*, **23**:2519–2524, 2002.

[248] C. D. Lloyd and P. M. Atkinson. Non-stationary approaches for mapping terrain and assessing uncertainty in predictions. *Transactions in GIS*, **6**:17–30, 2002.

[249] C. D. Lloyd and P. M. Atkinson. Archaeology and geostatistics. *Journal of Archaeological Science*, **31**:151–165, 2004.

[250] C. D. Lloyd, P. M. Atkinson, and P. Aplin. Characterising local spatial variation in land cover using geostatistical functions and the discrete wavelet transform. In P. Renard, H. Demougeot-Renard, and R. Froidevaux, editors, *Geostatistics for Environmental Applications: Proceedings of the Fifth European Conference on Geostatistics for Environmental Applications*, pages 391–402. Springer, Berlin, 2005.

[251] C. D. Lloyd, S. Berberoglu, P. J. Curran, and P. M. Atkinson. A comparison of texture measures for the per-field classification of Mediterranean land cover. *International Journal of Remote Sensing*, **25**:3943–3965, 2004.

[252] C. D. Lloyd and K. D. Lilley. Cartographic veracity in medieval mapping: analyzing geographical variation in the Gough Map of Great Britain. *Annals of the Association of American Geographers*, **99**:27–48, 2009.

[253] C. D. Lloyd and I. G. Shuttleworth. Analysing commuting using local regression techniques: scale, sensitivity and geographical patterning. *Environment and Planning A*, 37:81–103, 2005.

[254] A. Løland and G. Høst. Spatial covariance modelling in a complex coastal domain by multidimensional scaling. *Environmetrics*, **14**:307–321, 2003.

[255] D. J. Lunn, A. Thomas, N. Best, and D. Spiegelhalter. WinBUGS — a Bayesian modelling framework: concepts, structure, and extensibility. *Statistics and Computing*, **10**:325–337, 2000.

[256] S. Mallat. A theory for multiresolution signal decomposition: the wavelet representation. *IEEE Transactions on Pattern Analysis and Machine Intelligence*, **11**:674–693, 1989.

[257] S. Mallat. *A Wavelet Tour of Signal Processing*. Academic Press, San Diego, 1999. Second Edition.

[258] D. Martin. Mapping population data from zone centroid locations. *Transactions of the Institute of British Geographers*, **14**:90–97, 1989.

[259] D. Martin. An assessment of surface and zonal models of population. *International Journal of Geographical Information Systems*, **10**:973–989, 1996.

[260] D. Martin. *Geographic Information Systems: Socioeconomic Applications*. Routledge, London, 1996. Second Edition.

[261] D. Martin. Census population surfaces. In P. Rees, D. Martin, and P. Williamson, editors, *The Census Data System*, pages 139–148. John Wiley & Sons, Chichester, 2002.

[262] D. Martin and I. Bracken. Techniques for modelling population-related raster databases. *Environment and Planning A*, **23**:1069–1075, 1991.

[263] D. Martin, D. Dorling, and R. Mitchell. Linking censuses through time: problems and solutions. *Area*, **34**:82–91, 2002.

[264] P. M. Mather. *Computer Processing of Remotely-Sensed Images: An Introduction*. John Wiley & Sons, Chichester, 2004. Third Edition.

[265] G. Matheron. *Les Variables Régionalisées et leur Estimation*. Masson, Paris, 1965.

[266] G. Matheron. *The Theory of Regionalised Variables and its Applications*. Centre de Morphologie Mathématique de Fontainebleau, Fontainebleau, 1971.

[267] G. Matheron. The intrinsic random functions and their applications. *Advances in Applied Probability*, **5**:439–468, 1973.

[268] A. B. McBratney and R. Webster. Choosing functions for semivariograms of soil properties and fitting them to sampling estimates. *Journal of Soil Science*, **37**:617–639, 1986.

[269] J. M. McKinley, C. D. Lloyd, and A. H. Ruffell. Use of variography in permeability characterization of visually homogeneous sandstone reservoirs with examples from outcrop studies. *Mathematical Geology*, **36**:761–779, 2004.

[270] W. Meiring, P. Monestiez, P. D. Sampson, and P. Guttorp. Developments in the modelling of nonstationary spatial covariance structure from space-time monitoring data. In E. Y. Baafi and N. Schofield, editors, *Geostatistics Wallongong '96*, pages 162–173. Kluwer, Dordrecht, 1998.

[271] Y. Meyer. *Wavelets: Algorithms and Applications*. Society for Applied and Industrial Mathematics, Philadelphia, 1993.

[272] F. P. Miranda, L. E. N. Fonseca, J. R. Carr, and J. V. Taranik. Analysis of JERS-1 (fuyo-1) SAR data for vegetation discrimination in northwestern Brazil using the semivariogram textural classifier (STC). *International Journal of Remote Sensing*, **17**:3523–3529, 1996.

[273] F. P. Miranda, J. A. Macdonald, and J. R. Carr. Application of the semivariogram textural classifier (STC) for vegetation discrimination using SIR-B data of Borneo. *International Journal of Remote Sensing*, **13**:2349–2354, 1992.

[274] L. Mitás and H. Mitásová. General variational approach to the interpolation problem. *Computers and Mathematics with Applications*, **16**:983–992, 1988.

[275] L. Mitás and H. Mitásová. Spatial interpolation. In P. A. Longley, M. F. Goodchild, D. J. Maguire, and D. W. Rhind, editors, *Geographical Information Systems. Volume I: Principles and Technical Issues*, pages 481–492. John Wiley & Sons, New York, 1999. Second Edition.

[276] H. Mitásová, J. Hofierka, M. Zlocha, and L. R. Iverson. Modelling topographic potential for erosion and deposition using GIS. *International Journal of Geographical Information Systems*, **10**:629–641, 1996.

[277] H. Mitásová and L. Mitás. Interpolation by regularised spline with tension: I. Theory and implementation. *Mathematical Geology*, **25**:641–655, 1993.

[278] H. Mitásová, L. Mitás, W. M. Brown, D. P. Gerdes, I. Kosinovsky, and T. Baker. Modeling spatially and temporally distributed phenomena: new methods and tools for GRASS GIS. *International Journal of Geographical Information Systems*, **9**:433–446, 1995.

[279] A. J. Moffat, J. A. Catt, R. Webster, and E. H. Brown. A re-examination of the evidence for a Plio-Pleistocene marine transgression on the Chiltern Hills. I. Structures and surfaces. *Earth Surface Processes and Landforms*, **11**:95–106, 1986.

[280] P. Monestiez, D. Allard, I. Navarro Sanchez, and D. Courault. Kriging with categorical external drift: use of thematic maps in spatial prediction and application to local climate interpolation for agriculture. In J. Gómez-Hernández, A. Soares, and R. Froidevaux, editors, *GeoENV II: Geostatistics for Environmental Applications*, pages 163–174. Kluwer Academic Publishers, Dordrecht, 1999.

[281] I. D. Moore, R. B. Grayson, and A. R. Ladson. Digital terrain modelling: A review of hydrological, geomorphological, and biological applications. *Hydrological Processes*, **5**:3–30, 1991.

[282] D. J. Mulla. Using geostatistics and spectral analysis to study spatial patterns in the topography of southeastern Washington State, U.S.A. *Earth Surface Processes and Landforms*, **13**:389–405, 1988.

[283] D. E. Myers. To be or not to be ... stationary? That is the question. *Mathematical Geology*, **21**:347–362, 1989.

[284] R. H. Myers, D. C. Montgomery, and G. G. Vining. *Generalized Linear Models with Applications in Engineering and the Sciences*. John Wiley & Sons, New York, 2002.

[285] T. Nakaya. Local spatial interaction modelling based on the geographically weighted regression approach. *GeoJournal*, **53**:347–358, 2001.

[286] T. Nakaya, A. S. Fotheringham, C. Brunsdon, and M. Charlton. Geographically weighted Poisson regression for disease association mapping. *Statistics in Medicine*, **24**:2695–2717, 2005.

[287] S. P. Neuman and E. A. Jacobson. Analysis of nonintrinsic spatial variability by residual kriging with application to regional groundwater levels. *Mathematical Geology*, **16**:499–521, 1984.

[288] R. Nicolau, L. Ribeiro, R. R. Rodrigues, H. G. Pereira, and A. S. Câmara. Mapping the spatial distribution of rainfall in Portugal. In W. J. Kleingeld and D. G. Krige, editors, *Geostatistics 2000 Cape Town*, pages 548–558. Geostatistical Association of Southern Africa, South Africa, 2002.

[289] M. Nogami. Geomorphometric measures for digital elevation models. In R. J. Pike and R. Dikau, editors, *Advances in Geomorphometry — Proceedings of the Walter F. Wood Memorial Symposium*, pages 53–67. Gebruder Borntraegar, Berlin and Stuttgart, 1995.

[290] I. Ntzoufras. *Bayesian Modeling Using WinBUGS*. Wiley, Hoboken, 2009.

[291] R. A. Olea. *Optimum Mapping Techniques using Regionalized Variable Theory*. Kansas University Geological Survey Series on Spatial Analysis No. 2. University of Kansas, Lawrence, 1975.

[292] R. A. Olea. *Measuring Spatial Dependence with Semivariograms*. Kansas University Geological Survey Series on Spatial Analysis No. 3. University of Kansas, Lawrence, 1977.

[293] R. A. Olea. Sampling design optimization for spatial functions. *Mathematical Geology*, **16**:369–392, 1984.

[294] M. A. Oliver. Determining the spatial scale of variation in environmental properties using the variogram. In N. J. Tate and P. M. Atkinson, editors, *Modelling Scale in Geographical Information Science*, pages 193–219. John Wiley & Sons, Chichester, 2001.

[295] M. A. Oliver, E. Bosch, and K. Slocum. Wavelets and kriging for filtering and data reconstruction. In W. J. Kleingeld and D. G. Krige, editors, *Geostatistics 2000 Cape Town*, pages 571–580. Geostatistical Association of Southern Africa, South Africa, 2002.

[296] M. A. Oliver and R. Webster. A geostatistical basis for spatial weighting in multivariate classification. *Mathematical Geology*, **21**:15–35, 1989.

[297] M. A. Oliver and R. Webster. Kriging: a method of interpolation for geographical information systems. *International Journal of Geographical Information Systems*, **4**:313–332, 1990.

[298] M. A. Oliver, R. Webster, and J. Gerrard. Geostatistics in physical geography. Part I: theory. *Transactions of the Institute of British Geographers*, **14**:259–269, 1989.

[299] M. A. Oliver, R. Webster, and J. Gerrard. Geostatistics in physical geography. Part II: applications. *Transactions of the Institute of British Geographers*, **14**:270–286, 1989.

[300] S. Openshaw. *The Modifiable Areal Unit Problem*. Geobooks, Norwich, 1984. Concepts and Techniques in Modern Geography 38.

[301] S. Openshaw. Using a geographical analysis machine to detect the presence of spatial clustering and the location of clusters in synthetic data. In F. E. Alexander and P. Boyle, editors, *Methods for Investigating Localized Clustering of Disease*, pages 68–86. International Agency for Research on Cancer, Lyon, 1996. IARC Scientific Publications No. 135.

[302] S. Openshaw, M. E. Charlton, C. Wymer, and A. W. Craft. A mark I Geographical Analysis Machine for the automated analysis of point data sets. *International Journal of Geographical Information Systems*, **1**:359–377, 1993.

[303] S. Openshaw and P. J. Taylor. A million or so correlation coefficients: three experiments on the modifiable areal unit problem. In N. Wrigley, editor, *Statistical Applications in the Spatial Sciences*, pages 127–144. Pion, London, 1979.

[304] D. O'Sullivan and D. J. Unwin. *Geographic Information Analysis*. John Wiley & Sons, Hoboken, NJ, 2002.

[305] C.J. Paciorek and M.J. Schervish. Spatial modelling using a new class of nonstationary covariance functions. *Environmetrics*, **17**:483–506, 2006.

[306] A. Páez. Anisotropic variance functions in geographically weighted regression models. *Geographical Analysis*, **36**:299–314, 2002.

[307] Y. Pannatier. *VARIOWIN: Software for Spatial Analysis in 2D*. Springer-Verlag, New York, 1996.

[308] E. Pardo-Igúzquiza. GCINFE: a computer program for inference of polynomial generalized covariance functions. *Computers and Geosciences*, **23**:163–174, 1997.

[309] E. Pardo-Igúzquiza. MLREML: a computer program for the inference of spatial covariance parameters by maximum likelihood and restricted maximum likelihood. *Computers and Geosciences*, **23**:153–162, 1997.

[310] E. Pardo-Igúzquiza. VARFIT: a Fortran-77 program for fitting variogram models by weighted least squares. *Computers and Geosciences*, **25**:251–261, 1999.

[311] T. Pavlidis. *Algorithms for Graphics and Image Processing*. Computer Science Press, Rockville, MD, 1982.

[312] E. J. Pebesma. *Mapping Groundwater Quality in the Netherlands*. 1996. Unpublished Ph.D. Thesis, University of Utrecht.

[313] E. J. Pebesma and C. G. Wesseling. Gstat, a program for geostatistical modelling, prediction and simulation. *Computers and Geosciences*, **24**:17–31, 1998.

[314] D. B. Percival and A. T. Walden. *Wavelet Methods for Time Series Analysis*. Cambridge University Press, Cambridge, 2000.

[315] P. Petitgas. Reducing non-stationary random fields to stationarity and isotropy using a space deformation. *Fish and Fisheries*, **2**:231–249, 2001.

[316] T. K. Peuker, R. J. Fowler, J. J. Little, and D. M. Mark. The triangulated irregular network. In American Society of Photogrammetry, editor, *Proceedings of the Digital Terrain Models (DTM) Symposium. St. Louis, Missouri, May 9–11, 1978*, pages 516–540. American Society of Photogrammetry, Falls Church, VA, 1978.

[317] G. M. Philip and D. F. Watson. Matheronian geostatistics — quo vadis? *Mathematical Geology*, **18**:93–117, 1986.

[318] F. Potvin, B. Boots, and A. Dempster. Comparison among three approaches to evaluate winter habitat selection by white-tailed deer on Anticosti Island using occurrences from an aerial survey and forest vegetation maps. *Canadian Journal of Zoology*, **81**:1662–1670, 2003.

[319] W. H. Press, B. P. Flannery, S. A. Teukolsky, and W. T. Vetterling. *Numerical Recipes in Fortran 77: The Art of Scientific Computing*. Cambridge University Press, Cambridge, 1996.

[320] G. Ramstein and M. Raffy. Analysis of the structure of radiometric remotely-sensed images. *International Journal of Remote Sensing*, **10**:1049–1073, 1989.

[321] J.-M. Rendu. Disjunctive kriging: Comparison of theory with actual results. *Mathematical Geology*, **12**:305–320, 1980.

[322] B. D. Ripley. *Spatial Statistics*. John Wiley & Sons, New York, 1981. Wiley Series in Probability and Mathematical Statistics.

[323] J. Rivoirard. *Introduction to Disjunctive Kriging and Non-linear Geostatistics*. Clarendon Press, Oxford, 1994.

[324] A. H. Robinson, J. L. Morrison, P. C. Muehrcke, A. J. Kimerling, and S. C. Guptill. *Elements of Cartography*. John Wiley & Sons, New York, 1995. Sixth Edition.

[325] P. A. Rogerson. *Statistical Methods for Geography: A Student's Guide*. SAGE Publications, London, 2010. Third Edition.

[326] A. G. Royle and E. Hosgit. Local estimation of sand and gravel reserves by geostatistical methods. *Institution of Mining and Metallurgy*, pages A53–A62, 1974.

[327] Y. Sadahiro. Accuracy of count data transferred through the areal weighting interpolation method. *International Journal of Geographical Information Science*, **14**:25–50, 2000.

[328] P. D. Sampson. Constructions for nonstationary spatial processes. In A. E. Gelfand, P. J. Diggle, M. Fuentes, and P. Guttorp, editors, *Handbook of Spatial Statistics*, pages 119–130. CRC Press, Boca Raton, 2010.

[329] P. D. Sampson, D. Damien, and P. Guttorp. Advances in modelling and inference for environmental processes with nonstationary spatial covariance. In P. Monestiez, D. Allard, and R. Froidevaux, editors, *GeoENV III: Geostatistics for Environmental Applications*, pages 17–32. Kluwer Academic Publishers, Dordrecht, 2001.

[330] O. Schabenberger and C. A. Gotway. *Statistical Methods for Spatial Data Analysis*. Chapman and Hall/CRC Press, Boca Raton, FL, 2005.

[331] I. G. Shuttleworth and C. D. Lloyd. Are Northern Ireland's communities dividing? evidence from geographically consistent Census of Population data, 1971–2001. *Environment and Planning A*, 41:213–229, 2009.

[332] R. Sibson. A brief description of natural neighbour interpolation. In V. Barnett, editor, *Interpreting Multivariate Data*, pages 21–36. John Wiley & Sons, Chichester, 1981.

[333] B. W. Silverman. *Density Estimation for Statistics and Data Analysis*. Chapman and Hall, London, 1986.

[334] A. K. Skidmore. A comparison of techniques for calculating gradient and aspect from a gridded digital elevation model. *International Journal of Geographical Information Systems*, **3**:323–334, 1989.

[335] W. H. F. Smith and P. Wessel. Gridding with continuous curvature splines in tension. *Geophysics*, **55**:293–305, 1990.

[336] T. Snijders and R. Bosker. *Multilevel Analysis: An Introduction to Basic and Advanced Multilevel Modelling*. SAGE Publications, London, 1999.

[337] R. R. Sokal, N. L. Oden, and B. A. Thomson. Local spatial autocorrelation in biological variables. *Biological Journal of the Linnean Society*, **65**:41–62, 1998.

[338] M. Sonka, V. Hlavac, and R. Boyle. *Image Processing, Analysis and Machine Vision*. PWS Publishing, Pacific Grove, CA, 1999. Second Edition.

[339] D. Spiegelhalter, A. Thomas, N. Best, and D. Lunn. *WinBUGS User Manual: Version 1.4, January 2003.* 2003. http://www.mrc-bsu.cam.ac.uk/bugs.

[340] R. M. Srivastatva. Philip and Watson — quo vadunt? *Mathematical Geology,* **18**:119–140, 1986.

[341] R. M. Srivastava and H. M. Parker. Robust measures of spatial continuity. In M. Armstrong, editor, *Geostatistics,* pages 295–308. Reidel, Dordrecht, 1989.

[342] B. A. St-Onge and F. Cavayas. Automated forest structure mapping from high resolution imagery based on directional semivariogram estimates. *Remote Sensing of Environment,* **61**:82–95, 1997.

[343] J. L. Starck, F. Murtagh, and A. Bijaoui. *Image Processing and Data Analysis: The Multiscale Approach.* Cambridge University Press, Cambridge, 1998.

[344] T. H. Starks and J. H. Fang. The effect of drift on the experimental semivariogram. *Mathematical Geology,* **14**:309–319, 1982.

[345] A. Stein, I. G. Startitsky, J. Bouma, A. C Van Eijnsbergen, and A. K. Bregt. Simulation of moisture deficits and areal interpolation by universal co-kriging. *Water Resources Research,* **27**:1963–1973, 1991.

[346] E. J. Stollnitz, T. D. DeRose, and D. H. Salesin. Wavelets for computer graphics: A primer, part 1. *IEEE Computer Graphics and Applications,* **15**(3):76–84, 1995.

[347] E. J. Stollnitz, T. D. DeRose, and D. H. Salesin. *Wavelets for Computer Graphics: Theory and Applications.* Morgan Kaufmann Publishers, San Francisco, CA, 1996.

[348] G. Strang. Wavelets. *American Scientist,* **82**:250–255, 1994.

[349] S. V. Subramanian. Multilevel modeling. In M. M. Fischer and A. Getis, editors, *Handbook of Applied Spatial Analysis: Software Tools, Methods and Applications,* pages 507–525. Springer, Heidelberg, 2010.

[350] C. J. Swain. A Fortran IV program for interpolating irregularly spaced data using the difference equations for minimum curvature. *Computers and Geosciences,* **1**:231–240, 1976.

[351] J. L. Swall. *Non-Stationary Spatial Modeling using a Process Convolution Approach.* 1999. Unpublished Ph.D. Thesis, Institute of Statistics and Decision Sciences, Duke University.

[352] P. A. V. B. Swamy, R. K. Conway, and M. R. LeBlanc. The stochastic coefficients approach to econometric modelling. Part I: A critique of fixed coefficients models. *The Journal of Agricultural Economics Research,* **40**:2–10, 1988.

[353] P. A. V. B. Swamy, R. K. Conway, and M. R. LeBlanc. The stochastic coefficients approach to econometric modelling. Part II: Description and motivation. *The Journal of Agricultural Economics Research*, **40**:21–30, 1988.

[354] P. A. V. B. Swamy, R. K. Conway, and M. R. LeBlanc. The stochastic coefficients approach to econometric modelling. Part III: Estimation, stability testing, and prediction. *The Journal of Agricultural Economics Research*, **41**:4–20, 1988.

[355] M. J. Symons, R. C. Grimson, and Y. C. Yuan. Clustering of rare events. *Biometrics*, **39**:193–205, 1983.

[356] J. Testud and M. Chong. Three dimensional wind field analysis from dual-doppler radar data. Part I: filtering, interpolating and differentiating the raw data. *Journal of Climate and Applied Meteorology*, **22**:1204–1215, 1983.

[357] J. Theiler and G. Gisler. A contiguity-enhanced *k*-means clustering algorithm for unsupervised multispectral image segmentation. *Proceedings of the SPIE*, **3159**:108–118, 1997.

[358] I. L. Thomas, R. Howarth, A. Eggers, and A. D. W. Fowler. Textural enhancement of a circular geological feature. *Photogrammetric Engineering and Remote Sensing*, **47**:89–91, 1981.

[359] M. Tiefelsdorf. *Modelling Spatial Processes: The Identification and Analysis of Spatial Relationships in Regression Residuals by Means of Moran's I*. Springer-Verlag, Berlin, 2000. Lecture Notes in Earth Sciences 87.

[360] M. Tiefelsdorf. The saddlepoint approximation of Moran's *I* and local Moran's *I* reference distributions and their numerical approximations. *Geographical Analysis*, **34**:187–206, 2002.

[361] W. R. Tobler. A computer movie simulating urban growth in the Detroit region. *Economic Geography*, **46**:234–240, 1970.

[362] W. R. Tobler. Smooth pycnophylactic interpolation for geographical regions. *Journal of the American Statistical Association*, **74**:519–536, 1979.

[363] M. Tomczak. Spatial interpolation and its uncertainty using automated anisotropic inverse distance weighting (IDW) — cross-validation/jackknife approach. *Journal of Geographic Information and Decision Analysis*, **2**:18–30, 1998. ftp://ftp.geog.uwo.ca/GIDA/Tomczak.pdf.

[364] B. Tso and P. M. Mather. *Classification Methods for Remotely Sensed Data*. Taylor & Francis, London, 2001. CATMOG 12.

[365] J. W. Tukey. *Exploratory Data Analysis*. Addison-Wesley, Reading, MA, 1977.

[366] B. W. Turnbull, E.J. Iwano, W.S. Burnett, H.L. Howe, and L.C. Clark. Monitoring for clusters of disease: Application to leukemia incidence in upstate New York. *American Journal of Epidemiology*, **132**:S136–S143, 1990.

[367] A. Unwin and D. Unwin. Exploratory spatial data analysis with local statistics. *The Statistician*, **47**:415–421, 1998.

[368] A. R. Unwin. Exploratory spatial analysis and local statistics. *Computational Statistics*, **11**:387–400, 1996.

[369] C. Valens. A really friendly guide to wavelets, 2005. http://pagesperso-orange.fr/polyvalens/clemens/wavelets/wavelets.html; last accessed 02/10/2009.

[370] K. G. van den Boogaart and A. Brenning. Why is universal kriging better than IRFk-kriging: estimation of variograms in the presence of trend. In *Proceedings of the 6th Annual Conference of the International Association of Mathematical Geology, Cancún, México, September 6–12, 2001*. 2001.

[371] H. Wackernagel. *Multivariate Geostatistics: An Introduction with Applications*. Springer, Berlin, 2003. Third Edition.

[372] G. Wahba. *Spline Models for Observational Data*. Society for Industrial and Applied Mathematics, Philadelphia, Pennsylvania, 1990. CBMS-NSF Regional Conference Series in Applied Mathematics No. 59.

[373] G. Wahba. (Smoothing) splines in nonparametric regression. In A. El-Shaarawi and W. Piegorsch, editors, *Encyclopedia of Environmetrics, Volume 4*, pages 2099–2112. Wiley, Chichester, 2001.

[374] J. S. Walker. Fourier analysis and wavelet analysis. *Notices of the American Mathematical Society*, **44**:658–670, 1997.

[375] L. A. Waller and C. A. Gotway. *Applied Spatial Statistics for Public Health Data*. Wiley, Hoboken, NJ, 2004.

[376] L. A. Waller, L. Zhu, C. A. Gotway, D. M. Gorman, and P. J. Gruenewald. Quantifying geographic variations in associations between alcohol distribution and violence: a comparison of geographically weighted regression and spatially varying coefficient models. *Stochastic Environmental Research and Risk Assessment*, **21**:573–588, 2007.

[377] M. D. Ward and K. S. Gleditsch. *Spatial Regression Models*. Sage Publications, Los Angeles, 2008. Quantitative Applications in the Social Sciences 155.

[378] D. F. Watson. *Contouring: A Guide to the Analysis and Display of Spatial Data*. Pergamon Press, Oxford, 1992.

[379] D. F. Watson and G. M. Philip. Neighborhood-based interpolation. *Geobyte*, **2** (2):12–16, 1987.

[380] R. Webster and A. B. McBratney. On the Akaike information criterion for choosing models for variograms of soil properties. *Journal of Soil Science*, **40**:493–496, 1989.

[381] R. Webster and M. A. Oliver. Optimal interpolation and isarithmic mapping of soil properties. VI. Disjunctive kriging and mapping the conditional probability. *Journal of Soil Science*, **40**:497–512, 1989.

[382] R. Webster and M. A. Oliver. *Statistical Methods in Soil and Land Resource Survey*. Oxford University Press, Oxford, 1990.

[383] R. Webster and M. A. Oliver. Sample adequately to estimate variograms of soil properties. *Journal of Soil Science*, **43**:177–192, 1992.

[384] R. Webster and M. A. Oliver. *Geostatistics for Environmental Scientists*. John Wiley & Sons, Chichester, 2007. Second Edition.

[385] D. Wheeler. Diagnostic tools and a remedial method for collinearity in geographically weighted regression. *Environment and Planning A*, **39**:2464–2481, 2007.

[386] D. Wheeler. *Spatially Varying Coefficient Regression Models: Diagnostic and Remedial Method for Collinearity*. VDM Verlag Dr Müller, Saarbrücken, 2009.

[387] D. Wheeler and M. D. Tiefelsdorf. Multicollinearity and correlation among local regression coefficients in geographically weighted regression. *Journal of Geographical Systems*, **7**:161–187, 2005.

[388] D. C. Wheeler. Simultaneous coefficient penalization and model selection in geographically weighted regression: the geographically weighted lasso. *Environment and Planning A*, **41**:722–742, 2009.

[389] D. C. Wheeler and C. A. Calder. An assessment of coefficient accuracy in linear regression models with spatially varying coefficients. *Journal of Geographical Systems*, **9**:145–166, 2007.

[390] D. C. Wheeler and A. Páez. Geographically weighted regression. In M. M. Fischer and A. Getis, editors, *Handbook of Applied Spatial Analysis: Software Tools, Methods and Applications*, pages 461–486. Springer, Heidelberg, 2010.

[391] D. C. Wheeler and L. A. Waller. Comparing spatially varying coefficient models: a case study examining violent crime rates and their relationships to alcohol outlets and illegal drug arrests. *Journal of Geographical Systems*, **11**:1–22, 2009.

[392] B. M. Whelan, A. B. McBratney, and B. Minasny. Vesper 1.5 — spatial prediction software for precision agriculture. In P. C. Robert, R. H. Rust, and W. E. Larson, editors, *Precision Agriculture, Proceedings of the 6th International Conference on Precision Agriculture*. ASA/CSSA/SSSA, Madison, 2002. http://www.usyd.edu.au/agric/acpa/vesper/vesperpaper.pdf.

[393] A. Wilhelm and R. Steck. Exploring spatial data by using interactive graphics and local statistics. *The Statistician*, **47**:423–430, 1998.

[394] D. W. S. Wong. Modeling local segregation: A spatial interaction approach. *Geographical and Environmental Modelling*, **6**:81–97, 2002.

[395] S. N. Wood. Thin plate regression splines. *Journal of the Royal Statistical Society, Series B*, **65**:95–114, 2003.

[396] S. N. Wood. *Generalized Additive Models: An Introduction with R*. Chapman and Hall/CRC Press, Boca Raton, FL, 2006.

[397] S. N. Wood and N. H. Augustin. GAMs with integrated model selection using penalized regression splines and applications to environmental modelling. *Ecological Modelling*, **157**:157–177, 2002.

[398] I. Yamada and J.-C. Thill. Local indicators of network-constrained clusters in spatial point patterns. *Geographical Analysis*, **39**:268–292, 2006.

[399] S. R. Yates and A. W. Warrick. Estimating soil water content using co-kriging. *Soil Science Society of America Journal*, **51**:23–30, 1987.

[400] S. R. Yates, A. W. Warrick, and D. E. Myers. Disjunctive kriging. 1. Overview of estimation and conditional probability. *Water Resources Research*, **22**:615–621, 1986.

[401] S. R. Yates, A. W. Warrick, and D. E. Myers. Disjunctive kriging. 2. Examples. *Water Resources Research*, **22**:623–630, 1986.

[402] P. Zatelli and A. Antonello. New GRASS modules for Multiresolution Analysis with wavelets. In M. Ciolli and P. Zatelli, editors, *Proceedings of the Open Source Free Software GIS — GRASS users conference 2002*. 2002. http://www.ing.unitn.it/~grass/conferences/GRASS2002/home.html.

[403] L. W. Zevenbergen and C. R. Thorne. Quantitative analysis of land surface topography. *Earth Surface Processes and Landforms*, **12**:47–56, 1987.

[404] D. Zimmerman, C. Pavlik, A. Ruggles, and M. P. Armstrong. An experimental comparison of ordinary and universal kriging and inverse distance weighting. *Mathematical Geology*, **31**:375–390, 1999.

[405] D. L. Zimmerman and M. B. Zimmerman. A comparison of spatial semivariogram estimators and corresponding ordinary kriging predictors. *Technometrics*, **33**:77–91, 1991.

Index

A

Accuracy
  areal interpolation, 189
  cokriging vs. ordinary kriging,
    212
  local vs. global variograms,
    234
  spatially adaptive filtering, 110
  spatial prediction, 190
Acid deposition, 233
Acidification, 80
Adaptive approaches, 23–27
Adaptive bandwidth, 123, 259
Adaptive filtering, 110–111
Adaptive kernel estimation, 259
Adaptive window size, 47
Aggregation problem in areal in-
    terpolation, 173
Akaike Information Criterion (AIC),
    127, 136, 203
Altitude derivatives, 65–70
Analysis, scale of, See Scale of
    analysis
Anisotropy
  convolution approach for spa-
    tial prediction, 239
  geographically weighted regres-
    sion contexts, 124
  geometric and zonal, 202–203
  inverse distance weighting, 157
  variogram, 197, 202–203, 204,
    233
Anisotropy ratio, 29, 202
Anomalous point patterns, geograph-
    ical analysis machine, 265
ANUDEM, 172–173
Approximate interpolators, 145, 147

ArcGIS™, 124, 133, 142, 148
  ANUDEM algorithm (TOPOGRID),
    173
  Geostatistical Analyst exten-
    sion, 158, 275
  kernel estimation functions, 259
  Spatial Analyst, 258, 275
Areal data, 10, 13, 17–19
  centroid-based approaches, 10,
    81
  modifiable areal unit problem
    (MAUP), 3
  point predictions vs., 216
  surface model generation from,
    177–180
Areal interpolation, 18, 145, 173–
    189
  change of support problem, 173,
    216
  control zones, 184–186
  dasymetric mapping, 184–185
  intelligent interpolators, 184–
    185
  limitations, 189–190
  local models and local data,
    177
    population surface modelling,
     177–180
    use of prior knowledge, 184–
     189
    volume preservation, 181–
     184
  modifiable areal unit problem
    (MAUP), 3, 173
  overlay approach, 175–177
  pycnophylactic method, 171,
    181–184, 216

307

scale and aggregation problem, 173
spatial smoothing method, 178
target zones, See Target zones, areal interpolation
uncertainty in, 189
Areal weighting method, 175–176
ARIMA (autoregressive integrated moving average), 226
Arithmetic mean, 38
Artificial neural networks, 47, 240
Aspect derivative of altitude, 65–70
Atomistic fallacy, 108
At risk population, 10, 245, 260, 264–265, 267–269
Autocorrelation, spatial, See Spatial autocorrelation
Autocorrelation function (correlogram), 194, 196, 239
Automation, 273–274
  basis function selection, 61
  local point pattern analysis, 265
  nonstationary variogram model fitting, 233–234
  trend derivation for spatial prediction, 220
  variograms for seafloor classification, 240
Autoregression, See Spatial autoregressive models
Autoregressive integrated moving average (ARIMA), 226
Autovariograms, cokriging, 212
Azimuth, 65

B
Bandwidth, See also Window size
  adaptive, 259
  distance decay, 24
  Gaussian function, 74, 76
  kernels for intensity measurement, 258, 259
  kernels for ridge regression, 132

locally-adaptive kernels, 123
selection procedures, 126–127
Basis functions
  automated selection, 61
  completely regularised splines, 166
  orthogonality, 52
  thin plate splines, 161, 166, 215
  wavelets, 51–53
Bayesian Gaussian process model, 239
Bayesian geostatistical framework, 217
Bayesian regression approaches, 116–121
  geographically weighted regression, 124, 142, 276
  software tools, 142, 276
Bending energy function, 161
Best linear unbiased predictor (BLUP), 206
Biased variograms, 198, 221
  OLS residuals, 221–222
Bimodal distribution, segmentation techniques, 63
Biophysical relationships, spatial resolution, 7
Bi-square nearest neighbour scheme, 123, 132
Bivariate measures of spatial autocorrelation, 83
Block kriging, 206, 216
BLUP (best linear unbiased predictor), 206
Bounded variogram models, 201–202
British censuses, 178, 184

C
California Redwood seedling location dataset, 259
Cancer distribution, 119–121, 266–267

CAR (conditional autoregressive model), 107

Cartogram, 245

Categorical data and spatial association, 98

Census data, 9, 17–20, 76–81, 127, 146, 173, 178, 180, 184

Census enumeration districts, 184, 267

Census of quadrats, 247

Centroid-based areal data analysis, 10, 18, 81, 177–180

Centroid linkage segmentation, 64–65

Centroids, enumeration district, 267–268

Change of support problem, 173, 216

Chapter review, 271

Classification, spatially weighted, 138

Cluster detection, 35, 243, 264–269, See also Point pattern analysis
  births and infant death dataset, 19
  enumeration district centroids, 267–268
  geographical analysis machine, 265–268
  testing for local autocorrelation, 89–91

Cluster evaluation permutation procedure (CEPP), 268

Clustering
  contiguity enhanced k means algorithm, 138
  image segmentation, 62–65, 138
  Moran's I, 82
  point patterns, 246, See also Cluster detection

ClusterSeer® software, 268

Cluster size index, 249

CNSD (conditional negative semidefinite) variogram models, 199–201, 212

Coefficient of variation, geographically weighted, 78–79

Cokriging, 212–215
  accuracy, 212
  applying, 213–215
  linear model of coregionalisation, 212–213, 215
  prediction performance comparison, 215
  variograms and, 212–213

Collinearity, 113, 114, 124, 128–135

Colocated cokriging, ordinary (OCK), 215

Columbus, Ohio, crime data, 19, 112–113, 134–135

COMPASS, 18

Completely regularised spline (CRS), 166–167

Complete spatial randomness (CSR) model, 33, 246
  point pattern K function, 252–255, 264
  point pattern quadrat counts, 247–249

Compression, image, 61

Computational time, local interpolation methods, 145

Computerised Point Address Service (COMPASS), 18

Concavity derivative, 68

Conditional autoregressive model (CAR), 107

Conditional negative semidefinite (CNSD) variogram models, 199–201, 212

Conditional simulation, 215–216

Condition indexes, 133

Contiguity
  grid data, 39
  queen's case, 39, 81, 82, 90, 93
  rook's case, 39, 81, 82, 93
  spatial autocorrelation, 81

Contiguity enhanced k means clustering algorithm, 138

Contiguity ratio (Geary's c), 82, 94–97

Continuous wavelet transform (CWT), 49, 50

Contour values, DEM derivation, 157, 167, 171

Contrast enhancement, 38

Contrast stretching, 38, 59–60

Control process, population variations, 261

Control zones, areal interpolation, 184–186
    expectations maximum likelihood algorithm, 184–185, 187–189

Convexity derivatives of altitude, 65, 68

Convolution filters, 39

Convolution mask, 41–42

Convolution model, spatially varying regression effect, 119

Coocurrence matrix, 47, 240

Coordinate notation, 20

Coregionalisation, 212–213, 215

Correlogram, 194, 196

Covariance, 83, 195–196, See also Semivariance; Variance
    cokriging, 212–213
    generalised of order k, 226
    geographically weighted, 128–129, 138
    IRF-k kriging, 226
    kernel parameters, 239
    kriging, 207
    nonstationary models, 217, 233
    point-support, 216
    positive semidefinite, 200, 212
    second-order point pattern, 246
    spatial deformation models, 238
    spline, 215
    stationarity issues, 193–194
    stationary random function, 200

variance-covariance matrix, GLS model, 105–106

variogram, 195, 197, 233, 238

Covariance function, 194, 195–196
    nonstationary variograms, 238
    spatial deformation models, 238–239

Cressman decay function, 180

Crime data and spatial models, 19, 112, 134–135

Cross-correlation, 50

Cross K function, 31, 255, 264

Cross-validation (CV), 147, 190
    bandwidth selection, 126–127
    generalised (GCV), 130, 132, 160, 162–163
    inverse distance weighting performance assessment, 158
    neighbourhood size, 157
    ridge regression, 130, 132

Cross-validation prediction errors
    kriging with external drift model, 230
    moving window regression, 154
    nonstationary variograms, 234
    ordinary kriging spatial prediction, 210–211
    thin plate splines, 167–168

Cross-variogram, 198, 212

CRS (completely regularised splines), 166–167

CSR hypothesis, See Complete spatial randomness (CSR) model

Cubic variogram model, 201

Cumulative distribution function (cdf)
    indicator geostatistical approach, 216
    nearest neighbour point distances, 250
    random function models, 192

CWT (continuous wavelet transform), 49–50

D

Dasymetric mapping, 184–185

Data density and moving window size adaptation, 29, 30, 156

Data models, 9–10
  areal, See Areal data
  geostatistical, 10, See also Geostatistics
  grid, See Grid data analysis
  point patterns, 10, See also Point pattern analysis

Data segmentation, See Segmentation

Datasets (for illustrations of methods), 11–19

Datasets, local statistical model relationships, 32

Data transformation, local modelling, 24, 26, 30

DAUB4 wavelet, 54–55, 59

Daubechies basis function, 52

Daubechies wavelets, 49, 52, 54

Decimation, 54

Decomposable global statistics, 31

Decomposition, See also Wavelets
  Fourier transform, 48
  singular-value, 128–130
  standard deviation contrast stretching, 59–60
  two-dimensional wavelet transform, 58
  variance in GWR, 128–133
  wavelets, See Wavelet transforms

Deformation models, 26, 238–239

Delaunay triangulation, 151

DEM, See Digital elevation models

Density, population, 10, 18, 187–188

Density and window size adaptation, 29, 30, 156

Density-equalised point pattern, 245

Density function, point pattern intensity, 245

Dependence, spatial, See Spatial dependence

Detail, local, 54

Deterministic (trend) component, geostatistical model, 191
  separating deterministic and stochastic components, 220

Deterministic methods, spatial prediction, See Interpolation; Spatial prediction, deterministic methods

Detrending, 30, See also Trends
  intrinsic random functions of order k kriging, 225–226
  large scale trends, 2
  local modelling, 26, 30
  nonstationary variograms, 233
  OLS, 221–222
  trend-free variogram estimation, 217, 220

DFT (discrete Fourier transform), 49

D function, 255

Dichotomy of scale, 8

Differences across space, 1, 3, 4

Diffusion models, 189

Digital elevation models (DEM), 154
  contour line values, 157
  dataset, 11, 13
  derivatives of altitude, 65–70
  grid data analysis, 65–70
  kriging with external drift model, 231
  locally adaptive approaches, 171–173
  local terrain errors, 67
  PRISM and, 122
  wavelet applications, 62

Digital number (DN), 16, 37, 92, 93, 240

Digital orthophoto dataset, 11, 17, 60

Digital terrain models, 48

Digitised boundaries, 93, 240

Digitised images, 16
Directional variation, 124, See also
        Anisotropy
    inverse distance weighting, 157
    locally varying anisotropy, 233
    variogram estimation, 197
Directional variogram, 203–205, 230
Dirichlet polygons, 148
Dirichlet's integral, 182
Disaggregate behavior, 108
Disaggregated data, 7
Discrete Fourier transform (DFT),
        49
Discrete image transforms, 47, 48,
        See also Wavelet trans-
        forms
Discrete wavelet transform (DWT),
        49, 50–51
    implementation, 53–56
    standard deviation contrast stretch-
        ing, 59–60
    two-dimensional, 56–60
Disease distribution, 10, 119, 238,
        253, 260, 266–268
Disjunctive kriging, 217
Dispersion, index of, 249
Dissimilarity matrix, 138
Distance-based weighting, 24, 177–
        178, See also Inverse dis-
        tance weighting
Distance matrix, 24
Distance measures, non-Euclidean,
        239
Distance methods, point pattern,
        245, 249–255
DN (digital number), 16, 37, 92,
        93, 240
Double log variogram, 240
Downscale, 8
Drainage enforcement algorithm,
        171
Drainage networks, 68
Drift
        geostatistical terminology, 220

kriging with external, 215, 220,
        227–232
kriging with trend model, 220,
        223
DWT, See Discrete wavelet trans-
        form

E
Earthquake prediction, 61
Ecological fallacy, 7, 108, 173
Edge correction factor, 257
Edge detection, 39, 44
Edge effects, point patterns, 252,
        253
Edge preservation, 42
Edge sharpening, 44
Elevation, digital elevation mod-
        els, See Digital elevation
        models
Elevation and precipitation, See
        Precipitation-elevation re-
        lationship
Elevation derivatives, 65–70
Elevation measurement applications
    inverse distance weighting, 157
    locally-adaptive approaches for
        DEM construction, 171–
        173
Enumeration district (EM) centroids,
        267–268
Environmental heterogeneity, 260
Environmental processes, scale de-
        pendence, 8
Error
    cross-validation prediction, See
        Cross-validation prediction
        errors
    local errors in terrain, 67
    mean integrated square, 259
    measurement and nugget vari-
        ance, 199
    RMSE, See Root mean square
        error
    spatial error model, 311
Euler constant, 166

Event intensity measurement, 245–249

    local methods, 256–260

Exact interpolators, 145, 147, 154

Expectation maximum likelihood (EM) algorithm, 184–185, 187–189

Experimental variogram, 197–198

Exponential variogram model, 201

External control zones,' See Control zones, areal interpolation

External drift, See Drift; Kriging with external drift

Extrapolation, 224

F

Fast Fourier transform (FFT), 49

Fast wavelet transform (FWT), 53–56

Father wavelet, 49

FFT (fast Fourier transform), 49

F function, 250–251

Filtering, spatially adaptive, 110–111

Filters

    ARIMA, 226

    convolution, 39

    local approaches for grid data

        high pass, 42–44

        low pass, 41–42

        quadrature mirror filter pair, 53

        spatial filters, 39–41

        two-dimensional wavelet transform, 57

        wavelet coefficients, 53–54

    use of local models, 3

Finite difference methods, 171–173

Finite variance variograms, 199, See also Sill, variogram

First Law of Geography, 5

First-order properties, point patterns, 245–247

Fitted trend model, spatial prediction, 145

Fitting

    polynomial trend models, 30

    variogram models, 203

Fixed size windows, 47

Focal operators, 28, 39

Focused local space, 31

Forest stand identification, 240

Fourier transform, 47, 48–49

Fractal analysis, 8

Fractal dimension, 240

Frequency components, 42–43

Frequency localisation, Fourier functions and wavelets, 49

Freshwater acidification, 80

Friction surfaces, 68

Future developments, 273–274

FWT (fast wavelet transform), 53–56

G

Gaussian distribution, 124

    fitting variogram models, 203

    inappropriate assumptions, 217

Gaussian kernel and correlation function, 239

Gaussian prior distribution, 120

Gaussian process model, 239

Gaussian simulation, sequential, 216

Gaussian variogram model, 201

Gaussian weighting scheme, 42, 74, 123

Gauss-Seidel iteration, 172

GCV, See Generalised cross-validation

Geary's contiguity ratio (c), 82, 94–97

Generalised additive models (GAMs), 168

Generalised covariance (GCV) of order k, 226

Generalised cross-validation (GCV), 130, 132, 160, 162–163

Generalised least squares (GLS), 105–106, 203, 204, 222, 234

GeoDa™ software, 83, 86, 89, 90, 275

Geographical analysis machine (GAM), 265–268

Geographical applications, spatial scale in, 3, 8–9

Geographical Information Science (GIScience), 2

Geographical Information Systems (GISystems), 1, 3, 241, 272
    DEM analysis, 65–70
    geographical analysis machine, 265
    gridded data analysis, 38
    IDW applications, 157, See also Inverse distance weighting
    interpolation methods, 145
    measures of spatial autocorrelation, 81
    software tools, 273
    wavelets, 48, See also Wavelet transforms

Geographically weighted regression (GWR), 101, 105, 109, 122–138, 143
    alternative approaches, 142
    anisotropy, 124
    application illustration, 124–126
    bandwidth issues, 123
    bandwidth selection, 126–127
    Bayesian implementation, 124, 142
    case study, 135–138
    collinearity effects, 128–135
    condition indexes, 133
    goodness-of-fit, 124
    MWR and, 105, 154, See also Moving window regression

physical geography applications, 127
    random coefficient modelling vs., 116
    ridge regression (GWRR), 130–135
    significance testing, 127–128
    slopes, 137, 140
    spatial interaction model, 138
    variance decomposition, 128–130, 133

Geographical weighting schemes, 6, 24, 35, 74–80, See also Inverse distance weighting; Weighting
    coefficient of variation, 78
    common approaches, 24
    covariance, 128–129, 138
    differences of global and local mean, 77
    distance matrix, 24
    Gaussian function, 42, 74
    location notation, 25
    mean, 74–76, 156
    moving windows, 28
    moving window variogram, 238
    principal components analysis (GWPCA), 138
    skewness, 80–81
    spatial scale analysis, 8
    standard deviation, 76–77
    standard score, 77, 79
    variance-covariance matrix, 138

Geography, First Law of, 5

Geological dataset (minerals in rock), 11, 13, 258–260

Geometric anisotropy, 202–203

Geostatistical Analyst, 158, 275

Geostatistical data, 10

Geostatistical software, 241, 275

Geostatistical Software Library (GSLIB), 210, 227, 230, 275

Geostatistics, 10, 35, 241
    anchor algorithm of, 208, See also Ordinary kriging

anisotropy, 198, 202–203
change of support problem, 216
conditional simulation, 215–216
exploring spatial variation, 195–204
  covariance function, 195–196
  variogram, 197–204, See also Variograms
globally constant mean, 206–207
global models, 194–195
indicator approach, 216
kriging, 204–215, See also Kriging
limitations, 216
local approaches, nonstationary models, 194, 217–219
locally constant mean models, 208
model-based, 217
multidimensional scaling, 238–239
notation, 20
other approaches, 216–217
random function models, 191, 192
  stationarity, 193–194
regionalised variables, 191–192
separating deterministic and stochastic components, 220
stationarity, 193–194
texture analysis, 240
trend and drift terminology, 220
use of probabilistic models, 191
Getis and Ord statistic, 85, 93, 96–97
G function, 250–251
Gibbs sampling, 118
GIScience, See Geographical Information Science
GIS GRASS, 61, 275
GISystems, See Geographical Information Systems

Global 30 arc-second (GTOPO 30) digital elevation model, 11–13
Global centring, 130–131
Global estimate of intensity, 33
Global mean, geographical weighted difference from local mean, 77
Global mean and variance of nearest-neighbour distances, 33
Global measures, spatial autocorrelation, 80–84
Global methods, 1
  deterministic spatial prediction, See Spatial prediction, deterministic methods
  interpolation, 1, 145, See also Interpolation
  local methods vs., 2–4
  point interpolation, 147–148
  point patterns, 1, 243, 246–247
Global models
  decomposable global statistics, 31
  defining, 23
  detrending, 2, 30
  geostatistical spatial prediction, 194–195, See also Geostatistics
  limitations, 272, 274
  relationship between global and local statistics, 32–34
  spatial expansion method, 111
Global regression, 101–103, See also Regression
  accounting for spatial dependence, 103
  spatial prediction, 148
Global space, local space relationship, 32
Global trends, See also Large scale trends; Trends
  directional variogram, 230
  kriging with a trend, 224

removal, 2, 26
spatial expansion method, 101
Global univariate statistics, 38
GLS, See Generalised least squares
Goodness-of-fit
  Akaike Information Criterion,
    127, 203
  geographically weighted regres-
    sion, 124
  variogram models, 203
Gradient derivative of altitude, 65–
    69
Gradient operators, 39, 40, 44
Granite slab mineral positions, 11,
    13, 258–260
GRASS GIS, 61, 275
Great Britain DEM dataset, 13,
    See also Digital elevation
    models
Great Britain population data and
    models, 184
Great Britain precipitation dataset
    and models, 11–13, 135–
    138, See also Precipitation-
    elevation relationship
Grey levels
  cooccurrence matrix, 47, 240
  negative autocorrelation, 80
  texture, 44, 47
Grey-scale transformations, 38
Grid data analysis, 9–10, 34, 37, 44
  additional tools, 71
  contiguity arrangements, 39,
    See also Contiguity
  digital elevation models, 65–
    70
  edge detection, 39
  exploring spatial variation in
    single variables, 37
  finite difference-based spatial
    prediction, 172
  global univariate statistics, 38
  local univariate statistics, 38
  moving windows, 3, 10, 38,
    39–47, See also Moving

window methods
  multivariate methods, 37
  segmentation, 62–65
  standard deviation, 28
  wavelets, 48–62, See also Wavelet
    transforms
GSLIB, 210, 227, 230, 275
Gstat, 203
G statistic, local measure of auto-
    correlation, 85–86
GTOPO 30, 11–13
Guard zones, 252, 257
GWPCA (geographical weighted prin-
    cipal components analy-
    sis), 138
GWR, See Geographically weighted
    regression
GWRR (geographically-weighted ridge
    regression), 130–131

H
Haar wavelet, 51–53
Hat matrix, 127, 162, 164
Heterogeneity
  environmental, 261
  variables across zones, 176
Hierarchical models
  Markov chain Monte Carlo (MCMC)
    methods, 118
  multilevel modelling, 108–109,
    143
Hierarchies of data, 7
High frequencies, 2D wavelet trans-
    forms, 57
High pass filters, 42–44
Histogram, 38
Hole effect, 202
Homogeneity
  assumption for areal weight-
    ing, 176
  assumption for dasymetric method,
    185
  geometrically-weighted standard
    deviation, 77
  population at risk, 245

segmentation, 38, 62–64
spatial scale and, 7, 263
stationarity and, 193
Horizontal edges, 57, 59
Human vision applications, 61
Hybrid linkage region growing, 63,
64
Hydrological modelling application,
172
Hyperparameters, 117
Hyperprior distributions, 117

I
ICS (index of cluster size), 249
IDW, See Inverse distance weighting
Image analysis and processing
discrete transforms, 47
focal operators, 28, 39
frequency components, 42–43
local approaches for grid data,
39
local filters, 3
moving windows, See Moving
window methods
segmentation, 62–65, 138, See
also Segmentation
spatial resolution and, 5
standard deviation contrast stretching, 59–60
texture, 44–47, 240
Image compression, 61
Index of cluster size (ICS), 249
Index of dispersion, 249
Indicative Goodness of Fit, 203
Indicator approach, 216
Integral wavelet transform, 50–51
Intelligent interpolators, 184
Intensity, global estimate, 33
Intensity, point pattern, 243, 245–
249
density function, 245
kernel estimation, 247, 256–
257
local methods, 256–260

population at risk, 260
Intercept removal, ridge regression
in GWR, 130–131
Intercept variation, multilevel model,
108
Interpolation
areal, 145, 173–189, See also
Areal interpolation
change of support problem, 173,
216
exact and approximate interpolators, 145
finite difference methods, 171
geostatistical, See Kriging
global vs. local methods, 145
inverse distance, See Inverse
distance weighting
locally adaptive approaches for
DEM construction, 171–
173
location notation, 25
natural neighbours, 151, 153
point, 145, 146–173, See also
Point interpolation
prediction performance vs. krig-
ing/cokriging, 215
splines, 158–168, See also Thin
plate splines
weighting schemes, See Weighting
Intrinsic random functions of order
k (IRF-k) kriging, 220,
225–226
Intrinsic stationarity, 194
Inverse distance weighting (IDW),
6, 24–25, 145
exact interpolator, 154
local anisotropy, 157
nonstationary kriging models
vs., 225
performance assessment, 158
precipitation prediction application, 157–158
prediction performance comparison, 215

spatial prediction application,
        154–158
Inverse Fourier transform, 48
Invertibility, 181
ISATIS software, 226
Isotropic point process, 246
Isotropic weights, 47
Isotropy, trend-free variogram, 230
Isotropy assumption and spatial
        dependence, 246
Iterative finite difference interpola-
        tion, 171

J
Jackknifing, 147, 157
Joins count approach, 33, 81, 98

K
Kernel bandwidth, 123, 132, 258,
        259
Kernel estimation (KE)
    cluster analysis, 264
    comparing KEs, 259
    edge correction, 257
    event intensity measurement,
        247, 256–257
    naive estimator, 256–257
    software, 259
Kernel methods, See also Kernel
        estimation
    anisotropy ratio, 29
    convolution approach for spa-
        tial prediction, 239
    geographically-weighted regres-
        sion, 123
    locally-adaptive modelling, 26,
        27, 28–30
    local univariate statistics, 38
    parameter estimation, 239
    quartic kernel, 29, 257–258
    smoothing gridded data in mov-
        ing windows, 42
    spatially adaptive kernels, 123
    spatial scale analysis, 8
K function, 251–255, 259

cross *K*, 31, 255, 264
L function, 253
local, 261–264
local *L* function, 262
spatial scale analysis, 8
k-means clustering algorithm, 138
Kriging, 204–215, 241
    block, 206, 216
    BLUP, 206
    change of support problem, 216
    cokriging, 212–215
    conditional simulation, 215–
        216
    disjunctive, 217
    equivalence of splines, 215
    globally constant mean, 206–
        207
    intrinsic random functions of
        order *k*, 220, 225–226
    limitations, 216
    locally constant mean, 208
    locally-varying mean and, 222–
        233
    median polish, 220, 222
    modelling anisotropic structures,
        202
    nonstationary, 208
    optimal weights, 206, 207
    ordinary, 207–211
    punctual or point, 206
    segmentation and, 240
    simple, 206–207
    software, 241
    spatially-varying mean, 194–
        195
    unbiased, 206
    variogram, 198, 202, See also
        Variograms
        moving window approaches,
        233–234
    weight selection, 204
Kriging, nonstationary models for
        prediction, 222–233
    intrinsic random functions of
        order *k*, 220, 225–226

kriging with a trend, 220, 222

kriging with external drift, 215, 220, 227–232

median polish kriging, 220, 222

simple kriging with locally varying mean, 215, 220, 226–227

using secondary data, 226–231

Kriging with a trend (KT), 220, 222–225

IRF-k kriging prediction performance vs., 226

software applications, 241

Kriging with external drift (KED), 215, 220, 227–232

L

Lag, variograms, 197, 199, 233

Lagrange multipliers, 209, 214

Landcover data, 11, 13, 16

edge detector, 44

K function application, 264

multilevel modelling, 109

population estimation using, 185–187

texture, 45–46

Landform spatial variation, 198

Landsat Thematic Mapper (TM), 11, 13, 47

Land use

areal interpolation, change of support problem, 173

multilevel modelling, 109

spatial autocorrelation, 93

spatial prediction application, 187

Laplace equation, 171, 182

Laplacian differential operator, 44

Large scale trends, See also Global trends; Trends

kriging with a trend model, See Kriging with a trend

modelling and removing, 220–222

removal, 2, 30

spatial expansion method, 101

unbounded variogram model, 202

Least squares methods

detrending data, 221

drift in geostatistical contexts, 220

generalised, 105–106, 203, 222, See also Kriging

OLS, 101–103, 105–106, 203, 221, 222

outliers and, 221

polynomial fitting and trend surface analysis, 147

thin plate smoothing splines, 160

weighted, for fitting variogram models, 203

L function, 253, 262

LiDAR, 157

Light Detection And Ranging (LiDAR) sensor, 157

Linear geostatistics, See Geostatistics

Linearly decomposable global statistics, 31

Linear model of coregionalisation, 212–213, 215

Linear regression methods, See Regression

Linear weighted moving average, 208

LISA (local indicators of spatial association), 83, 86, 97

Local, defining, 4–5

Local anisotropy, See Anisotropy

Local centring, 130–131

Local cluster detection, See Cluster detection

Local detail, 54

Local filters, See Filters

Local indicators of spatial association (LISA), 83, 86, 97

Localised models, 29–30

Local K function, 261–264

Local L function, 262
Locally-adaptive Bayesian modelling, 116–121
Locally-adaptive geostatistical models, 194
Locally-adaptive kernel bandwidth, 123
Locally-adaptive window, 29
Locally-varying mean, simple kriging with, 215, 220, 226–227
Locally varying model parameters, 30
Local measures, spatial autocorrelation, 85–97, See also Spatial autocorrelation
Local methods
    deterministic spatial prediction, See Spatial prediction, deterministic methods
    event intensity measurement, 256–260
    global methods vs., 2–4
    interpolation, 145, 148–160, See also Interpolation; specific methods
    locally-adaptive approaches, 23–27
    point patterns, See Point pattern analysis
    types, 4
Local models and methods, structure of the book, 34–35
Local models and modelling
    applications, 3
    categorisation, 25–26, 30–34
        global and local statistic relationship, 32–34
        relationship between global and local spaces, 32
        spatial subsets, 31
    data models, 9–10
    data transformation, 24
    defining, 23–24
    interest in, 2

localised models, 29–30
locally-adaptive approaches, 23–27, See also specific approaches
locally-varying parameters, 30
moving window/kernel methods, 26, 28–30, See also Kernel methods; Moving window methods
need for, 3
nonstationarity, See Nonstationarity
nonstationary geostatistical, 217–219
problematic issues, 5
quantitative geography developments, 4
relationship between global and local statistics, 32–34
remit of this book, 2
selecting, 274
stationarity, See Stationarity
stratification/segmentation of spatial data, 26, 27, See also Segmentation
transforming and detrending spatial data, 26, 30, See also Detrending
Local Moran's I, 87–89, 92–93, 256
Local nearest-neighbours distance, 33–34
Local point interpolation, See Point interpolation
Local point pattern analysis, cluster detection, 264–269
Local regression, 109–111, See also Regression
    advantages and limitations, 142
    alternative approaches, 142
    GWR, See Geographically weighted regression
    kriging with trend model
    MWR, See Moving window regression

random coefficient modelling, 116–121
significance testing, 127–128
spatial prediction application, 153–154
variogram estimated from residuals
Local segregation indices, 98
Local space, 4–5
focused and unfocused, 31
global space relationship, 32
Local spatial autocorrelation, See Spatial autocorrelation
Local spatial statistical analysis (LoSSA), 31, 34, 264
Local standard deviation, texture characterisation, 45–46
Local summary statistics, 73
Local univariate statistics, grid data, 38
Local variation, spatial dependence, See Spatial dependence
Local variation, standard approaches for assessing, 23
Location, notation, 20, 25
Log-ratio data, 80, 85
Look-up tables, 174
LoSSA (local spatial statistical analysis), 31, 34, 264
Low pass filtering, 41–42

M
Marked point process, 255
Markov chain Monte Carlo (MCMC) methods, 118, 217
Masks, convolution, 41–42
Mass preservation, 181, 184, 216
Matrices, 1
cokriging, 213–214
cooccurrence, 47, 240
coregionalisation, 213
digital elevation models, 67
dissimilarity, 138
Fourier transform, 49

geographically weighted regression, 123, 127, 129–131
geographical weighting, 24, 138
global regression, 102
hat, 127, 162, 164
kriging with a trend model, 224
ordinary kriging, 209
pycnophylactic method, 183
random coefficient modelling, 119–120
simple kriging, 207
spatial autoregression, 106
spatial proximity, 83
thin plate splines, 161–164
variance-covariance, 105–106, 138
wavelet transforms, 53–56
MAUP (modifiable areal unit problem), 3
Maximum likelihood (ML)
expectations maximum likelihood algorithm, 184–185, 187–189
fitting variogram models, 203
nonstationary variograms, 234
per-field variogram inputs, 240
simultaneous autoregressive model estimation, 107
texture characterisation, 47
Mean, geographically weighted, 74–76, 156
Mean, spatial variation, 39, See also Moving average
kriging, 194–195, 206–208
globally constant, 206–207
locally constant, 208
nonstationary, 220–222
nonstationary models, 222–233
ordinary kriging applications, 208
simple kriging with locally varying mean, 215, 220, 226–227

nonstationarity, 233
nonstationary geostatistical mod-
    els, 217–219
nonstationary mean, 220–222
nonstationary model, 194
    simple kriging with locally vary-
        ing mean, 215
Mean integrated squared error (MISE),
    259
Measurement error, nugget vari-
    ance, 199
Measurement scales, 7
Median computation in moving win-
    dow, 42
Median polish kriging, 220, 222
Mediterranean landcover, 27
Meyer basis functions, 52
Mineral location, 11, 13, 258–260
Minimum error variance, 207
MISE (mean integrated squared er-
    ror), 259
Missing data, expectation maxi-
    mum likelihood approach,
    187
ML, See Maximum likelihood
MLM, See Multilevel modelling
MLwiN, 109
Mode computation in moving win-
    dow, 42
Model-based geostatistics, 217
Modifiable areal unit problem (MAUP),
    3, 173
Monte Carlo methods
    areal interpolation accuracy as-
        sessment, 189
    cluster evaluation permutation
        procedure, 269
    cluster significance assessment,
        266
    Markov chain Monte Carlo (MCMC)
        methods, 118, 217
    point pattern applications, 253,
        264
    spatial autocorrelation, random
        permutation test, 84

stationarity significance test-
    ing, 128
Moran scatterplot, 23, 83, 90
Moran's *I*
    case study, 93–94
    global, 82–83
    local, 87–89, 92–93, 256
Mother wavelet, 49–52, 59
Moving average, 38
    ARIMA, 226
    weighted, 154, 208–210, 215
Moving window methods, 26, 28,
        35, 41–42, 44–47, 272
    detrending, 2
    edge detection, 44
    edge preservation, 42
    focal operators, 28
    Fourier transforms, 49
    fractal dimension estimation,
        240
    grid data analysis, 3, 10, 38,
        39–47
    image texture, 44–47
    locally-adaptive approaches, 29
    local modelling, 26, 28–30
    low pass filtering, 41–42
    median and mode computa-
        tion, 42
    ordinary kriging and constant
        mean, 208
    point pattern quadrat counts,
        247
    rotating windows, 47
    spatial filters, 39–41
    standard deviation, 28
    summary statistics, 73
    texture analysis, 240
    variogram, 233, 238, 272
    weighting schemes, 28, See also
        Weighting
    window size, 29, 30, 39, 156,
        189
    adaptive, 47
    Fourier transform, 49

Moving window regression (MWR), 105, 122
   application illustration, 124–126
   case study, 135–138
   simple kriging with locally varying mean, 227
   spatial prediction application, 154
Multicollinearity, See Collinearity
Multidimensional scaling, 238–239
Multilevel modelling, 105, 108–109, 143
   random coefficient modelling, 116
Multiresolution analysis, 51, 53
Multivariate data analysis, 4, See also Regression; Spatial relations
   gridded data analysis, 37
MWR, See Moving window regression

N

Naive estimator, 256–257
Natural neighbour interpolation, 151, 153
Nearest neighbour methods, 148, 250
   bi-square scheme, 123, 132
   digital elevation models, 13
   geographically-weighted regression, 123, 132
   geographical weighted schemes, 24
   global and local models, 33–34
   inverse distance weighting, 156
   local Moran's I, 87
   point pattern analysis, 250–251
   spatially adaptive filtering, 110
   Thiessen polygons, 148, 151, 153
Negative autocorrelation, 5

Neighbour contiguity arrangements, 39, 81, 82
Neighbourhood size identification, 157
Nested spatial structures, 105
Neural networks, 47, 240
Newark, New Jersey, digital orthophoto, 11, 17, 60
Noise
   high pass filters and, 42
   reducing using smoothing operators, 40
   wavelet applications, 61
   white, Gaussian process, 239
   white, thin plate smoothing splines, 161
Non-Euclidean distance measures, 239
Nonstationarity, 3, 15, See also Nonstationary models; Stationarity
   mean, 220–222, 233
Nonstationary mean, 220–222, 233
Nonstationary models, 9, 15, 24, 194, 208
   geostatistics, 194, 217–219, 220, See also Ordinary kriging; Random function (RF) model
   significance testing, 127–128
   spatial prediction, 9, 222–233, See also Kriging, nonstationary models for prediction
   variogram, 233–240, 272
Normal distribution, See also Gaussian distribution
   kriged prediction reliability and, 210
   spatial autocorrelation issues, 88
Normalised Haar basis, 52
Normality
   orthonormal basis function, 52
   variogram issues, 216

North Carolina births and SIDS
    dataset, 19, 268
Northern Ireland population data,
    11, 17–19, 76–81, 127, 146,
    180
Notation, 20, 25
Nugget effect, 199, 201, 221, 233
Nugget variance, 199, 221
Null hypothesis of complete spatial
    randomness, 248

O
OK, See Ordinary kriging
OLS, See Ordinary least squares
    (OLS) methods
Omnidirectional variogram, 203–
    204
One-dimensional filters, 57
Online tools, 264
Ordinary colocated cokriging (OCK),
    215
Ordinary kriging (OK), 207–211
    accuracy, local vs. global, 234
    case study, 210
    cokriging equations, 213
    kriging with trend model vs.,
        224–225
    limitations, 216
    reliability, 210
    software applications, 241
Ordinary least squares (OLS) meth-
        ods, 101–103, 105–106, 203,
        221–222
Orthogonal basis, wavelet, 52
Orthonormal basis, 52
Orthophoto, digital, 11, 17, 60
Oscillating function, 60
Outliers
    local nearest-neighbours dis-
        tance, 34
    Moran scatterplot, 23
    OLS and, 221
    smoothing operators, 40
    variograms and, 198

Overlay approach, areal interpola-
    tion, 175–177
Overview of the book, 21

P
Paired observations
    experimental covariance func-
        tion, 196
    nonstationary variograms, 233
    second-order point pattern in-
        tensity, 246
    trend-free variogram estima-
        tion, 230
    variogram, 197
    variogram cloud, 195
Parabolic behavior, variograms, 201
Partial thin plate splines, 165
Penalised regression splines, 168
Per-field image analysis approaches,
    27, 47
Per-field variograms, 240
Periodicity, 202
Permeability in sedimentary struc-
    tures, 198
Permutation-based tests for spatial
    autocorrelation, 84, 88
Permutation procedure, cluster eval-
    uation, 269
Physical geography, 4, 8, 73, 139,
    192, 244
Pits, 153, 157, 167, 171
Pixel operations, global univariate
    statistics, 38
Plan convexity, 65
Point interpolation, 145, 146–173
    areal data vs., 216
    exact and approximate inter-
        polators, 145, 147
    finite difference methods, 171–
        173
    global methods, 147–148
    inverse distance weighting, 154–
        158
    limitations, 189–190

local methods, 148–160, See also specific methods

local regression, 153–154

natural neighbours, 151, 153

partial thin plate splines, 165

polynomial fitting and trend surface analysis, 153

predictor performance assessment, 147

splines, 158–168, See also Thin plate splines

Thiessen polygons, 148, 151, 153

triangulation, 151

Point kriging, 206

Point operations, global univariate statistics, 38

Point pattern analysis, 10, 35, 243–244

cluster detection, 243, 264–269

complete spatial randomness, 246

distance methods, 245, 249–255

cross K function, 255

K function, 251–255, 259

nearest neighbour, 250–251

global approach, 1, 243

intensity estimation, 243, 245–249

kernel estimation, 247, 256–257

local methods, 256–260

population at risk, 260

local K function, 261–264

mineral dataset, 13

online tools, 264

population at risk, 10, 245, 264–265, 267–269

quadrat count methods, 247–249

second-order, 33

visual examination, 244–245

Point patterns, 10, 243–244, See also Point pattern analysis

edge effects, 252, 253

first- and second-order properties, 245–247

marked point process, 255

random variable representation, 245

scale characterisation, 252

statistical test, 246

Point separation, 245

Point-support covariance model, 216

Poisson distribution, 119–120

point pattern quadrat counts, 248–249

point patterns, 269

Poisson parameter estimation, 187

Poisson process, complete spatial randomness model, 246

Poisson regression

geographically weighted regression, 124

spatial prediction application, 186–187

Polygons, Thiessen, 148, 151, 153, See also Dirichlet polygons; Voronoi polygons

Polynomial fitting

digital elevation models, 65

trend surface analysis, 147, 153

Polynomial modelling, local regression, 109

Polynomial trend model, 23, 30, 221

Population at risk, 10, 245, 260, 264–265, 267–269

Population census temporal linkage, 173, 184–185

Population density, 10, 18, 187–188

Population estimation, 185–189

Population intensity, 260, 264

Population mapping from zone centroids, 177–179

Population of Northern Ireland, 11, 17–19, 127, 146, 180

Population surface modelling, 177–180

Population variations, control process, 261

Population-weighted centroids, 18

Portugal, 122, 215

Positive semidefinite covariance, 200, 212

Positive spatial autocorrelation, 5

Postal addresses, spatially-referenced, 18

Posterior distribution, 116, 118

Power model for variograms, 202

Precipitation-elevation Regressions on Independent Slopes Model (PRISM), 122

Precipitation-elevation relationship British dataset, 11–13
    digital elevation models, See Digital elevation models
    global regression, 101–103
    kriging and cokriging, 215
    local variogram modelling, 234, 237
    PRISM, 122
    regression applications, 122, 124–126, 135–141, 153–154
    spatial autocorrelation, 80
    variogram and SKlm, 227–230

Precipitation prediction or modelling
    inverse distance weighting, 157–158
    kriging with a trend model, 221
    kriging with external drift model, 238
    ordinary kriging, 210–211
    thin plate splines, 163–165, 167–170
    variogram, 203–204, 234

Prediction, spatial, See Spatial prediction

Prediction errors, See Error

Predictor-corrector, 110–111

Prewitt operator, 44

Principal components analysis (PCA), 138

Prior distribution, 116–117, 120

Prior knowledge, areal interpolation, 184–189

PRISM, 122

Probabilistic models, 191, See also Geostatistics

Probability distribution
    geographical analysis machine, 268
    local Moran's I, 88
    quadrat counts, 248–249

Profile convexity, 65

Punctual kriging, 206

Pure nugget effect, 199

Pycnophylactic areal interpolation method, 171, 181–184, 216

Q

Quadrat analysis, 247–249, 256

Quadratic surface, 67

Quadrature mirror filter pair, 53

Quantitative geography, 4

Quartic kernel, 29, 257–258

Quasi-intrinsic stationarity, 194

Queen's case contiguity, 39, 81, 82, 90, 93

R

Random (stochastic) component, geostatistics, 191

Random coefficient modelling (RCM), 116–121

Random function (RF) model, 191, 192
    intrinsic random functions of order k kriging, 220, 225–226
    stationarity, 193–194, 220

Random permutation test, 84, 88

Random sampling, Monte Carlo methods, 266

Random sampling variations, 4

Random slopes and intercepts model, 108

Random variable (RV) representation, point patterns, 245

Raster data, 9, See also Grid data analysis

RCM (random coefficient modelling), 116–121

Regionalised variable (ReV), 191–192

Region-growing segmentation, 27, 64, 240

Regression, 127–128

  accounting for spatially varying relationships, 103

  accuracy assessment, 189

  autoregressive models, 103, 106–107, See also Spatial autoregressive models

  Bayesian spatially varying coefficient modelling, 116–121

  collinearity effects, 113, 114, 124, 128–135

  Gaussian distribution assumption, 124

  geographically-weighted, See Geographically weighted regression

  global, 101–103

  GLS, 105–106

  local, 109–111, See also specific methods

    GWR, See Geographically weighted regression

    MWR, See Moving window regression

    random coefficient modelling, 116–121

  mapping residuals, 103

  moving window, See Moving window regression

  multilevel modelling, 108

OLS, See Ordinary least squares (OLS) methods

penalised regression splines, 168

Poisson regression, spatial prediction application, 186–187

polynomial modelling, 109

prediction performance vs. kriging/cokriging, 215

ridge, 113–114, 130–135

spatial data and, 105–106

spatial interaction modelling, 4

spatial prediction application, 153–154

using spatial subsets, 109

variance-covariance matrix, 105–106

Regression residuals, See Residuals

Regularised splines, 166–167

Remotely-sensed imagery, 3, See also Image analysis and processing

  areal interpolation, 173

  change of support problem, 216

  edge detection, 44

  geostatistics, 192, 240

  grey scale autocorrelation, 80

  gridded data analysis, 37, 38

  grid operations, 10, See also Grid data analysis

  negatively autocorrelated grey scales, 80

  nugget variance in estimating measurement error, 199

  spatial resolution, 7, 45

  texture, 45–47

  wavelet applications, 48, 61, See also Wavelet transforms

Residuals

  autocorrelation, 105

  global detrending, 2

  global regression applications, 103

identifying local variations, 23

trend-free variogram estimation, 220–222

Residual sum of squares (RSS), 234

Resolution, See Spatial resolution

Restricted maximum likelihood (REML) estimation, 203

Ridge regression, 113–114, 130–135

RMSE, See Root mean square error

Roberts operator, 44

Rook's case contiguity, 39, 81, 82, 93

Root mean square error (RMSE), See also Cross-validation prediction errors

    inverse distance weighting performance assessment, 158

    kriging spatial prediction, 210–211

    kriging with external drift model, 230

    moving window regression for spatial prediction, 154

    nonstationary variograms, 234

    thin plate spline applications, 167–168

Rotating windows, 47

Roughness penalty, 171–172

R programming language, 80, 273, 275

    spatial autocorrelation application, 80

R squared (r^2), 112–113, 124, 126, 128–129, 136–137, 141

R statistic, point patterns, 251

S

S+ SpatialStats™ software, 253

S+ Wavelets® software, 59

Sample spacing

    IDW cross-validation errors and, 157

    nugget effect, 199, 201

prediction accuracy and, 190

spatial scale issues, 8, 272

variogram, 199

Sampling, Gibbs, 118

Sampling configuration, inverse distance weighting, 157, 225

Sampling density variation, triangulation, 151

Sampling distribution, 83–84, 98, 253

Sampling strategy, 8

    variogram, 198–199, 202

Sampling variations, 4

SAR (simultaneous autoregressive model), 107

SaTScan™ software, 269

Scale, dichotomy of, 8

Scale, spatial, See Spatial scale

Scale-dependent models and processes, 7–8

Scale of analysis, 3, 29, 127

    point patterns, 251–252

Scale of spatial variation, 7, See also Spatial scale

Scale problem in areal interpolation, 173

Scales of measurement, 7

Scaling, multidimensional, 238–239

Scaling function, wavelet, 51–53, 56–59

    DAUB4, 55

    Fourier transform, 49

    Haar, 51

    scaling factor, 57

    two-dimensional wavelet transform, 56

Scatterplot, 23, 83, 90, 244

Scotland, 119

Seafloor classification, 240

Secondary data, 147, 207, 226, 230

Secondary variables, spatial prediction, 148, 154, 167, 212, 215, 220, 226–228, 230, 231

Second-order effects, spatial dependence, 105

Second-order point pattern analysis, 33

Second-order properties, point patterns, 246

Second-order stationarity, 193

Sedimentary structures, 198

Segmentation, 62–65
  centroid linkage, 64–65
  edge detection, 44
  local modelling, 26, 27
  region growing, 27, 64, 240
  spatial clustering, 138
  stratified kriging, 240
  variogram application, 240

Segregation indices, 98

Semivariance
  cokriging, 214
  kriging with trend model, 224
  ordinary kriging, 209
  variograms, 197, 199, 202, 204

Semivariogram, 197

Sequential Gaussian simulation (SGS), 216

Sharpening, 3, 44

Significance testing, geographically-weighted regression, 127–128

Sill, variogram, 199, 201, 202, 230, 233–234

Simple kriging (SK), 206–207
  kriging with external drift model, 228

Simple kriging with locally varying mean (SKlm), 215, 220, 226–227

Simulation, conditional, 215–216

Simulation, Monte Carlo, See Monte Carlo methods

Simultaneous autoregressive model (SAR), 107

Singular-value decomposition (SVD), 128–130

SK (simple kriging), 206–207

Skewed data, ordinary kriging reliability and, 210

Skewed distribution, variogram bias and, 198

Skewness, geographically weighted, 80–81

Slope
  geographically weighted regression, 137, 140
  local variation models, 108, 118
  random intercepts and slope model, 108

Slope derivatives of altitude, digital elevation models, 65–69
  errors, 67
  PRISM, 122

Smoothing, 40
  kriging and, 215
  local approaches for grid data, 39–42
    low pass filtering, 41–42
    spatial filters, 39–41
  population surface modelling, 178
  spline approaches for spatial interpolation, 158–168, See also Thin plate splines
  wavelet coefficients, 53–54

Sobel operator, 44

Socioeconomic contexts, 5, 7, 109

Socioeconomic data, 73

Software, 1, 272–273, 275–276, See also specific programs
  geostatistical, 241
  wavelet transforms, 61
  web addresses, 276

Soil erosion, 65

Source zones, 173–176, 181–189

Spaces, differences across, 1, 3, 4

Spatial Analyst, 258

Spatial autocorrelation, 4, 5–6, 73
  allowing for second-order effects, 105

bivariate measures, 83
global measures, 80–84
identifying local clusters, 89–91
joins-count statistic, 81
local cluster identification, 89–91
local measures, 81, 85–97
    case study, 93–94
    categorical data, 98
    Geary's contiguity ratio, 94–97
    G statistic, 85–86
    local indicators of spatial association, 83, 86, 97
    Moran's I, 87–89, 92–94
Moran scatterplot, 23, 83, 90
Moran's I, global, 82–83
Moran's I, local, 87–89, 92–94
positive and negative, 5
quadrat counts, 249
software, 275
spatial dependence, 80
testing for, 83–84
    random permutation test, 84, 88
vector boundary data, 16
Spatial autoregressive models, 103, 106–107
    conditional autoregressive model, 107
    simultaneous autoregressive model, 107
Spatial clustering, image segmentation, 62–65, 138
Spatial coordinate notation, 20
Spatial data, 3
Spatial data, transforming and detrending, 3
Spatial data and regression, 105–106, See also Regression
Spatial data models, See Data models
Spatial data sources, 274

Spatial deformation models, 26, 238–239
Spatial dependence, 5–6
    cross-variogram, 198
    "First Law of Geography", 5
    global regression modelling, 103
    measure of spatial autocorrelation, 80
    second-order effects, 105
    strong and weak examples, 6
    weighting-based approaches, 6, See also Geographical weighting schemes; Weighting
Spatial error model, 106–107
Spatial expansion method (SEM), 110, 111–115, 142
Spatial filters, 39–41, See also Filters
    collinearity effects, 128
Spatial hierarchies, multilevel modelling, 108–109
Spatial hypothesis generator, 265
Spatial interaction modelling, 4
    geographical weighted, 138
Spatial interpolation, See Interpolation
Spatial law of random function, 192
Spatial location notation, 20, 25
Spatially adaptive filtering, 110–111
Spatially adaptive kernels, 123, See also Kernel methods
Spatially lagged dependent variable model, 106
Spatially-referenced postal addresses, 18
Spatially-referenced variable, geostatistical random function model, 191
Spatially varying coefficient processes, 116–121, 142
Spatially weighted classification, 138
Spatial nonstationarity, See Nonstationarity

Spatial patterning, 73
  areal data, See Areal data
  autocorrelation, See Spatial autocorrelation
  categorical data, 98
  point patterns, See Point pattern analysis
  weighting schemes, See Weighting
Spatial prediction, 35, 241
  accuracy assessment, 189, 190
  change of support problem, 173, 216
  comparing stationary and nonstationary models, 9
  geostatistics, 191–241, See also Geostatistics
  global regression modelling, 148
  kriging, 204–215, See also Kriging
  limitations of conventional approaches, 216
  local approaches, nonstationary models, 217–219
  nonstationary kriging models, 222–233
  secondary variables, 148, 154, 167, 212, 215, 220, 226–228, 230, 231
  separating deterministic and stochastic components, 220
  spatial deformation models, 238–239
  spatial interpolation, See Interpolation
  stationarity, 193–194
  uncertainty in, 189, 191
  variogram, See Variograms
Spatial prediction, deterministic methods, 35, 145–190, See also Interpolation; specific methods
  areal interpolation, See Areal interpolation
  finite difference, 171–173

  generalised additive model using, 168
  global methods, 147–148
  limitations, 189–190, 191
  local point interpolation, See Point interpolation
  local regression, 153–154
  performance assessment, 147
  separating deterministic and stochastic components, 220
  splines, 158–168, See also Thin plate splines
  using prior knowledge, 184–189
Spatial proximity matrix, 82
Spatial relations, 4, 35, 101, 142–143, See also Regression
  cross-variogram, 198
  GWR, See Geographically weighted regression
  local indicators of spatial association, 83, 86
  local regression, See Local regression
  measures of spatial autocorrelation, 83, 86, See also Spatial autocorrelation
  multilevel modelling, 105, 108–109, 143
  MWR, See Moving window regression
  spatially varying coefficient processes, 116–121
Spatial resolution, 5, 272, See also Spatial scale
  ecological fallacy, 7
  texture characterisation, 45–46
  upscale and downscale, 8
  wavelet transforms, 53–54
Spatial scale, 7–9, 272
  analytical approaches, 8
  defining, 7
  dichotomy of scale, 8
  ecological fallacy, 7

in geographic applications, 3, 8–9

scale-dependent models and processes, 7–8

spatial deformation models, 238–239

spatial structure variations and, 6

Spatial structure, texture, 44–45

Spatial structure and global methods, 147

Spatial structure and scale, 6, See also Spatial scale

Spatial structure variation, variograms, See Variograms

Spatial subsets, 31

Spatial variation

　exploring, 195–203

　　covariance and correlogram, 195–196

　　variogram, 197–204, See also Variograms

　geostatistical spatial prediction, 10, 195–204, See also Geostatistics

　mean, See Mean, spatial variation

　in multivariate relations, See Multivariate data analysis; Regression; Spatial relations

　nonstationarity, See Nonstationarity

　predicting, See Spatial prediction

　separation of stochastic and deterministic components, 220

　single variables, gridded data, See Grid data analysis

　single variables, point patterns, See Point pattern analysis

Spatial weights, See Geographical weighting schemes; Weighting

Spectral information, 47, 65

Spherical variogram model, 201

Spline covariance model, 215

Splines, 158–168, See also Thin plate splines

　completely regularised, 166–167

　equivalence of kriging, 215

　local regression, 109

　penalised regression, 168

Split and merge segmentation, 64

Standard deviation

　contrast stretch for wavelet transform image, 59–60

　geographically-weighted, 76–77

　local, texture characterisation, 45–46

　moving windows

　　focal operators, 28

　　gridded data analysis, 28

Stationarity, 3, 9, 24, See also Non-stationarity; Nonstationary models

　geostatistical context, 193–194

　geostatistical prediction issues, 241

　intrinsic, 194

　problematic issues, 3

　quasi-intrinsic, 194

　second-order, 193

　significance testing, 128

　strict, 193

　testing for, 9

　variogram and, 194

Statistics, See Geostatistics; specific applications, methods, statistics

Steerable wavelets, 59

Stochastic (random) component, geostatistical model, 191

　separating deterministic and stochastic components, 220

Stochastic simulation, 215–216

Stratification, 26, 27

Stratified kriging, 240
Strict stationarity, 193
Structure, texture, 44–45
Sudden infant death syndrome (SIDS)
    data, 19, 268
Summary statistics, local, 73
Support, 173, 216
Surface model generation, areal in-
    terpolation, 177–180
Surface water flow, 172

T

Target zones, areal interpolation,
    173–175
    control zone approaches, 184–
        187
    EM algorithm, 187–188
    overlay approach, 175
    pycnophylactic method, 181
    uncertainty, 189
Temperature mapping, 230
Tension, thin plate splines with,
    166
Terrain errors in DEMs, 67
Tessellation, 32, 151
Texture, 44–47, 240
Theory of Regionalised Variables,
    191
Thiessen polygons, 148, 151, 153
Thin plate splines (TPS), 158–168
    basis function, 161, 166, 215
    completely regularised splines
        (CRS), 166–167
    computational demand, 168
    generalised additive models, 168
    generalised cross-validation, 160,
        162–163
    kriging and, 215
    partial TPS, 165
    penalised regression splines, 168
    precipitation prediction appli-
        cation, 163–165, 167–170
    smoothing parameter, 160
    software, 275
    TPS with tension, 166

weights, 162
Thresholding, 63
Tobler's pycnophylactic method, See
    Pycnophylactic areal in-
        terpolation method
Tone, texture, 44
Topographic fractal dimension, 240
TPS, See Thin plate splines
Transformation, 49
    Fourier, 47
    grid data and moving windows,
        47
    local modelling, 24, 26, 30
    wavelet, 47–62
    wavelets, 47–62, See also Wavelet
        transforms
Transitive variogram model, 199
Trend-free variogram, 217, 220–
    225, 230
Trends, See also Detrending; Drift;
    Large scale trends
    deterministic components, geo-
        statistical models, 191
    geostatistical covariance func-
        tion, 196
    geostatistical terminology, 220
    global detrending, 2
    kriging with trend model, 220,
        222–223
    nonstationary mean, 220–222
    nonstationary spatial predic-
        tion models, 217
    polynomial model and detrend-
        ing, 29
    spatial expansion method, 110,
        111–115
    spatially adaptive filtering, 110–
        111
    trend-free variogram estima-
        tion, 217, 220–222, 230
Trend surface analysis, 145, 147,
    153
Triangulated irregular network (TIN)
    model, 65, 151
Tukey box and whiskers plot, 34

Turkish landcover dataset, 13, 16
    texture, 45–46
Two-dimensional (2D) variogram,
    195
Two-dimensional (2D) wavelet trans-
    forms, 56–60

U
Unbiased kriging, 206
Unbounded variogram model, 199,
    202
Uncertainty in spatial prediction,
    189, 191
Unfocused local space, 31
Union operator overlay procedure,
    175
Univariate data analysis, 4
Universal kriging, 220, 222, See
    also Kriging with a trend
Upscale, 8

V
Variance, 39, 197, See also Covari-
    ance; Mean, spatial vari-
    ation
    anisotropic functions, 124
    cokriging, 212–214
    collinearity effects, 128, See
        also Collinearity
    decomposition, 128–133
    geographically weighted, 138
    IRF-k kriging, 226
    kriging, 206–210, See also Semi-
        variance
    kriging with a trend model,
        223–226
    nugget, 199, 221
    null distribution, 128
    spatial resolution and, 45–46
    spline methods, 162
    stationarity issues, 3, 9, 24,
        128, 194
    variogram models, 199–201, 221,
        226, See also Semivari-
        ance

Variance-covariance matrix, 105–
    106, 138
Variance inflation factors (VIFs),
    128–129, 133
Variance/mean ratio (VMR), fit-
    ting Poisson distribution,
    249
Variogram cloud, 23, 198, 203–204
Variogram models, 198–204
    anisotropic, 202–203
    bounded, 201–202
    conditional negative semidefi-
        nite, 199–201
    fitting, 203
    nonstationary, 194
    nugget effect, 199, 201
    unbounded, 199, 202
Variograms, 197–204
    accuracy, local vs. global, 234
    anisotropy, 197, 202–203, 204,
        233
    bias, 198, 221
    cokriging and, 212–213
    conditional negative semidefi-
        nite, 212
    cross-variogram, 198, 212
    digital number, 240
    double log, 240
    estimation and modelling ex-
        ample, 203–204
    experimental, 197–198
    kriging and, 198, 202, See also
        Kriging
    moving window approaches,
        233–234
    simple kriging with locally
        varying mean, 227
    limitations, 216
    nonstationary models, 233–240
    convolution approach, 239–
        240
    locally varying anisotropy,
        233
    moving window approaches,
        233, 238, 272

spatial deformation models, 238–239

nugget effect, 199, 201, 221, 233

observations needed to estimate, 198

omnidirectional, 203–204

outliers and, 198

separating deterministic and stochastic components, 220

spatially weighted classification, 138

spatial scale analysis, 8

stationarity issues, 194

texture analysis, 240

trend-free estimation, 217, 220–225, 230

two-dimensional, 195

Vector contour lines, 157

Vector field boundary dataset, 11, 16

Vertical edges, 57, 59

Very Important Points (VIP) algorithm, 151

Vesper software, 234, 275

Vision applications, 61

Volume preservation, 181–184

Voronoi polygons, 148

Voronoi tessellation, 151

W

Water runoff, 65

Wavelet packets, 60

Wavelets, 4, 48, See also Wavelet transforms

applications, 61–62

coefficients, 53–54

Daubechies, 49–50, 52, 54

digital orthophoto data, 11, 17

frequency localisation, 49

Haar, 51–53

Mother, 49–52

notation, 20

spatial scale analysis, 8

steerable, 59

Wavelet transforms, 47–62, 272

basis functions, 51–53

continuous, 49, 50

discrete, 49, 50–51

2D data, 56–60

implementation, 53–56

fast (FWT), 53–56

Fourier transforms, 48–49

packets, 60

software, 61

two-dimensional, 56–60

Weighted least squares (WLS), 203

Weighted moving average, 154, 208–210, 215, See also Inverse distance weighting

Weighting, 6, 74–80, See also Geographical weighting schemes

adaptive window size, 47

areal weighting method, 175–176

bi-square nearest neighbour scheme, 123

convolution mask for grid data, 41–42

distance-based, 24, 177–178

Gaussian function, 42, 74, 123

geographically weighted statistics, 138

global Moran's I, 82

GWR, See Geographically weighted regression

inverse distance, See Inverse distance weighting

kernel-based methods, 28–29, See also Kernel methods

kriging, 204, 206

kriging with external drift model, 228

simple kriging, 207

variograms, 198

moving windows, 28

population-weighted centroids for areal interpolation from zone, 18, 177

spatially weighted classification,
    138
surface model generation from
    areal data, 177
thin plate splines, 162
variogram, 198
White noise, 161, 239
WinBUGS software, 119–121, 142,
    276
Windowed Fourier transform (WFT),
    49
Windows, locally-adaptive, 29
Windows, rotating, 47
Window size, 29, 30, 39, 156, 189,
        272, See also Bandwidth;
        Moving window methods
    adaptive, 47
    texture analysis, 240
    variogram estimation, 233
Wishart prior distribution, 120
WLS (weighted least squares), 203

Z
Zonal anisotropy, 202–203, See also
        Anisotropy
Zonal overlay approach, 175–177
Zone centroids, 177–179, 261, 266–
    267
Zones, source, 173–176, 181–189
Zones, target, See Target zones,
        areal interpolation
Zoning problem in areal interpola-
        tion, 173

Printed in the United States
by Baker & Taylor Publisher Services